FIFTY YEARS OF ELECTRON DIFFRACTION

FIFTY YEARS OF
Electron Diffraction

In recognition of
fifty years of achievement
by the Crystallographers and
Gas Diffractionists in the field
of Electron Diffraction

General Editor

P. GOODMAN

Regional and Special Subject Editors

M. J. WHELAN *United Kingdom*

R. UYEDA *Japan*

S. A. SEMILETOV *USSR*

K. HEDBERG *Gas Diffraction Editor*

G. A. SOMORJAI *Low Energy Electron Diffraction Editor*

Published for the International Union of Crystallography by

D. REIDEL PUBLISHING COMPANY

DORDRECHT : HOLLAND / BOSTON : U.S.A.
LONDON : ENGLAND

Library of Congress Cataloging in Publication Data

Main entry under title:

Fifty years of electron diffraction.

Includes bibliographies and indexes.
1. Electrons—Diffraction—History—Addresses, essays, lectures.
I. Goodman, P. (Peter), 1928–. II. International Union of Crystallography.

QC793.5.E628F53   530,4'1   81-2069
ISBN 90-277-1246-8   AACR2

Published by D. Reidel Publishing Company,
P.O. Box 17, 3300 AA Dordrecht, Holland

Sold and distributed in the U.S.A. and Canada
by Kluwer Boston Inc.,
190 Old Derby Street, Hingham, MA 02043, U.S.A.

In all other countries, sold and distributed
by Kluwer Academic Publishers Group,
P.O. Box 322, 3300 AH Dordrecht, Holland.

D. Reidel Publishing Company is a member of the Kluwer Group

Printed in the Netherlands

Table of Contents

Period 3: 1945–Present

PART III – THE PRESENT SUBJECT

ACCESS LISTING OF GAS DIFFRACTION ARTICLES
edited by KENNETH HEDBERG

Introduction

The idea of a '50 Years of Electron Diffraction' volume was bound to occur to many people as the year 1977 approached, as a consequence of the success of P. P. Ewald's volume on the history of X-ray diffraction. There were many reasons for wanting to produce such a volume, and for believing that it would in no way be a mere shadow volume for the X-ray history. For electron diffraction had had its own exciting beginning which, incidentally, it shared with the whole of quantum mechanics, and it has now its own exciting and successful present. This is an essential point. A mere 50 years of existence is not in itself sufficient reason for a volume, simply as a dutiful gesture; but it is, in fact, a success story. Between the beginning and the present there came a long, difficult, and often fruitless search for truth and recognition, for much of that time laboring under the knowledge that the older more established techniques of X-ray diffraction were a necessary reference point for new results. It is possible to say today that electron diffraction has found its own fields of application, often outside the reach of other techniques. It is, then, more important than ever that the techniques of X-ray, neutron and electron diffraction become mutually understandable and that results from one discipline be discussed and evaluated by workers from others. This situation is one which the IUCr expressly wishes to establish and one to which a 50 years' assessment might hopefully contribute.

The fact that this volume came into existence is due to the enthusiasm of a few people and the encouraging support of the IUCr Executive Committee. The idea was discussed at the London Meeting in 1977, organised by Professor Blackman at Imperial College for the 50 years' Anniversary; it was here that Professor Ryozi Uyeda became involved, and played a central role in further planning. The need for writing the early international history had become clear in the stimulating atmosphere of the anniversary year. Professor Uyeda was asked how it came about that the experiments in Japan followed so quickly those of Thomson, and Davisson and Germer. What means of contact did they have? Were they direct, or did they have independent access to the French theoretical school of de Broglie? It was then clear that many of the people who knew the history, in Japan, Europe and America, were around and willing to help.

A formal submission was made by the Commission on Electron Diffraction after their 1978 meeting. The unanimous proposal of the Commission, that a 50 year volume be produced as a Commission project was generously endorsed by the Executive Committee in the following year. The detailed planning of the volume was however left very much in the control of the Commission. It was due to the skill of the Commission chairman, Professor Kuchitsu, that the initial debate within the Commission resulted in a successful submission, and that the various responsibilities were allocated. I was equally fortunate, as Editor, in finding in Melbourne a group of local crystallographers who were willing at very short notice to act as interim advisory committee. This group suggested the basic plan of a three-part volume, and that an international history should be written around centers of research, in whatever countries these should occur. Both ideas survived, but were only made practicable by the hard work both of Commission members and invited contributors. In particular a great deal of responsibility fell to contributors to Parts I and III, where long historical or survey articles were required.

The three-part division of the volume was planned as follows: Part I would contain all the history up to 1928, and include an account of the interplay between the esoteric circle of French scientists in de Broglie's laboratory and the extremely practical and industrially related research workers at Davisson's laboratories in America, as well as the exchanges between these, Thomson, and Kikuchi. Apart from the short note by de Broglie himself this section was necessarily written by scientific historians, sufficiently detached from the actual events to give a rounded history. Part II was to deal with the developments from 1928, when Bethe's theory was formulated, to the present day, in the form of personal memoirs. Part III was intended as a concise, self-contained report on the present state of the subject. The six authors chosen here, would, we hoped, be able to put the historical development of their specialist fields in a current perspective, and give some indication of the capabilities of present-day electron diffraction.

Part II turned out to be by far the largest section and contains most of the history. Here 36 authors with a variety of approaches to electron diffraction give their personal accounts. The compilation is as near as possible chronological, and includes gas diffraction, a subject which started in 1930, as well as diffraction from the solid state. It must be said that the gas diffraction workers, Hedberg, and Hargittai in Hungary, had initially planned a separate history. I am happy to say that I have enjoyed a friendly collaboration with Ken Hedberg and that we came to a general agreement that we should have the joint history in its present form. Hedberg's preface will follow mine.

Part II fell conveniently into three periods, one of the dividing lines being 1945–the end of the war. The periods appeared to fall into place on their own account, distinguished at first by an advancing technology. It is worth mentioning here the role planned for photographs, other than in-text illustrations, for this volume. As with Part I, the individual periods in Part II are introduced by picture pages displaying equipment or, where appropriate, laboratories or personnel evocative of their period. The contrast between typical laboratory equipment in use in the early '30s and the early '40s, for example, can be quickly assessed from these pages. Then the final division (preceding period 3) contains photographs of some equipment, but also of some of the people most responsible for the more recent developments, limited only slightly by the collection made available to us. For one thing which is clear from this history is that there was a dramatic advance in the subject between 1955 and the early 1960s. This advance is easy to detect and was due as much to a greater theoretical understanding as to the advance in technology and instrumentation. It was, I believe, the intensive development of this time which has transformed the subject from a relatively minor branch of science, with its own private terminology and private procedures, into a highly developed and important one. In this respect it must be mentioned that the period classification used may be far from clear-cut with respect to individual participating authors. Some leading contributors have had a life-time association with electron diffraction and it would be unfair to suggest, for example, that some of the authors listed in an earlier period did not contribute substantially, both directly and indirectly, to the recent subject.

Of the very many points to emerge from this history I mention only one, which I believe to be a key issue in transmission studies. It was once held by many that thin-film structures discovered by electron diffraction had their own meaning and very little if anything to do with bulk structures. If this were the case, there would have developed an interesting but independent study of thin-film and surface structures. There are two places in this volume at which this once commonly held belief is found to be not quite true. Ogawa recalls that during his career, by comparing X-ray and electron diffraction data, he found, much to his excitement, that structures determined by electron diffraction had relevance to the bulk structure. Elsewhere in the volume we can read Cowley's statement on the same subject, made during his investigation of short-range-order in binary alloys: "Lately", he wrote, "we have become more bold in our interpretation of these patterns. In this approach we have been encouraged by continued success." (Cowley, 1965). If this single understanding is conveyed by this volume, then it will have gone at least some way towards fulfilling its obligations as a history.

That the subject is now growing very rapidly is evidenced by Part III. If I have any disappointment at all, it is that we still need six independent authorities to write about six diverse topics. It is clear, however, that such divisions are slowly disappearing. Today we find, for example, surface studies being made by fast as well as by slow electrons, and inelastic scattering being analysed by electron microscopists as well as by spectroscopists. It is furthermore encouraging to think that our subject may be falling into line in terminology, particularly in the theory, with the rest of solid state and scattering physics. If I have a single wish for this volume, therefore, it is that it will fall into the hands of some scientists outside the field of electron diffraction, and that they will find it understandable, and, to a certain extent, entertaining.

In acknowledgement, I wish to thank Professor Uyeda for his enthusiasm in organizing the Japanese contributions, the Executive Committee of the International Union of Crystallography for generously supporting our Commission project, and Professor K. Kuchitsu our Commission Chairman, and individual Commission members, for their collaboration. I particularly thank IUCr. Secretary Dr Jim King for his direct assistance and advice during the course of the production, and Professors S. Rassmussen and N. Kato, of the

Executive Committee for their continued support and advice. I also wish to thank laboratory colleagues for helping in various ways; Dr and Mrs Jeff Wunderlich for supervising contributions from France, my laboratory Chief, Dr L. T. Chadderton for generously consenting to my absence from normal duties for this period, and CSIRO for its financial support. I have furthermore to thank members of my family, and in particular my wife Pat, who for several months now has lived with this volume, correcting manuscripts and generally behaving as unpaid co-editor. Finally, any success this volume might have is due directly to the many contributors who gave up much valuable time, and who for the most part have shown great patience and tolerance to editorial intrusion.

P. GOODMAN

28th October 1980

Foreword to Gas Diffraction Articles

Gas-phase electron diffraction had its beginning over fifty years ago with the pioneering work of Mark and Wierl on carbon tetrachloride. Those workers had foreseen some of the advantages of electron scattering over X-ray scattering for obtaining structural information from gas molecules, but they doubtless did not foresee the evolution of their experiment into the powerful method for structure analysis that we have today. That evolution was the consequence of work by many people in several different countries, work that has included improvements in the experiment, development of a more complete theory, and the introduction of versatile, high-speed computational methods. These advances permit gas electron diffraction to be applied to a wide variety of species over astonishingly large temperature ranges. Kozo Kuchitsu's excellent article in Part III of this volume traces the historical development of the method, describes important aspects of contemporary theory and practice, and summarizes a few of the important results.

The articles in Part II of the volume are accounts of the birth and development (and in some cases the death) of several gas electron-diffraction laboratories as recalled by some of the older workers in the field. The intent of these articles is to convey some sense of the excitement and challenge experienced by those who worked in the earlier days. As a group, the articles are a valuable history of these early days, but one necessarily limited by the scope and plan of the volume. Some day, perhaps, a more complete history of gas electron diffraction will be written, one that includes as well the important advances made by the many people in the newer laboratories of England, France, Germany, Holland, Hungary, Japan, the United States, and the Soviet Union. An access-listing of the gas diffraction articles is included in the Table of Contents.

This occasion must not pass without mention of the very special place occupied by Lawrence Brockway in the development of our science. More than anyone else, he may be said to be the father of the gas diffraction method. His contributions need no detailed documentation here: the literature speaks for itself, and as Larry Bartell notes, his professional descendents are everywhere. But those of us who knew Lawrence Brockway think of him as more than a colleague. He was a generous and thoughtful person, with a zest for life and a never-failing sense of humor that made him a friend to all. When he was asked, shortly before he became ill, if he would contribute an article to this book, he responded with characteristic enthusiasm. Unfortunately, he was unable to do more than to write part of a preliminary draft, which, edited by Larry Bartell, appears in Part II. Lawrence Brockway died in October, 1979. He will be missed.

K. HEDBERG

October 1980

PART ONE

The Start

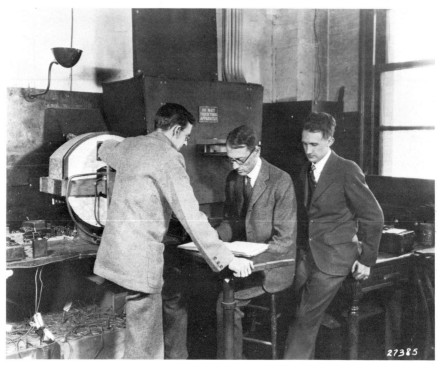

The pioneers of experimental electron diffraction. *Upper Left:* Sir George Thomson F.R.S. (supplied by P. B. Moon, F.R.S.) *Upper Right:* Seishi Kikuchi, in 1970. (supplied by S. Miyake). *Lower:* Davisson, Germer, and Calbick in 1927. Grouped in the laboratory at 463 West Street, New York City, are, left to right, Davisson (age 46), Germer (age 31), and Calbick (age 23). Note the apparatus in the background; Germer is seated at the observer's desk, ready to read and record the electron current from the galvanometer (the large 'box' beside his head); the banks of dry cells behind Davisson supplied the DC voltage for the experiments. Photo courtesy of Bell Labs.

Seminar in the library of the de Broglie X-Ray Physics Laboratory in 1929. From left to right: Jean Thibaud, Maurice de Broglie, Louis de Broglie, Alexandre Dauvillier, and Jacques Trillat. (Photograph supplied by J. J. Trillat.)

LOUIS DE BROGLIE

I.1. A General View of My Scientific Works

I presented the basic principles of Wave Mechanics in three letters to *Comptes Rendus de l'Académie des Sciences de Paris* in September and October 1923, and subsequently in a more developed fashion in my Doctoral Thesis which was submitted on 29 November, 1924. My idea essentially being to extend to all particles the coexistence of waves and particles discovered by Einstein in 1908 for the particular case of light and photons. Consistent with the known ideas of classical physics, I tried to represent a real physical wave carrying the small particles which was localized in space and travelling in time.

Two methods for representing the waves occurred to me. The first, which I considered the more fundamental, was based in part on the difference between the relativistic transformation of the frequency of a wave and the frequency of a clock, and allowed particles to *behave* like a clock pendulum moving constantly in phase with its associated wave. This idea led me to write the formula $p = h/\lambda$, where p is the particle momentum, λ the wavelength of the wave carrying the particle, and h is Planck's constant. This hypothesis permits the identification of the Principle of Least Action for a particle with the Principle of Fermat applied to its carrying wave.

My ideas, in my view falsely interpreted, led to the development of a purely wavelike theory of the electron, namely the theory of the wavefunction ψ which overlooks the localized nature of the particle. It is this misconception which mars the otherwise quite remarkable work of Schrödinger, who made the mistake of abandoning the concept of a localized particle on its wave.

In 1928 I was called to teaching duties and resigned myself to teaching the prevailing ideas bearing the name Quantum Mechanics and, for a long time, gave up developing my initial ideas. But, for about the last 30 years, I have been once again convinced that it is necessary to consider the particle as a very small object localized in space at each instant and describing a trajectory in time.

As I have shown in a more and more fundamental series of works, it is possible completely to conserve the statistical significance of the wavefunction ψ, which is only a representation of probability and supports my initial concept of a localized particle carried by a real physical wave. I have been able to complete this theory by postulating a hidden thermodynamic concept whose introduction opens quite new perspectives. A very important consequence of this thermodynamic concept appears to be the following: the Principle of Least Action, the basis of mechanics, becomes only an aspect of the Second Law of Thermodynamics.

I conclude this article by noting that the fundamental idea, in which the motion of a particle is determined by the propagation of the wave which carries it, leads one directly to a prediction of the remarkable phenomenon of electron diffraction.

Photograph of the author, Louis de Broglie, taken in 1978. (Photograph kindly supplied by J. J. Trillat.)

HEINRICH A. MEDICUS

I.2. The Origin of de Broglie's Concept

No historical document on electron diffraction can be complete without some examination of events surrounding de Broglie's revolutionary concept. The equation for matter waves is usually baldly attributed to de Broglie's thesis of 1924. The following analysis by Medicus* gives the background to his theory and an analysis of its content.

1. INTRODUCTION

Early in 1922 de Broglie wrote an article about black-body radiation in which he derived Wien's radiation law by means of thermodynamics, kinetic gas theory and quantum theory, without using electromagnetic theory (de Broglie, 1922a). Here he says, "The hypothesis of quanta of light is adopted." (This article was written one year before the explanation of the Compton effect, which settled definitively the existence of light quanta.) He treats the photons as particles, or 'atoms of light' with mass $h\nu/c^2$ and momentum $h\nu/c$. On assuming a mixture of photon gas 'molecules,' each consisting of 1, 2, 3, . . . atoms of light, he obtains Planck's law. In a sense, this paper was a precursor of the Bose statistics. Yet another article (de Broglie, 1922b), written in the same year, deals with the energy fluctuations in black body radiation.

In a 1963 interview in Paris with Thomas S. Kuhn, André George and Théo Kahan, de Broglie indicated that this paper on blackbody radiation was the point of departure for his later work: "I began to have the idea – it was not yet born. I probably would not have dared to tell about it, but I began to have it in my mind." This interest in the properties of quanta motivated his search for a theory that would unify the wave and particle aspects. In addition, the occurrence of integral quantum numbers in descriptions of the movement of electrons in an atom reminded him of wave theory, where integers appear in interference phenomena and at many other places.

Suddenly, in the summer of 1923, the idea occurred to de Broglie to generalize this wave–particle duality to include material corpuscles as well. (He was thinking of electrons in particular.) This idea, to be more fully developed the following year in his thesis, initially came to light in 1923–in three short articles published in the 10 and 24 September and 8 October issues of the *Comptes Rendus*, (de Broglie, 1923a), and in a very short note submitted to *Nature* on 12

* This article is an extract from a more complete article entitled '50 Years of Matter Waves' in *Physics Today*, of February 1974, and is reproduced here by kind permission of the author and the American Institute of Physics.

September, in which he sketches some of the main points and refers for details to the first two *Comptes Rendus* articles (de Broglie, 1923b).

These preliminary notes, which contained the essentials of his new theory on a total of seven or eight pages, were written with the thought of developing these ideas more completely in a thesis. In October of the same year he also submitted a paper to the *Philosophical Magazine*, which however, did not appear until 1924 (de Broglie, 1924).

Thus, de Broglie's discovery actually took place in 1923. The year of his thesis, 1924, has gained common acceptance, particularly outside France, as the date of the discovery. But de Broglie himself, in the volume published on the occasion of his 60th birthday in 1952, said that 1923 would be a more accurate date. The thesis itself was published as an article of over 100 pages in the *Annales de Physique* (de Broglie, 1925).

It should be noted in passing that, contrary to popular understanding, de Broglie was not a young, unknown student when he wrote the thesis for which he won the Nobel Prize in 1929. Because he had spent six years in the military, stationed at the military radio-telegraphic station on the Eiffel tower, he was 32 years old when he received his doctorate and had already published about two dozen scientific papers on electron, atomic and X-ray physics. Upon discharge from the military, de Broglie resumed his theoretical studies, at the same time working in the private laboratory of his considerably older brother, the Duc Maurice de Broglie, a highly respected X-ray physicist, with whom he had many conversations about wave–particle duality in X rays.

2. DE BROGLIE'S THEORY IN BRIEF

It will be helpful to sketch briefly the thought behind de Broglie's discovery. Basically, his idea was an extension, to include all particles, of Einstein's 1905 theory about the wave–particle duality of photons: Corpuscles are accompanied by waves. Whereas the equation $E = h\nu$ applied for photons, and $E = mc^2$ for material particles, de Broglie assigns a frequency ν_0 as the frequency of an internal vibration as measured in the reference system fixed to the particle of rest mass m_0. However, if a stationary observer sees the particle passing with a certain velocity, he will see the frequency of the internal vibration decrease to $\nu_1 = \nu_0 (1 - \beta^2)^{1/2} = (m_0 c^2/h)(1 - \beta^2)^{1/2}$, because moving clocks go slowly. On the other hand, the energy of the moving particle is $m_0 c^2 (1 - \beta^2)^{-1/2}$, which corresponds, according to the quantum relation, to a frequency $\nu = (m_0 c^2/h)(1 - \beta^2)^{-1/2}$. The frequencies ν_1 and ν are different. De Broglie overcomes this apparent difficulty by stating that the periodic phenomenon that is inherent in the moving particle and that for an observer at rest has the frequency ν_1, appears to this observer to be constantly in phase with a wave having the frequency ν and propagating with a velocity $V = c/\beta$ in the same direction as the particle. This velocity V is thus the phase velocity of the wave.

Therefore, de Broglie speaks of phase waves and concordance, or harmony, of phases. The particle itself, and hence also the energy, moves with the group velocity v. These considerations are of relativistic nature because they are based on the difference between the relativistic transformations for waves and for a moving clock, which represents the particle. De Broglie also shows that, in the case where the propagation of the wave can be described in the approximation of geometrical optics, one has to identify Fermat's principle with the principle of least action of Maupertuis, and this leads again to the same phase waves. He also expresses his belief that the new dynamics of mass points exhibits a relationship to the classical one (including relativity) similar to the relationship existing between wave optics and geometrical optics.

As early as the *Comptes Rendus* article of 10 September 1922 de Broglie applied his electron-wave hypothesis to electron orbits in an atom, requiring that the wave be in phase with itself, that the circumference be an integral multiple of the wavelength. In the summary concluding his thesis he comments, "We believe that this is the first physically plausible explanation for the Bohr–Sommerfeld stability rules."

It is noteworthy that the famous formula $\lambda = h/mv$ is found in this explicit form only once in de Broglie's thesis, namely in the chapter on statistical mechanics, where he calculates the momentum of molecules in an enclosure forming standing waves. For de Broglie, it is not the wavelength of the particle that is in the foreground, but its frequency. There is no essential difference between photons and particles. In order to avoid difficulties in his theory, de Broglie at first assumes that light quanta, or atoms of light, "have an extremely small mass (not infinitely small in the mathematical sense). . . . It seems that m_0 [the rest mass of the quanta] should be at most of the order of 10^{-50} g."

Thus, the titles of all but one of his early papers dealing with matter waves make no explicit mention of the concept of matter waves. The first two *Comptes Rendus* articles have the headings 'Ondes et quanta', 'Quanta de lumière, diffraction et interférences'; the thesis and the article in *Annales de Physique* are entitled 'Recherches sur la théorie des quanta'; the note in *Nature* is called 'Waves and quanta,' and the article in the *Philosophical Magazine* 'A tentative theory of light quanta.' Only the title of the third *Comptes Rendus* paper, 'Les quanta, la théorie cinétique des gaz et le principe de Fermat', promises perhaps more than a discussion of light. In a recent article (de Broglie, 1973), de Broglie has provided an illuminating summary of his early ideas and his current interpretation of matter waves.

3. EXPERIMENTAL PROOF

How did de Broglie feel about experimental proof of his theory? In the *Comptes Rendus* note of 24 September 1923 he writes, "A beam of electrons passing through a very small opening could present diffraction phenomena. This is perhaps the direction in which one may search for an experimental confirmation of our ideas." The letter to *Nature*, written at the same time, likewise implies this possibility. However, the article in the *Philosophical Magazine* of 1924 contains nothing of that sort. Even the thesis, although much longer and more detailed, is silent about potential experiments. De Broglie said in an interview with Fritz Kubli in November 1968 (R. Kubli, 1970) that his brother Maurice had suggested that the thesis should also include an experimental part. Louis declined, saying that he was not an experimentalist. Maurice de Broglie's attitude appears to indicate that he considered an experimental proof of the existence of matter waves within the realm of possibility.

De Broglie's examination committee, composed of Jean Perrin as chairman, Paul Langevin as thesis advisor, Elie J. Cartan and Charles Mauguin, was impressed by the candidate's work. First-hand knowledge of the opinion of one of its members is available. In 1952, Mauguin, who was a crystallographer, recalled de Broglie's examination: "Today I have difficulty understanding my state of mind [in 1924] when I accepted the explanation of the facts [the Bohr–Sommerfeld quantization rules] without believing in the physical reality of the entities that provided this explanation." (A. Michel, 1953). Today, interestingly enough, we have no such inhibitions about accepting the quark model, for example, even though no one has ever found a quark! Of the other members of the committee, Langevin was sufficiently enthusiastic to send a copy of the manuscript (before the exam) to his old friend Einstein.

4. EINSTEIN'S INTERCESSION

De Broglie's theoretical discovery did not become widely known, nor did it win immediate credence, for several reasons. Although the *Comptes Rendus* were widely circulated in Europe, they were not intensively read, and thus de Broglie's articles had little impact.

De Broglie's ideas found a sympathetic reception with Einstein, as he himself had gone through an extended struggle to convince his colleagues of the wave–particle duality of photons. Einstein had a liking for symmetry arguments in physics, and de Broglie's theory established such symmetry between photons and material particles (Klein, 1964). Einstein enthusiastically gave Langevin his judgment, writing that de Broglie had "lifted a corner of the great veil" (de Broglie, 1962) and he in turn alerted others to the importance of this far-reaching work.

Professor of Physics
Rensselaer Polytechnic Institute
Troy, New York 12181, U.S.A.

REFERENCES

de Broglie, L.: 1922a, *J. de Physique* IV, **3**, 422.
de Broglie, L.: 1922b, *Compt. Rendus* **175**, 811.
de Broglie, L.: 1923a, *Compt. Rendus* **177**, 507, 548, 630.
de Broglie, L.: 1923b, *Nature* **112**, 540.
de Broglie, L.: 1924, *Thèse de doctorat*, Masson, Paris.
de Broglie, L.: 1925, *Ann. de Physique* **3**, 22.

de Broglie, L.: 1962, *New Perspectives in Physics* (Oliver and Boyd, Edinburgh), p. 139.
de Broglie, L.: 1973, *Compt. Rendus* B **277**, 71.
Klein, M. J.: 1964, 'Einstein and Wave Particle Duality', in *The Natural Philosopher* (Xerox Publishing Company).
Kubli, R.: 1970, *Arch. His. Exact Sci.* **7**, 26.
Michel, A.: 1953 *Louis de Broglie – Physicien et Penseur* (A. George, Ed.)

R. K. GEHRENBECK

I.3. Davisson and Germer

1. DAVISSON: THE EARLY YEARS

Clinton Joseph Davisson was born on 22 October 1881 in the small agricultural town of Bloomington, Illinois, about 100 miles from Chicago. His father, Joseph Davisson (1841–1906) a native of Ohio and the descendant of early French and Dutch settlers in Virginia, was a contract painter and paperhanger by trade. Davisson's mother, Mary Calvery Davisson (1861–1954), was of English and Scottish parentage and a native of Erie, Pennsylvania. The only other child, Carrie Louise, two years younger than Davisson, remembered the Davisson home as "a happy, congenial one – plenty of love but short on money."

Davisson graduated from Bloomington High School in 1902 at age 20. He entered the University of Chicago in the fall of 1902 on a full tuition scholarship which lasted three quarters. Each quarter he studied physics with R. A. Millikan, and it was from Millikan that Davisson "was delighted to find that physics was the concise, orderly science he had imagined it to be, and that a physicist (Professor Millikan) could be so openly and earnestly concerned about such matters as colliding bodies." Davisson further appreciated the fact that Millikan introduced "only as many concepts, definitions, laws and principles as could be completely absorbed by earnest application," and that there were "no loose ends and no emphasis on 'practical applications'."

Davisson's education at the University of Chicago was abruptly interrupted at the end of his first year for lack of funds. Unable to earn enough money to return to school that fall, he worked at the telephone company in Bloomington until Millikan secured for him a temporary position at Purdue University in the spring of 1904. He returned to Chicago that summer as an assistant to Millikan, and was able to remain for five quarters. Again funds ran out, and this time Millikan obtained a part-time post for him at Princeton, from where he returned each summer to Chicago, until finally obtaining his bachelor's degree there in 1908. One of the most brilliant undergraduates in the physics department, he took six years to get his degree.

While at Princeton Davisson came under the influence of another electron physicist, Owen W. Richardson, who had come over from England as research professor in 1906. By 1908, when Davisson completed his studies at Chicago, an old basement laboratory at Princeton had been replaced by an ultra modern and specialized laboratory (the Palmer Physical Laboratory), and it was here that Davisson began his research apprenticeship in electron physics.

During his research fellowship year (1910–1911) he completed a doctoral dissertation on positive thermions, a field of special interest to Richardson. Davisson later attributed his success as a physicist to "having caught the physicist's point of view – his habit of mind – his way of looking at things" from men such as Millikan and Richardson. A further link with Richardson which was later to prove of critical importance had occurred during his Princeton years. It was then that he first met Richardson's sister, Charlotte Sara, who had come over from England to visit her brother, and in August of 1911 they were married.

For the next few years (1912–1917) he worked as a physics instructor at the Carnegie Institute of Technology in Pittsburgh, Pennsylvania. The eighteen class hours per week plus individual tutoring assignments were not particularly appealing to such a shy, withdrawn person. One of his students, A. D. Moore, remembered him as a lecturer who "had all the weak points. He was a wispy little nervous man with a prominent Adam's apple, and a stammer – but we respected him." With his heavy teaching load, he managed to publish only three short theoretical notes during his six years at Carnegie. When World War I came along Davisson was refused enlistment in the U.S. Army because of his frailty, but he found a Summer position in war-related research in the Engineering Department of the Western Electric Company in New York City. He remained there throughout the war, and turned down a university promotion at the war's end in order to stay with the company. His wartime work had been concerned primarily with the testing and production of vacuum tubes: Davisson was given more flexible research opportunities in 1919, at which time he was joined by a young man just discharged from active service.

2. ENTER: GERMER

The new partner was Lester Halbert Germer, born in Chicago on 10 October 1896 when fifteen-year-old Davisson was about to enter high school in nearby Bloomington. Germer's parents were of German descent, from a professional background; his father was a local physician. In 1898 the Germers moved to Canastota, a small busy town on the Erie Canal in upstate New York, where Mrs. Germer had lived as a child. Germer attended public schools there, "shining both as a student and as an athlete, one of the finest young men Canastota ever produced." Excelling in algebra, geometry, physics, and taking four years of German, Germer won a four-year scholarship to Cornell University where he and two fellow students, finding themselves "unsatisfied with the course in electricity and magnetism given . . . bought a more advanced text and met regularly in the vacant classroom . . . and really learned something."

He graduated from Cornell in 1917, six weeks early due to the entry of the United States into World War I, and obtained a research position at Western Electric. Two months later he volunteered for the Army and trained as a pilot. He was sent to France early in 1918, and was officially credited with having brought down four German warplanes. Following his discharge on 5 February

1919, he was treated for severe headache, nervousness, restlessness, and loss of sleep, symptoms attributed to the strain of the campaigns, but he refused to file for compensation feeling that "others were worse off." Upon returning to Western Electric he was duly assigned to work with Davisson.

That same fall he married his Cornell sweetheart, Ruth Woodard.

3. COMPANY RESEARCH

Davisson's research in 1919 was related to one of the primary concerns of the Western Electric company's parent organization, the American Telephone and Telegraph Company (AT&T), then actively engaged in extending its long-distance telephone lines throughout the North American continent, and aware of the need for a reliable, distortionless repeater, or amplifier. In fact, the company was involved in the famous Arnold–Langmuir patent suit concerning the development and employment of high-vacuum triode electron tubes. Although, contrary to Germer's later recollection, his and Davisson's work at this time was not directly related to this particular legal battle, their work in the emission of secondary electrons from oxide-coated cathodes was a natural outgrowth of questions raised during that controversy. It was certainly as a part of that involvement that Western Electric had procured the pumps which Davisson and Germer found so necessary in their later work. During their initial collaboration Davisson and Germer tried reversing the tube voltages, as a sideline interest, to measure electron emission under electron, rather than positive ion, bombardment of the test cathodes. (See also C. Calbick's article, later in this volume.) They then discovered much to everyone's surprise (J. J. Thomson, who was visiting the laboratory at about this time, was almost ready to "stake his reputation" that this could not be so), that a small fraction (about 1%) of the incident electrons were coming back off the target with the same energy with which they struck it, i.e., that they were being elastically scattered. This discovery was the beginning of the sequence of experiments that led ultimately to the discovery of electron diffraction.

4. DAVISSON AND KUNSMAN: 1920–1923

At about this time (1920), Germer was assigned to a separate project on thermionic emission, and Davisson took on a new assistant, Charles H. Kunsman, a recent Ph.D. from the University of California, to help with the electron scattering experiments. It was Davisson's vision to use the elastically scattered electrons as probes of the extranuclear structure of the atom, much as Rutherford and his colleagues had been using alpha-particles to explore the atomic nucleus. It will be remembered that there was much perplexity about the electronic structure of the atom at this time; Bohr's orbital model had enjoyed marked success in the case of hydrogen, but heavier atoms were still hard to envisage theoretically. Davisson and Kunsman threw themselves into this task with great enthusiasm, hoping to supply the needed experimental data. That

Western Electric was willing to allow so much time, equipment, and money to be spent on basic research which had little hope of being commercially profitable is somewhat surprising; the company's reservations may be illustrated by Germer's tale of how one day H. D. Arnold, Davisson's supervisor, startled Davisson by asking him: "Do you realize that your loaded salary on this project has already cost over $2000 – with no end in sight?" Imagine being concerned about $2000, a figure which included shop-time and equipment costs, in these days of megabuck budgets!

In a series of experiments lasting approximately two and a half years (mid-1921 to late 1923), Davisson and Kunsman had built – with the excellent help of the company machinist and glassblower, George Reitter – seven different specialized vacuum tubes, each designed to test a specific feature of the theory of electron–electron scattering then developing in Davisson's mind. A general schematic of these tubes is shown in Figure 1. In the accompanying series of papers describing the work of this period, Davisson and Kunsman were able to demonstrate a fairly limited and semiquantitative correspondence between their assumed configuration of the electron shells in the atom and the observed

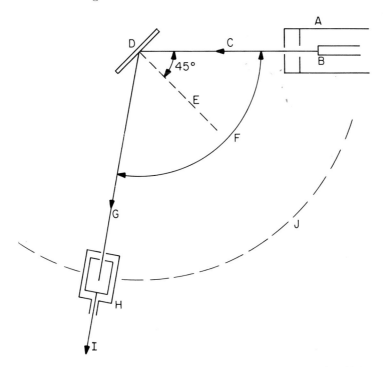

Fig. 1. Scheme of the first scattering tube, identifying the scattering angle which was later called the colatitude. This tube served as the prototype for all subsequent tubes, some of which included mechanisms for rotating the target 360° around the beam axis (azimuth) and for changing the angle of the incident beam with respect to the plane of the target. *Code*: A: Electron Gun. B: Filament. C: Incident Beam. D: Nickel Target. E: Target Normal. F: Scattering Angle (later called colatitude). G: Scattered Electrons. H: Faraday Collector Box. I: To Galvanometer. J: Arc on which Faraday Box Rotates.

direction and intensity of the scattered electron beams. Besides the original target material, nickel, they tried five others in the hope of being able to detect differences in scattering due to different electron shell arrangements. Despite high hopes and dedicated effort, however, their project appeared to founder, and the results of their last three series of experiments were not even published. Kunsman left the project at this time, and Davisson also abandoned it and returned to thermionic emission studies. Perhaps the best that can be said regarding the value of these scattering studies is that they enabled Davisson to perfect the techniques needed to obtain reliable measures of the direction and intensity of the electron beams scattered off metal targets under high vacuum (10^{-8} mm Hg and better). Davisson had no idea that these early Davisson–Kunsman experiments were to be heralded later by others as evidence of electron waves. He furthermore completely rejected the idea when he eventually learned of it.

5. DAVISSON AND GERMER: 1924–1926

About a year later Davisson was ready to have another go at it. Germer was just returning from a fifteen-month illness, and hospitalization for severe mental depression. He had already carried out several thermionic studies under Davisson's direction, and had pressed ahead with his graduate studies on a part-time basis at Columbia University. It appears to have been the heavy schedule of work, graduate study, and family responsibility that had put him in hospital, but at this time he was ready again for work. Regarding his graduate education, Germer had this to say: "I learned relatively little at Columbia . . . but was nevertheless fortunate in working . . . with Dr C. J. Davisson. I learned a simply enormous amount from him. This included how to do experiments, how to think about them, how to write them up, how even to learn what other people had previously done in the field . . . I do really owe to Dr Davisson much the best part of my education, and I am not really convinced that it is so inferior to that obtained in more conventional ways. It is certainly different." Davisson, recognizing Germer's development as a research physicist by this time, asked him to re-activate the abandoned electron scattering project.

The work was just getting underway in early 1925 when an apparent tragedy occurred. As Germer was attempting to heat up the tube and outgas it prior to a new set of experiments, the charcoal trap cracked, exposing the heated nickel target to air and oxidizing it badly. Instead of scrapping the tube and starting from scratch, Davisson decided to try to repair it by heating the cathode in a vacuum and in hydrogen, thereby reducing the oxide, a method which had been tried without success after an earlier break. This time the attempt worked, however, and two months later the experiments were resumed.

The results obtained with the repaired tube during early May 1925 were quite similar to those that had been found by Davisson and Kunsman during the initial scattering experiments back in 1921–23. These results changed rather abruptly in mid-May, however, when very sharp peaks began to appear in the

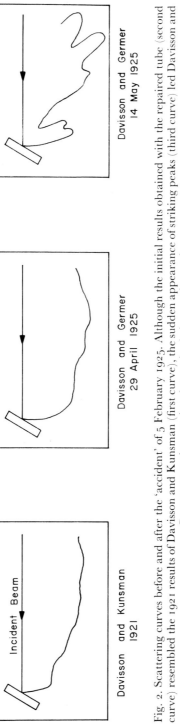

Fig. 2. Scattering curves before and after the 'accident' of 5 February 1925. Although the initial results obtained with the repaired tube (second curve) resembled the 1921 results of Davisson and Kunsman (first curve), the sudden appearance of striking peaks (third curve) led Davisson and Germer to make a major change in their experimental program.

scattered beams, A typical curve, with comparison curves for 1921 and early
May 1925, is shown in Figure 2. These results so confounded Davisson and
Germer that they immediately cut the tube open to see what had caused the
change. With the help of the company's microscopist, F. F. Lucas, they found
that the nickel target had been transformed during the repair process; what had
been a highly-polished polycrystalline surface was now seen to be an array of
about ten distinct crystal facets. Puzzling over this discovery, they concluded
that the changed pattern of scattered electron beams had been caused by the

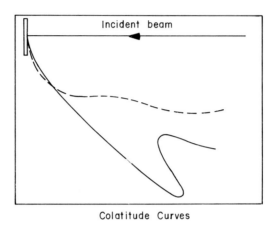

Colatitude Curves

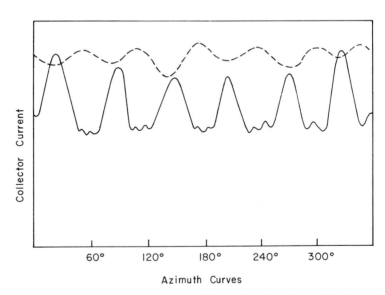

Azimuth Curves

Fig. 3. Colatitude and azimuth curves. The dotted lines show how the colatitude and azimuth
curves looked before the first 'quantum bump' was found (i.e., how they looked when Davisson
took them to England in 1926). The solid lines show the situation after the first 'quantum bump'
was observed on 6 January 1927.

rearrangement of the atoms into a crystalline array, or in other words, that it was not the arrangement of the electrons within the atom that determined the scattering pattern, but rather it was the arrangement of the atoms within the crystalline structure that was significant.

This conclusion led Davisson and Germer to modify their experimental program. Realizing that a random arrangement of small crystal facets was too complicated a target for systematic study, they decided to prepare a large single crystal of nickel from which they hoped to cut a new target of known orientation. Neither of them knew much about such matters, however, so it was almost a year before they were able to prepare the necessary target and a new experimental tube. It is interesting to note that during this period they made many X-ray diffraction photographs of the various nickel crystals with which they were working in order to become familiar with their structure, yet they made no connection between the spatially-arranged electron intensity peaks which they had found in their experiments and the X-ray intensity peaks which were a part of their X-ray diffraction studies. (Of course, hindsight is always easier and wiser than foresight.)

Finally, in April 1926, the scattering experiments were again resumed, this time with a nickel single crystal mounted as a target, and a means of rotating the beam collector in azimuth (i.e., the 360° angle encircling the beam axis) as well as in colatitude as in the earlier tubes. This modification in equipment reflected the new theoretical approach; it was expected that the electron beams would emerge along certain 'transparent directions' characteristic of the crystal, and a three-fold azimuthal symmetry was expected. Davisson and Germer had switched from trying to probe atomic structure to trying to probe crystalline structure.

The results of these experiments, so long in preparation, were unfortunately no more satisfying than the earlier ones. Once again the colatitude curves showed little detail, looking very much like the earliest Davisson–Kunsman results of 1921–23. They certainly had much less detail than the post-accident curves of 1925, and the azimuth curves were only slightly more encouraging. A vague three-fold symmetry could be seen in them with a little imagination. Representative curves from this period are shown in Figure 3 (dotted lines).

6. THE EUROPEAN CONNECTION: AUGUST 1926

At this time Davisson and his wife decided to 'get away' for a few weeks in the summer of 1926* in order to visit Mrs. Davisson's relatives in England. This break must have seemed particularly welcome to Davisson for it was a chance to escape from this last set of disappointing experiments. Davisson, nevertheless,

* It may seem trivial to note that the reason the Davissons were making the journey at this time was because Mrs. Davisson's sister May and her husband were available as baby-sitters. In view of the central role which this visit was to assume in the discoveries of 1927 it is worthwhile, however, recalling the casual manner in which the visit was planned. In a letter to his wife at the time Davisson makes this clear: "It seems impossible that we will be in Oxford a month from now . . . We should have a lovely time . . . a second honeymoon . . . even sweeter than the first!" This relaxed feeling was no longer evident after August 10!

carried a few of his most recent experimental results along with him, undoubtedly in the hope that his brother-in-law Richardson might be able to shed some light on them.

The Davissons sailed for England on 17 July 1926 and, after motoring around Devonshire and Cornwall for several days, drove to Oxford in early August, where Davisson and Richardson attended the annual meeting of the British Association for the Advancement of Science (BAAS). Imagine Davisson's surprise when, upon attending the meeting of 10 August, he heard Max Born state that "the experiments of Davisson and Kunsman . . . on the reflection of electrons from metallic surfaces" are an "indication" of the diffraction of electrons as predicted by the wave theory of Louis de Broglie and subsequently developed more fully by Erwin Schrödinger. Born went on to state that certain experiments by E. G. Dymond on the collisions of electrons in helium were "a complete verification of this radical hypothesis."

After the meeting Davisson and Richardson engaged Born and several other 'continental physicists' in conversation regarding the new wave ideas and the relationship of Davisson's electron scattering experiments to them. Davisson showed them his most recent experimental results, and what seemed to him to have been rather disappointing in terms of his current theory of 'transparent directions' in the crystal lattice was interpreted by others as being of great interest in terms of the new wave theory of the electron. As a result of this conversation, Davisson spent "the whole of the westward transatlantic voyage . . . trying to understand Schrödinger's papers, as he then had an inkling . . . that the explanation might reside in them."

7. NEW EXPERIMENTS

When Davisson returned to Bell Labs (as the Western Electric laboratories had been renamed in 1925) he immediately enlisted Germer's assistance in making sense of the German language edition of Schrödinger's papers which he had borrowed from Richardson. Armed with this knowledge they re-examined their recent experiments to see if any of the scattering peaks could be interpreted in the light of the new theory. Germer's notebook entry of 9 October 1926, under the heading "Attempt to Explain Scattering by the Assumption of Elsasser . . . that Each Electron is an Electromagnetic Wave of Wave Length $\lambda = h/mv$", includes calculations of what diffraction peaks were to be expected according to the given orientation of the scattering planes in the nickel crystal; it ends with the statement "[Colatitude] $= 41°$ as compared with $31\frac{1}{2}°$ found in [the curve of 26 July 1926]." This discrepancy was too much to be acceptable; Davisson's account of this event says that he "calculated where some of the beams ought to be, looked for them and did not find them." Note Germer's difficulty in grasping the nature of the electron waves; he referred to them as 'electromagnetic waves', and attributed their characteristic wavelength to Walter Elsasser rather than to de Broglie or Schrödinger. Note further that, whereas Born and Elsasser had been willing to accept the Davisson–Kunsman results as evidence for electron waves, Davisson and Germer rejected not only that but also their own later work. At this point Davisson and Germer 'laid out a program of thorough search' to see if they might yet find evidence for electron waves.

It was December 1926 before the 'thorough search' was finally underway. By that time the old experimental tube had been cut open, examined, and found to have been sufficiently out of alignment to explain the 10° discrepancy in colatitude. Davisson had also perservered with Schrödinger's papers and, as he wrote to Richardson, finally thought he knew "the sort of experiment we should make with our scattering apparatus to test the theory." These experiments, carried out with a new tube, had a new feature; they examined the effect of varying the electron speed (and hence wavelength) in addition to the traditional variations in colatitude and azimuth. The new tube is shown in Figure 4.

Fig. 4. The experimental tube of 1926–27. The main glass envelope was about 12 inches long, and the principal electrical and mechanical parts are diagrammed below it. Note the charcoal trap appended at the right end; it was used for pumping out residual gases in order to obtain the high vacuum needed for the experiment. It was a trap such as this that cracked, causing the famous 'accident' of 5 February 1925. Photo courtesy of Bell Labs.

8. A THIRD MAN: CALBICK

The sense of excitement that seems to have pervaded the laboratory in late 1926 can be seen in the fact that a third man, Chester H. Calbick, recently graduated from the University of Washington as an electrical engineer, was added to this non-commerical project at this time. The three men are shown together in the photograph reproduced at the beginning of this volume. Within a few days Calbick was doing practically all of the actual laboratory work, as the change in handwriting in the notebooks clearly indicates, while Davisson and Germer seem to have become fully occupied with calculations and interpretation.

The new sense of excitement and commitment of personnel was not misplaced. In the early days of 1927, after a few weeks of preliminary runs a dramatic event took place. In the midst of an experiment in which the intensity of the scattered beam was measured as a function of the incident electron energy, a sudden spike was observed which had never been seen before. Since this spike was only a single isolated data point, and it occurred at 65 volts instead of the calculated 78 volts, a second run was made for confirmation. It was unmistakably there. Calbick noted neatly in the margin: "Attempt to show 'quantum bump' at an intermediate (colatitude)-angle.

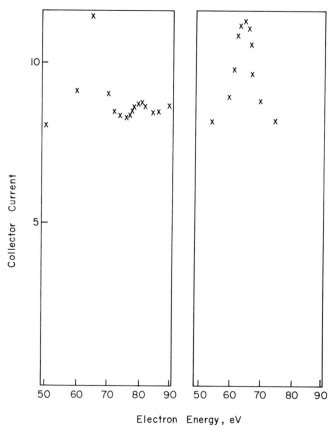

Fig. 5. The first 'quantum bump' of 6 January 1927. These curves represent what Davisson called "the sort of experiments we should make . . . to test the theory" in which a peak corresponding to the de Broglie wavelength $\lambda = h/mv = (150/V)^{1/2}$ was expected (V in volts, λ in Å). Looking for a peak at 78 volts, they took one-volt steps in that region, as seen in the first curve. Luckily, they extended their data below that region in five-volt intervals, and happened across a spike at 65 volts. They immediately repeated the experiment using one-volt intervals in this region, obtaining a clear peak (second curve) and the first 'quantum bump'.

Bump develops at 65 volts compared with calculated value for 'quantum bump' of $V = 78$ volts." But Germer was less reserved; he wrote "First Appearance of Electron Beam" in unmistakable, bold letters across the bottom of the page. The curves for this experiment are plotted in Figure 5.

In the next few days, follow-up curves of scattered beam intensity versus colatitude and azimuth were obtained, and it became immediately apparent that electrons of a specified energy were definitely being scattered in specified directions in space, similar to what one observes when a beam of X-rays is diffracted from the atomic planes of various crystals. Typical curves are shown in Figure 3 (solid lines). Was this indeed positive evidence for electron diffraction? Were electrons really behaving as waves?.

9. A NOTE TO *NATURE*

For the next two months, Davisson, Germer, and Calbick experimented and calculated at a furious pace. By 3 March 1927 the experiments were halted, Calbick took a brief leave to attend to family business back in Washington State, and Davisson and Germer submitted a note to *Nature* outlining their results. To the article Davisson attached a covering letter, addressed to Richardson, as follows: "I hope you will be willing, if you think it at all desirable, to get in touch with the editor with the idea of securing early publication. We know of three other attempts that have been made to do this same job, and naturally we are somewhat fearful that someone may cut in ahead of us." Who were the others? We can't say for sure, but the evidence I have seen points to at least five people who had made some attempt at one time or another: Walter Elsasser, E. G. Dymond, P. M. S. Blackett, James Chadwick, and Charles Ellis. Of these five I think the first three are the most likely, based on who attended the Oxford meeting of the BAAS. As it turned out, of course, none of these others ever completed any experiments in electron diffraction (Dymond's earlier results being eventually withdrawn as spurious). I can find no evidence to suggest that Davisson had any idea that G. P. Thomson was beginning experiments on electron diffraction at this time. Thomson, on the other hand, had certainly heard of Davisson's work, indirectly through E. G. Dymond if in no other way, although he probably was more aware of electron scattering in general than of Davisson's role in particular. Curiously enough, there was yet another person doing experiments on electron scattering, very similar to those of Davisson and Germer, about whom they knew nothing, even though he was virtually next door to them. This was H. E. Farnsworth, at Brown University in nearby Providence, Rhode Island. Unlike Davisson, however, Farnsworth did not happen to take a trip to England in the summer of 1926, attend the BAAS meeting, and learn about the possibility of interpreting his results in terms of electron diffraction; he therefore missed out on sharing in this important discovery. It was Farnsworth, however, who continued with his work on low-energy electron diffraction long after Davisson and Germer had turned to other fields.

Davisson and Germer's brief note in *Nature* was so low-key that, unless one already knew what electron diffraction was all about, the point could have easily been missed. Even its title, 'The Scattering of Electrons by a Single Crystal of Nickel', gave no hint as to its real content, and it seemed to be just one more paper in the series begun by Davisson and Kunsman back in 1921. Near the end of the article however there was a short statement to the effect that the results were "highly suggestive . . . of the ideas underlying the theory of wave mechanics," and there was an accompanying table comparing observed peaks with those expected on the basis of the computed de Broglie wavelengths. One of the problems that may have prevented Davisson and Germer from being more direct in their assertions was the fact that their results were consistently at variance with the predicted values; they suggested an *ad hoc* 'contraction factor' in the crystal spacing to account for this. That Davisson and Germer were not really doubtful of their own results can be seen in the short article published in the *Bell Labs Record* at the same time. Entitled 'Are Electrons Waves?' it argued, within the comparative safety of an in-house publication, that electrons were as wave-like as were X-rays, even though it contained not a single numerical result.

A third article, essentially an abstract of the paper Davisson gave at the Washington, D.C., meeting of the American Physical Society in April 1927, even went so far as to suggest that certain 'anomalous beams' mentioned in their *Nature* article might be due to diffraction of the electron waves from contaminants on the crystal surface. The conceptual shift represented by this comment is striking; it seems that almost before Davisson and Germer had got used to thinking that they might be able to use the 'known' structure of the single crystal target to learn about the 'unknown' nature of electron waves, they were already starting to use the 'known' structure of electron waves to investigate the 'unknown' character of the crystal's surface.

10. THE DEFINITIVE STUDY

Yet Davisson and Germer realized they had much more to do before the idea of electron waves was conclusively demonstrated. Consistent with Davisson's thorough and methodical approach, they spent five months (April to August) carefully repeating and extending all their experimental results. The comprehensive paper which they submitted to the *Physical Review* at the end of this period, published in December 1927, did indeed resolve most of the problems inherent in their first effort, and in so doing established the concept of electron waves on a sure footing. The 'anomalous beams' were found to be time-dependent and hence attributed to occluded gases contaminating the crystal surface; they were found to fit well a 'gas crystal' model in which the gas atoms fit into the crystal surface in a regular pattern. But the *ad hoc* 'contraction factor' could not be definitely settled, though an 'index of refraction' proposed by Carl Eckart and Fritz Zwicky was found to serve as a reasonable solution to this problem. It wasn't until 1928, when Hans Bethe proposed a theoretically sound understanding of this index of refraction, that this problem was convincingly settled. By that time Davisson and Germer had conducted further experiments, including an investigation of the law of reflection, and published several other papers which served solidly to further the idea that electrons behaved as waves. There seemed to be little room for doubt. As Davisson stated it: *"Are electrons waves? The easiest way of answering this question is to ask another. Are X-rays waves? If X-rays are waves, then so also are electrons."* Within the next year or two there were many additional experimental studies supporting the electron wave hypothesis; besides the work of A. Reid and G. P. Thomson there was the work of E. Rupp (some of which became highly suspect and was eventually withdrawn) and L. Szilard, D. C. Rose, S. Szczeniewski, R. T. Cox, C. G. McIlwraith, B. Kurrelmeyer, A.F. Joffé, A. N. Arsenieva, F. Wolf, S. Kikuchi, S. Nishikawa, R. Ironside, and H. E. Farnsworth.

One almost wonders, however, if there was any real need for experimental evidence to establish the idea of electron waves at this time. Reading the statements of some of the leading physicists of the day, such as Neils Bohr, Max Born, Werner Heisenberg, Irving Langmuir, Louis de Broglie, and Erwin Schrödinger, one gets the distinct impression that electron waves were practically a foregone conclusion even without any experimental evidence. Perhaps the most remarkable example of this state of affairs is the address given by Max Planck at the Franklin Institute on 18 May 1927, before he had even seen Davisson and Germer's *Nature* article: "[The motion of the electron in the atom] resembles . . . the vibrations of a standing wave . . . [Thanks] to the ideas introduced into science by L. de Broglie and E. Schrödinger, these principles have already established a solid foundation." Yet in this same paper Planck went on to say how he still had reservations about the corpuscular properties of electromagnetic radiation implied by his own quantum theory! (In other words, Planck was willing to accept electron waves *without experimental evidence* even though he was still hesitant to accept radiation corpuscles four years after the discovery of the Compton effect and long after others had used energy quanta to explain the photoelectric effect, phosphorescence, specific heats, the Franck–Hertz experiment, and a host of other phenomena.) In spite of this readiness to accept the wave nature of the electron, I am certain the physics community welcomed such elegant and convincing experimental evidence for it. Davisson and Germer were recognized for the importance of their work with numerous honors, the most noteworthy being the awarding of the 1937 Nobel Prize for Physics to Davisson (shared by G.P. Thomson).

11. ELECTRON OPTICS AND LEED

In the years immediately following their discoveries of 1927–28 Davisson and Germer found themselves doing as much writing and talking about their past work, as advancing their studies in new directions. Gradually, however, their interests turned to new subjects. Davisson, assisted by Calbick, became interested in electron optics; Germer, on the other hand, shifted his attention first to the study of high-energy electron diffraction (*à la* G. P. Thomson), and

later to the study of electrical sparks. Davisson retired in 1946 and spent his last
years at the University of Virginia. In 1959–60 Germer's interests returned to
low-energy electron diffraction (LEED) when he and several colleagues
perfected the 'post-acceleration technique' which had been originally proposed
by Wilhelm Ehrenberg in 1934 but never developed. Retiring from Bell Labs in
1961, Germer remained extremely active in this 'new–old' field, advancing it
commercially as well as theoretically and experimentally; he held a research
appointment at his alma mater Cornell and traveled all over the world in the
interests of LEED. He died while mountain climbing in the Shawangunk
Mountains of New York on 3 October 1971.

12. PERSONAL NOTES

In closing I should like to add a few personal remarks regarding my search for
the information in this article, which I feel will add something to the
appreciation and understanding of the personal characteristics of both Davisson
and Germer and the circumstances surrounding their historic work.

Germer, whom I visited in New Jersey at his home near Bell Labs, was
extremely helpful in obtaining the research notebooks from the Bell Labs'
archives and in assisting me in finding and interpreting the experimental details.
I was astonished to find that Davisson made no notebook entries whatsoever;
they were all made by Germer, Kunsman, and later Calbick. I enjoyed the
expression on Germer's face when he said to me, "Oh, Davisson never wrote in
the notebooks." Yet Germer also made it clear that Davisson always knew
exactly what went into the notebooks! It was fascinating to watch Germer
reconstruct the experiments in his mind as he sketched them out for me on paper
– how quickly and incisively he made the dusty pages come alive. I came to
appreciate how Davisson must have been impressed by this bright young man
from Cornell. There was just one difficulty I encountered in working with
Germer: he constantly kept shifting his attention to his current research on
LEED and the many projects with which he was then engaged and trips he had
just completed, and he never seemed to understand why I was interested in the
1920s when there were such exciting new things happening in the 1970s. He also
seemed much more genuinely interested in telling of his recent mountain
climbing excursions and mushroom hunting expeditions than in dwelling on
'ancient history'. Yet he always good-naturedly, if somewhat reluctantly, came
back to the task at hand, if I was simply patient enough. After Germer died, Mrs.
Germer was very generous in sharing with me many helpful details of family
history and personal memories.

My first visit with Mrs. Davisson took place at the cottage in Maine which
Davisson himself had built. Beginning in 1914 the Davissons spent every summer
there, Mrs. Davisson and the children going up in June with 'Davy' joining them
in August. This circumstance proved to be especially valuable for me; the daily
letters that Davisson wrote to his family in Maine during June and July provided
many details about his life and work that I would not have had otherwise,

especially since all his company papers (two packing cases) were either lost or destroyed when Bell Labs abandoned its old headquarters at 463 West Street in New York City. Many of the insights and quoted material in this article came from these delightful letters, and I am grateful to Mrs. Davisson for sharing them with me.

Davisson and Germer: two very different individuals who, through a fascinating combination of circumstances, love of physics, and patient hard work, made one of the truly fundamental discoveries of modern physics, still live on in the lives of scores of people whom they knew and influenced personally and on the pages of virtually every physics text published since 1930. It has been my extreme pleasure to have come to know something of them as men and of their accomplishments as physicists.

Associate Professor of Physical Science
Rhode Island College, U.S.A.

SOURCES

References to correspondence, personal remarks, archival material, and a complete bibliography of all published material used in this study are documented in Richard K. Gehrenbeck, *C.J. Davisson, L.H. Germer, and the Discovery of Electron Diffraction* (New York: Arno Press, forthcoming). Of the many articles documenting the early work of Davisson and Germer and Davisson and Kunsman, the following may be of special interest (in chronological order): C. J. Davisson, 'The Scattering of Electrons by a Positive Nucleus of Limited Field', *Phys. Rev.* **21**, 637 (1923); C. J. Davisson and C. H. Kunsman, 'The Scattering of Low Speed Electrons by Platinum and Magnesium', *Phys. Rev.* **22**, 245 (1923); C. J. Davisson and L. H. Germer, 'The Scattering of Electrons by a Single Crystal of Nickel', *Nature*, **119**, 558 (1927); C. J. Davisson, 'Are Electrons Waves?', *Bell Labs Record* **4**, 257 (1927); C. J. Davisson and L. H. Germer, 'Diffraction of Electrons by a Crystal of Nickel', *Phys. Rev.* **30**, 705 (1927). Of these five articles, the last is the one referred to by all subsequent workers as the definitive article on electron diffraction.

P. B. MOON

I.4. *George Paget Thomson**

1. INTRODUCTION

George Paget Thomson was born in Cambridge on 3 May 1892 and died there on 10 September 1975. His father, Joseph John Thomson, had been Cavendish Professor of Physics at Cambridge University for seven years when he was born, while his mother (Rose Paget) was the daughter of another very distinguished Cambridge Professor, and before marrying J. J. Thomson had worked as one of his students in the Cavendish Laboratory, Cambridge. Physics and mathematics were in George's blood and he greatly enriched the first two of these.

Best known for the discovery of the diffraction of electrons, he was a substantial contributor to the early stages of the study of neutrons and their use in the uranium chain reaction; he also independently initiated work on controlled nuclear fusion. Largely, though not originally, through his service in World War I, aerodynamics was also an important interest for Thomson. In this his mathematical skill was joined with a spirit of practical enquiry, for he flew aircraft as well as theorizing about them.

After professorships in Aberdeen and London, during which the bulk of his scientific and public work was done, he returned to Cambridge as Master of Corpus Christi College, where he had been a young Fellow after graduating from Trinity College.

2. ANCESTRY AND CHILDHOOD

His family background is exceptionally interesting, and the recurrence of 'George' and 'Paget' give a reason, if one were needed, for using initials. Sir Joseph Thomson was known to all physicists as 'J. J.'; the habit of calling his son 'G. P.' came naturally and persisted until he himself was knighted, and indeed often afterwards. J. J.'s forbears were well-respected Lancashire people, his father being a bookseller and publisher in Manchester, but he was the first of the family to win academic or public distinction. He went to Owens College, Manchester, at 14, even in those days a very early age, and was intending to become an engineer. Two years later his father died and there was not enough money to provide the premium for an apprenticeship, so he turned to physics.

* This article is an abridged version of the Memoir on G. P. Thomson by the same author [Moon (1977). *Biographical Memoirs of Fellows of the Royal Society*, **23**, 529–556]. Grateful acknowledgement is made to the Royal Society for permission to base this article on the Memoir, and to Dr M. J. Whelan for undertaking much of the abridgment and associated work.

Rose Paget's family had been more prosperous and distinguished; her grandfather Samuel Paget was a Yarmouth shipowner and brewer. Though in the end Samuel's business failed, his sons James and George set themselves the task of paying the debts while building their own careers with great success: Sir James Paget was a famous surgeon while Sir George Paget became Regius Professor of Physics at Cambridge, where James Clerk Maxwell was one of his patients. Sir George, who was G.P.'s grandfather, apparently passed the family talents, and certainly passed the tradition of academic interests and connections, to his daughters. Thus G. P. had on his mother's side a strong and abundant medical tradition; but his father's footsteps were, from the beginning, those he wished to follow, and J. J.'s care of his education put his feet early and firmly on the ladder of mathematics and physics.

He was born in Scroope Terrace, then on the fringe of the town of Cambridge and still bordered by the University Botanical Garden, a location that may or may not have been connected with J. J.'s keen interest in botany. Virtually an only child (his sister was born when he was 11 years old), he must have had more attention than would have been possible in a larger family, but this certainly did not give him the 'only child' mentality. The family situation, with his mother's teaching in his earliest years and with many lively and talented visitors to the Thomson home as he grew up, brought him into contact with a wide range of ideas as well as opportunity and encouragement to develop his own interests. His first recorded scientific study was of the twisting motion of a swing that hung in his nursery; he understood that when the seat was turned it rose slightly and would tend to drop again by untwisting the suspension, and he wrote a little 'paper' about this.

G. P. was left-handed and wrote with the words running downwards on the paper, which he placed almost at right angles to its ordinary position. Few left-handers do this, but it is clearly a very logical transformation of the right-handed method, though it cannot be applied on the blackboard. He was always grateful to his parents for not attempting to make him conform to right-handedness. His colour-vision, too, appears to have been non-standard, though his general vision was acute and he was interested in effects of optical contrast, an interest that became significant in the interpretation of his first electron diffraction results.

He was physically active but team games were unattractive to him, except for Rugby football which he much enjoyed when he went on to the Perse School in Cambridge. He gained much from the breadth of teaching at Perse; the headmaster even recommended that he should go on to read classics but this was firmly rejected by J. J. and G. P., who at that time had thoughts of naval architecture. An early interest in soldiers had brought him by way of the Vikings to an intense preoccupation with warships. So, when given a set of tools by his father, he immediately picked up a thread of enjoyment that ran in diverse forms to the end of his life; he launched into model ship-building on a grand scale, first in the numbers and later in the details of the vessels. These models developed his manual skill, gave scope for his ingenuity, and as he grew up through boyhood, became an increasing scientific and aesthetic delight.

Real ships as well as model ones early engaged G. P.'s attention; he became seriously interested in their past and their future, particularly in war. At the age of 14 he won the annual prize of the Navy League with an essay on 'The Effect on Naval Warfare of the Substitution of Steam for Sail'. Ships and guns go naturally together and G. P., who was later to make a vital contribution to the commencement of the British nuclear programme in World War II, went further than most boys with explosive weapons. He had been delighted by a gift of brass cannon from his father's friend, J. H. Poynting, Professor of Physics at Birmingham, and he cast some for himself with the help of W. G. Pye, the head of the Cavendish Laboratory workshop. He made live cartridges, too, and found many years later that one of them was in good explosive order.

3. CAMBRIDGE UNIVERSITY

In October 1910 Thomson entered Trinity College, Cambridge, as a Scholar, as his father had done, and as his two sons were later to do. His self-confidence was shown at the first meeting with his Director of Studies, who advised him to go to the lectures of G. H. Hardy. He demurred and was told drily that if he did not go to hear the greatest mathematician of the day he would later regret it – as indeed he did.

G. P. gained first-class honours and a 'B star' in mathematics after two years, and a first in physics in his third year. He acknowledged the advantages he had at this stage, not only from his father's teaching but also by the head-start of attending Dr Alex Wood's first-year University lectures while still at school. He also noted that C. T. R. Wilson's lectures were bad, but that in practical teaching Wilson had the good method of going through the plan of an experiment with the student and then leaving him entirely to his own resources.

Wilson's influence on physics was greater than is often appreciated; he is remembered as the inventor and the patient developer of the cloud-chamber, but his deep understanding of principles was communicated to younger men, particularly in connection with the properties of waves. The classic innovations of W. L. Bragg and E. V. Appleton were in the tradition of this understanding, which Rayleigh's brief tenure of the Cavendish professorship had first implanted. G. P.'s alertness to the possibility of electron waves was, he had recorded, partly due to an idea of Bragg's that had remained in his mind.

Thomson had become a Major Scholar of Trinity in 1911 and, with this status continuing after graduation, he started research in the Cavendish under his father's supervision, working on radicals such as CH and CH_2 produced as positive rays in discharge tubes. He also did theoretical work on the stability of aeroplanes in circular flight which, together with work done at Farnborough later, won him the Smith Prize in 1916. In 1914 he was elected Fellow and Mathematical Lecturer of Corpus Christi College, Cambridge – positions that he held until 1922 though he was absent on war service from 1914 to 1919.

He was commissioned as 2nd Lieutenant, The Queen's Regiment, in September 1914 and served in France until 1915, when he was attached to the

Royal Flying Corps for duty at the Royal Aircraft Factory (now the Royal Aircraft Establishment) at Farnborough. He worked there on problems of stability and performance of aircraft, learned to fly at the Central Flying School, and had a spell of attachment to the British War Mission in the U.S.A. which included working under R. A. Millikan. After returning to Farnborough, he was employed on navigational problems, and at the end of the war he worked briefly for the Aircraft Manufacturing Company, during which time he wrote his first textbook, *Applied Aerodynamics* (Thomson, 1919).

G. P. returned to Cambridge in 1919, continuing his work with positive rays and with anode rays; a distinction was made between these in the sense that 'positive rays' are generated in the body of the electric discharge but 'anode rays' come from the surface of the anode. At sufficiently low pressure these will have gained kinetic energy corresponding to the total voltage applied across the tube. Passing, along with the positive rays, through a hole in the cathode, they form a beam for experimental study. With them he discovered, simultaneously with F. W. Aston, who used a different method, that lithium comprises two isotopes of masses 6 and 7.

4. ABERDEEN UNIVERSITY AND ELECTRON DIFFRACTION

In 1922, Thomson was appointed Professor of Natural Philosophy in the University of Aberdeen, where he met and married Kathleen, daughter of Sir George Adam Smith, then Principal of the University. The marriage brought them great happiness until Kathleen Thomson died in 1941, leaving four children: John, David, Clare and two-year-old Rose, whom he cared for devotedly. John married the daughter of an American college president; David continued the Paget tradition of academic unions even more closely by marrying Patience Bragg, daughter of Sir Lawrence Bragg, his father's close friend and his grandfather's successor (Rutherford intervening) in the Cavendish Chair of Experimental Physics.

Aberdeen was the scene of G. P.'s greatest contribution to pure physics. He was continuing some experimental work on positive rays, but realizing that his apparatus could be quickly adapted to search for diffraction patterns with electrons, he clinched simply and in a few months what Clinton J. Davisson and Lester H. Germer of the Bell Telephone Laboratories in New York had just established after several years of superbly executed experimentation and difficult, though in the end convincing, interpretation. The following account is based on Thomson's own description given over the perspective of forty years in a lecture to the Institute of Physics and the Physical Society in 1967 at a conference at the Imperial College of Science and Technology celebrating the fortieth anniversary of the discovery of electron diffraction. The text, published in *Contemporary Physics*, is a fascinating contribution to the history of physics as well as an illuminating example of G. P.'s ways of thought, work and writing.

He describes Davisson's early experiments with C. H. Kunsman on the scattering of low-energy electrons by a target of polycrystalline nickel, in which they found more electrons to be scattered through about 70° than at other angles. Using classical theory of scattering by electron distributions in single atoms, they could explain their results by supposing the 28 electrons of nickel to be arranged in two shells. In 1923, Davisson and Kunsman obtained peaks from a platinum target.

Next there follows Davisson's own summary of the famous accident of 1925 when the target was oxidized by an inrush of air and was given prolonged heating to restore its surface. The angular distribution of scattered electrons was now different because the originally polycrystalline surface contained only a few large single crystals. Also in 1925, Elsasser had suggested that evidence for the wave nature of electrons could be found by studying their interactions with crystals and had instanced Davisson and Kunsman's work with platinum as a possible example.

Davisson and Thomson were both present at the meeting of the British Association for the Advancement of Science at Oxford in September 1926. Davisson certainly discussed his findings with Born and Hartree, and probably also with Blackett and Franck. After these discussions, Davisson spent all his time on the voyage back to New York studying Schrödinger's papers on wave mechanics. He calculated the angles of the principal beams on the supposition that they were diffraction peaks from the two-dimensional array of atoms in the crystal surface. He and Lester H. Germer looked for these beams but did not at first find them. In January 1927, after a more thorough search, they got strong beams corresponding in their relative angles to this assumption. The actual angles were different, which could be ascribed to a change of layer-separation near the surface or to a refractive index of the metal for electron waves.

With the above background, G. P.'s own account is now given to the stage where his experiments with A. Reid had led to conclusive results (Thomson, 1968).*

I also attended the Oxford meeting but did not, I think, meet Davisson; certainly I did not discuss his experiments with him. There was, however, a good deal of talk about de Broglie. If Elsasser's name was mentioned I took no note of it and did not read his paper till years later, but it may have influenced people with whom I talked. On the way back I saw Dymond in Cambridge who was working on the scattering of electrons by helium. He had got some curves which might have been some sort of diffraction and we discussed de Broglie. He sent a note about this time to *Nature* [1926] on results he had got a little before in Princeton; what he showed me was probably the same. Afterwards it occurred to me that if there was anything in this possibility – and I thought there was – it would be easier to do with solids and remembered the Young's 'eriometer' experiments I had done as an undergraduate in C. T. R. Wilson's laboratory. This worked with textile fibres and ordinary light, so organic molecules and waves of 10^{-9} cm seemed reasonable, which would allow one to use ordinary cathode rays, much easier to handle than Dymond's 100-volt electrons. Indeed Dymond later had unfortunately to withdraw these early results as erroneous. In fact, helium shows no striking effects as a scatterer of electrons in the range of energy used. He, like Elsasser, had been influenced by Davisson and Kunsman's paper.

It happened that one of the students working at Aberdeen on the problem of the scattering of positive rays had an apparatus which could be used for the purpose, with hardly more alteration than changing the polarity of the discharge, which came from an induction coil. So I asked Alexander Reid to try the experiment, using thin films of celluloid.

* The author is grateful to Taylor and Francis Ltd. for permission to quote from this paper.

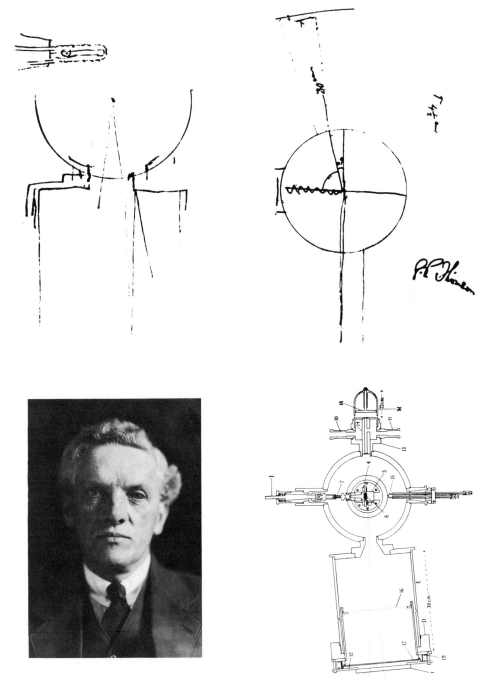

Fig. 1. Thomson's original sketches for a diffraction apparatus (above) as presented to Fraser, and a drawing of the actual construction (below: here reproduced from *Proc. Roy. Soc.* A **128**, 643; [1930]). Thomson's chief laboratory mechanic, Mr. C. G. Fraser, the 'mechanic of the old school' who brought about this transformation, is shown in inset.

Almost at once Reid got photographs apparently showing diffuse halos. I was a little sceptical, as photographs of scattered positive rays which we had been working on had apparently shown sharp edges which on examination merely turned out to be contrast effects. However, the photometer confirmed the reality of the electronic halos. They were what one might expect from molecules of definite size oriented at random, and the diameter varied with the energy of the electrons in the way to be expected from the de Broglie formula. The size was also about what one would expect for a long-chain carbon compound, but that of course was only rough. We published a note in *Nature* under our joint names in June 1927, two or three months after Davisson and Germer's first publication.* The structure of celluloid was then unknown, and for a proper test it was necessary to use crystal diffraction by a known lattice. Now metals suggested themselves; their structure was well known and simple, but this was before the techniques of sputtering and evaporation for making very thin films had been developed, or at least before I knew how to do it. The films which I used, this time working by myself, were of aluminum, gold and platinum, and were prepared by the chief mechanic of the laboratory, the late Mr C. G. Fraser to whom the success of the experiment is entirely owing, because this is the difficult part. He was a mechanic of the old school; he had been trained as a watchmaker, and he had that astonishing gift (which later came out) of being able to produce a complicated and perfectly working piece of apparatus after one had just talked to him and made some scribbles on dirty pieces of paper on his bench. Those of you who have had the immense good fortune to work with such a man will know that, whatever may be true of wives, the price of such mechanics is above rubies.

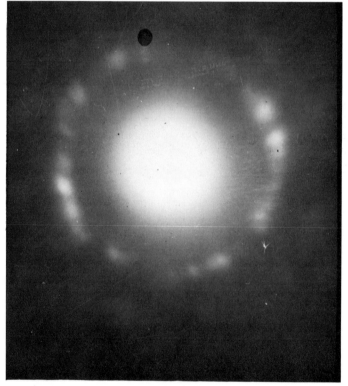

Fig. 2. Early electron diffraction pattern of aluminium showing spotty ring due to reflexions from discrete grains.

* Reid was killed in a motor cycle accident shortly after the experiment.

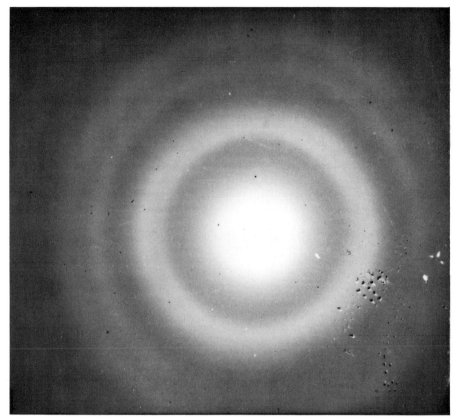

Fig. 3. Early electron diffraction pattern of gold leaf showing poorly resolved 111 and 200 rings (inner rings) together with 220 and 311 (outer rings).

The first films we tested were of aluminium, then of gold, and both gave rings which agreed quantitatively and qualitatively with the theoretical expectation for a face-centred cubic lattice of side about 4.06Å. Later we used platinum.

These three metals all agreed in detail with the predictions of de Broglie. That is to say, the rings were all of the right sizes, both in relative size as agreeing with the structure of the metal and varying with the electron energy as they ought and in absolute size as calculated from de Broglie's formula. I published a note in *Nature* in December 1927 on the platinum ones, and the Royal Society paper on the experiments with gold and aluminium appeared in February 1928 – it was two or three months later than Davisson and Germer's paper which appeared in the *Physical Review* late in 1927. The objection was raised to these experiments that they might in some way (it was not indeed explained how) be connected with the production of Bremsstrahlen, that is X-rays, by the collision of the electrons with the films. I was able to disprove this in a second paper by deflecting the electrons after they had passed through the films and before they had reached the photographic plate. I simply put on a magnetic field and showed (as indeed I had observed qualitatively before) that the electrons which formed the rings (the diffraction electrons as I called them) were equally bent with the quite numerous electrons that had gone through holes in the film which, although carefully made, was so heavily pitted with holes that nine-tenths of the electrons went straight through. That disposed of the view that they could be due to X-rays, which would not, of course, have deflected.

Though Thomson goes on to describe later experiments by himself and by others, this account can best be closed with two epitomizing paragraphs:

I think that the real genius of de Broglie comes from his acceptance of the waves as a concomitant of all particles. It would have been so easy to keep the idea just to the bodies in which he was originally interested, namely quanta of radiation. Then he would only have had another of the devices for getting over the radiation paradox and one which required the somewhat unsatisfactory postulate of a non-zero rest-mass for these entities.

   If one compares the two sets of experiments it is clear that those of Davisson and Germer were great triumphs of experimentation, among the greatest ever made. Those at Aberdeen were singularly simple and easy, the only serious difficulty being to prepare good specimens. It is not hard to see the reason for this difference. Davisson had discovered an effect which he thought might be important in a particular way. He had to study this effect more or less as he had found it. If he did a wholly different experiment it would neither prove nor disprove this explanation of what he had found. I was trying to confirm the truth (or otherwise) of an idea based on a very general theory and was entitled to try the easiest experiment which could reasonably be expected to do so.

When the British Association for the Advancement of Science met in Glasgow in 1928, de Broglie and Davisson were there and Thomson had the pleasure of their company, not only at the meeting where they all read papers, but afterwards in his home for a few days. Davisson and Thomson shared the 1937 Nobel Prize for Physics; de Broglie had become a Nobel Laureate eight years earlier.

   The fiftieth anniversary of the discovery was celebrated by an international conference at the Imperial College of Science and Technology in September 1977.

5. IMPERIAL COLLEGE

In 1930 Thomson was appointed to succeed H. L. Callendar as Professor of Physics at Imperial College, London. In contrast to his department at Aberdeen, the deparment of which he now became head was a large one with several sub-departments. In his early years at the College he took an interest in the subject of electron microscopy, then in its infancy, and encouraged the researches in L. H. Martin's group, which as a result acquired in 1936 the first commercially developed electron microscope in the world.

   Electron diffraction was also a burgeoning field, the applications of which were engaging the attention of numerous researchers both in Thomson's department and later in the nearby Department of Chemical Technology of Imperial College under the leadership of Professor G. I. Finch. These developments are reviewed by Blackman and Wilman in Part II of this volume. A further major area of interest of Thomson while at Imperial College, both in pre- and post-war days was the subject of nuclear physics, particularly during the Second World War in the early stages of the nuclear fission weapons project, and later in post-war days in researches into the controlled release of nuclear fusion energy.

6. LATER YEARS IN CAMBRIDGE

In 1952 Thomson returned to Cambridge as Master of Corpus Christi College, where he had been a young Fellow. There were many developments during the ten years of his Mastership, but the one by which he will be most remembered is the postgraduate establishment associated with the College at Leckhampton House. Taking root in a Victorian house away from the main College building, the establishment grew from a few Fellows and students to about a dozen Fellows and a hundred students residing in a new building which bears Thomson's name.

In 1962 he relinquished the Mastership and had eleven years of retirement in Cambridge. He long remained physically and mentally active, keeping in close touch both with his Colleges and with many friends, and strong family ties contributed much to his contentment.

7. HONOURS AWARDED

Thomson was elected to the Fellowship of the Royal Society in 1930, awarded the Nobel Prize in 1937, knighted in 1943, and had many other recognitions of his work for science and the public good. He received the Royal Society's Hughes Medal in 1939 and the Royal Medal in 1949, and he was a medallist of the Franklin Institute, Philadelphia. In 1960 he was awarded the Faraday Medal by the Institution of Electrical Engineers; the citation reads:

... in recognition of his outstanding contribution to the advancement of electrical science, of the leadership in research which he has given to others in this field, and of his inspiring services to education in science and technology.

He was a Foreign Member of the American Academy of Arts and Sciences and of the Lisbon Academy, and a Corresponding Member of the Austrian Academy of Sciences.

He was an honorary graduate of the Universities of Aberdeen, Dublin (Trinity College), Lisbon, Sheffield and Wales, an Honorary Fellow of both Trinity and Corpus Christi Colleges, Cambridge, of Imperial College, London, and of the Institute of Physics, having been President of the Institute from 1958 to 1960. He was President of the British Association for the Advancement of Science in 1960.

Emeritus Professor of Physics
University of Birmingham
England

REFERENCES

Literature cited directly in this article:

Thomson, G. P.: 1919, *Applied Aerodynamics*. Hodder and Stoughton, London.
Thomson, G. P. and Reid, A.: 1927, *Nature* **119**, 890.
Thomson, G. P.: 1927, *Nature* **120**, 802.
Thomson, G. P.: 1928, *Proc. Roy. Soc.* **A117**, 600–609.
Thomson, G. P.: 1968, *Contemp. Physics* **9**, 1–15.

BIBLIOGRAPHY

(a) *Papers in major scientific journals*

1920   A note on the nature of the carriers of the anode rays. *Proc. Camb. Phil. Soc.* **20**, 210–211.
        The spectrum of hydrogen positive rays. *Phil Mag.* **40**, 240–247.

1921   The application of anode rays to the investigation of isotopes. *Phil. Mag.* **42**, 857–867.

1922   The scattering of hydrogen positive rays, and the existence of a powerful field of force in the
        hydrogen molecule. *Proc. R. Soc. Lond.* A **102**, 197–209.

1923   Test of a theory of radiation. *Proc. R. Soc. Lond.* A **104**, 115–120.
        The cathode fall of potential in a high-voltage discharge. *Proc. R. Soc. Edinb.* **44**, 129–139.

1925   A physical interpretation of Bohr's stationary states. *Phil. Mag.* **1**, 163–164.

1926   An optical illusion due to contrast. *Proc. Camb. Phil. Soc.* **23**, 419–421.
        Bemerkungen zu der Arbeit von Herrn Richard Conrad: Über die Streuungs-absorption
        von Wasserstoff-Kanalstrahlen beim Durchgang durch Wasserstoff. *Zeitschr. Phys.* **40**, 652.
        The scattering of positive rays of hydrogen. *Phil. Mag.* **1**, 961–977.
        The scattering of positive rays by gases. *Phil. Mag.* **2**, 1076–1084.

1927   Über die Streuung von Kanalstrahlen durch Wasserstoff. *Zeitschr. Phys.* **46**, 93–105.
        The process of quantization. *Phil. Mag.* **2**, 1294–1305.
        (With A. Reid) Diffraction of cathode rays by a thin film. *Nature, Lond.* **119**, 890.
        Diffraction of cathode rays by thin films of platinum. *Nature, Lond.* **120**, 802.

1928   Waves of an electron. *Nature, Lond.* **122**, 279–282.
        Experiments on the diffraction of cathode rays. *Proc. R. Soc. Lond.* A **117**, 600–609.
        Experiments on the diffraction of cathode rays, II. *Proc. R. Soc. Lond.* A **119**, 651–653.
        The effect of refraction on electron diffraction. *Phil. Mag.* **6**, 939–942.

1929   On the wave associated with beta-rays, and the relation between free electrons and their
        waves. *Phil. Mag.* **7**, 405–417.
        Diffraction of cathode rays, III. *Proc. R. Soc. Lond.* A **125**, 352–370.

1930   (With C. G. Fraser) A camera for electron diffraction. *Proc. R. Soc. Lond.* A **128**, 641–648.
        The analysis of surface layers by electron diffraction. *Proc. R. Soc. Lond.* A **128**, 649–661.
        Electron diffraction by 'forbidden' planes. *Nature, Lond.* **126**, 55–56.
        Optical experiments with electrons. *Trans. Opt. Soc.* **32**, 159–160.

1931   The diffraction of electrons by single crystals. *Proc. R. Soc. Lond.* A **133**, 1–24.

1933   (With N. Stuart and C. A. Murison) The crystalline state of thin sputtered films of
        platinum. *Proc. Phys. Soc.* **45**, 381–387.

1934   Experiments on the polarization of electrons. *Phil. Mag.* **17**, 1058–1071.
        The small-scale structure of surfaces. *Phil. Mag.* **18**, 640–656.

1935   Attempts to produce helium 3 in quantity. *Nature, Lond.* **136**, 334.
        An apparatus for electron diffraction at high voltages. *Trans. Faraday Soc.* **31**, 1049–1051.

1937   (With J. A. Saxton) Attempt to detect radioactivity produced by positrons. *Phil. Mag.* **23**,
        241–246.

1938   With G. E. F. Fertel, P. B. Moon and C. E. Wynn-williams) Velocity distribution of
        thermal neutrons. *Nature, Lond.* **142**, 829.

1939   (With J. L. Michiels and G. Parry) Production of neutrons by the fission of uranium.
        Nature, Lond. **143**, 760.
        Possible delay in the emission of neutrons from uranium. *Nature, Lond.* **144**, 202.
        (With M. Blackman) Theory of the width of electron diffraction rings. *Proc. Phys. Soc.* **51**,
        485.

1948   The growth of crystals (Guthrie Lecture). *Proc. Phys. Soc.* **61**, 403–416.

1949   Nuclear explosions (Bakerian Lecture). *Proc. R. Soc. Lond.* A **196**, 311–324.
        The production of cosmic ray stars. *Phil. Mag.* **40**, 589.

1951   The cascade production of cosmic ray stars and the relative number of charged and
        uncharged particles. *Phil. Mag.* **43**, 978.

(b) *Books*

1919   *Applied aerodynamics.* London: Hodder & Stoughton.
1928   (With J. J. Thomson) *Conduction of electricity through gases*, vols I and II (being the 3rd edition
        of J. J. Thomson's book). Cambridge University Press. (2nd edition 1933.)
1930   *Wave mechanics of the free electron.* New York: Cayuga Press.
        The atom. Oxford University Press. (6th edition in the press.) (Translations: German, 1952;
        Swedish, 1952; Italian, 1956; Polish, 1957; Spanish, 1958.)
1939   (With W. Cochrane) *Theory and practice of electron diffraction.* London: Macmillan.
1947   Contribution to *Atomic challenge.* Winchester Publications.
1955   *The foreseeable future.* Cambridge University Press. (Readers' Union edition, 1957;
        American edition 1959.) (Translations: Swiss, 1956; Spanish, 1956; Dutch, 1957; Swedish,
        1957; Italian, 1957; Russian, 1958; French, 1958.)

SHIZUO MIYAKE

I.5. Seishi Kikuchi
(1902–1974)

1. APPROACH TO PHYSICS

The Kikuchis are a well-known family belonging to a famous lineage from which many distinguished scholars, including physicians, experts in foreign languages, jurists, historians, natural scientists, etc. – mostly of occidental disciplines – appeared over several generations in the revolutionary period of Japan: that is, before and after the Meiji Restoration in 1868. Among the Kikuchi family, Dairoku Kikuchi (1855–1917) was a particularly eminent figure. Having been recognised as a child of outstanding promise, Dairoku was sent to England by the Shōgunate in 1866 when he was a boy of only eleven years old. He returned home in 1868 but was sent again to England in 1870, this time by the Meiji Government, and studied in Cambridge. He graduated from the University in mathematics as a wrangler in 1877, and in the same year he was appointed Professor of Mathematics in the University of Tokyo, which had just opened as the first Western-style university in Japan. Dairoku was the first Japanese professor in mathematics there. Later he became President of the University of Tokyo, and also of Kyoto University. He also served as Minister of Education in the inner cabinet of the government from 1901 to 1903. Dairoku naturally played a leading part in the establishment of a modern educational system as demanded by the new Japan of the Meiji Era.

Seishi Kikuchi was born in Tokyo as the third son of Dairoku on 25 August, 1902. His mother, Tazu, was from the family of a retainer of high class belonging to the Shōgunate. Seishi's family was exceptionally fortunate in having a good environment in economical as well as intellectual aspects by comparison with the average Japanese at the time. No doubt his straight and open character must be largely attributed to the warm and free atmosphere of his home. In primary and middle schools, he was a cheerful boy beloved of all of his classmates though, according to reminiscences of his old friends, he had by then not yet stood out above the others. However, concentration and enthusiasm, two of his characteristic features, were to emerge a little later when he was a student in the First High School, Tokyo. There his interests increasingly turned to physics, and in 1923 he entered the Physics Department of the University of Tokyo.

Fortunately Kikuchi left us a personal reminiscence entitled 'My Early Days with Electron Diffraction' (Kikuchi, 1970), to which the present article is greatly indebted. In it he writes: "Undoubtedly physics was my favourite branch of science, but the choice of it as my line of speciality was to a certain extent due to the influence of my late elder brother, Taiji." Taiji Kikuchi (1893–1921) had graduated in physics from Tokyo University in 1915 and was expected to have a bright future.

In 1920 he went to England and began a study on radioactivity in the Cavendish Laboratory, Cambridge, under the guidance of Rutherford. However, it was a matter of deep regret that he died in 1921 quite unexpectedly of an infectious disease. Taiji's sudden and untimely death was naturally a great shock to the Kikuchi family, especially to Seishi who had respected his brother so deeply. Possibly, the unforgettable impression thereby left on him turned out later to have affected his choice of the field of study.

In the final year – the third year – of the undergraduate course, Seishi was placed under the guidance of Professor S. Kinoshita, well known for his nuclear emulsion studies of α-ray tracks. His research project involved investigation of the β-ray decay of RaD, E and F using a Wilson cloud chamber. The next year he continued in his graduate course with the same topic. He confirmed that the β-rays from RaE form exclusively single tracks, in conformity with the interpretation that the continuous energy spectrum of β-rays emitted at a β-ray disintegration is intrinsic in nature, a view which had not yet been considered as conclusive at the time. The full account of his study was published in 1927 (Kikuchi, 1927). In response to this paper, he received to his great delight a letter from Rutherford expressing his pleasure at having this interesting contribution done by a younger brother of his former pupil, Taiji.

Kikuchi's activities thus seemed to have had a happy start foreshadowing his later career as a nuclear physicist. Actually, however, he was not completely satisfied with his current project. He writes:

Although radioactivity was a frontier topic of physics at the time, I had a feeling that problems which were philosophically more rewarding would continue to be found 'outside the nucleus'. When I was an undergraduate student, a series of stimulating works appeared, including the discovery of the Compton effect in 1923, a suggestive theoretical paper by Bohr, Kramers and Slater in 1924, and the experimental study by Bothe and Geiger in 1925 proving that energy conservation holds in elementary processes. Influenced by these papers I was deeply concerned by the problem of the dual nature of light.

In the meantime, it had been revealed that dualism also held for matter. He knew de Broglie's and Schrödinger's papers, read at the Physics Colloquium held every week in the Physics Department, but initially thought that matter waves would not be so readily accessible to direct experiment. Therefore, he was surprised in 1927 to hear of Davisson and Germer's work, in which the new theory had been so straightforwardly approached by experiments. Having been enlightened by this work, he realized the necessity of more advanced studies of the wave nature of electrons, and became very anxious to conduct an electron diffraction experiment of his own. Fortunately he was supported in his enthusiasm by S. Nishikawa, who was professor in the Physics Department and at the same time head of a laboratory in the Institute of Physical and Chemical

Research. In the early spring of 1928, Kikuchi became a research student of the Nishikawa Laboratory, in order to devote himself exclusively to this new project.

2. INSTITUTE OF PHYSICAL AND CHEMICAL RESEARCH AND NISHIKAWA LABORATORY

In the beginning of the Meiji Era, the Japanese Government invited many teachers of science and technology from the Western countries to Japan, and also sent abroad a large number of young students. These measures of the government's successfully contributed to the rapid modernization of Japan, and the Japanese students efficiently absorbed Western science and technology. In the meantime, original work accomplished by Japanese scientists began to appear, from about 1890 onwards, and some of this proved to be of international standard. In the field of pure physics, we may quote as examples the study of magnetism by H. Nagaoka and K. Honda, the proposal of an atomic model by Nagaoka, and the study of X-ray diffraction and crystal structures by T. Terada and Nishikawa.

However, the general standard of Japanese science around 1915 was still poor, with relatively few research workers and largely obsolete facilities. In the meantime, the 1914–1918 War broke out and Japanese industry immediately suffered from an abrupt interruption in the supply of pure chemical materials, precision machinery, etc. from Europe. As a result the leaders of the country came to realize the pressing need for self-sufficiency in Japanese industry, and to recognise the fundamental importance of promoting basic science.

The Institute for Physical and Chemical Research (IPCR), Tokyo, was a semi-governmental research organization established in 1917 in response to this need, and was designed to cover the fields of physics and chemistry, both pure and applied. After an initial period of organization, the IPCR began its real activities from 1922 as an advanced and modernized research institute, splendidly equipped, well organized, and managed with a degree of flexibility which encouraged and promoted free research. It possessed a complete workshop on a large scale to support the laboratory work. The department of pure physics consisted of laboratories headed by Nagaoka, Terada, Nishikawa and others, and in 1931 Nishina's laboratory joined. The IPCR played a significant and timely role, effectively filling many of the gaps in science and technology, which had occurred due to the speed with which occidental science had been introduced. The IPCR contributed to the development of science not only by its own achievements, but also by acting as a centre for supplying scientific brains. It sent a number of scientists to universities, and many of them later became leading figures – such as Kikuchi. Incidentally, after the Second World War the IPCR ceased to be a research institute, but was reorganized in 1958 and has since recovered its former activity.

S. Nishikawa (1894–1952) is famous as one of the pioneers in the field of X-ray crystallography (Nitta, 1962). In his laboratory in the IPCR, a number of active young workers gathered around him. For example, I. Nitta had been there as a

Fig. 1. Professor Shoji Nishikawa (1884–1952).

research student when Kikuchi entered this laboratory. Nitta came from the field of chemistry and has since become an international figure in X-ray crystallography. Kikuchi writes:

I was exceedingly fortunate to have studied in the Nishikawa Laboratory which was by then a top-ranking X-ray diffraction laboratory. The smooth progress I could make in my work on electron diffraction was largely due to the team in this laboratory. Professor Nishikawa's profound knowledge of and penetrating insight into physics was indeed a surprise to me; and in particular, it was nothing but a wonder to see him produce just the right interpretation immediately after he was introduced to a new electron diffraction pattern full of difficult features.

3. SUCCESS OF THE MICA EXPERIMENT

Kikuchi's challenge began with the construction of a new apparatus for low-energy electron diffraction consisting of an open vacuum system, accompanied by a number of ground joints, like the vacuum X-ray spectrometers which were then popular in physics laboratories. He expected thereby that, because the crystal specimens and their geometrical setting could be easily changed, the new apparatus would adduce the necessary information much more efficiently than the sealed-off apparatus originally used by Davisson and Germer. The parts of the apparatus appear to have been carried into the laboratory from the workshop in February or March, 1928. However, needless to say this initial plan was doomed to fail, inasmuch as the apparatus constructed had been incapable of achieving ultra-high-vacuum.

Kikuchi had to struggle with a lot of seemingly avoidable troubles due to innumerable leaks at ground joints and in the cast metal body. He writes:

> I wasted almost a couple of months in this way, and became increasingly irritated. I eventually realized that experiments using electrons of higher energy of 20 ~ 30 keV and a sample of thin single crystal, such as a thin layer of mica, might be technically easier, and more likely to bring about a fruitful result. I discussed this idea with Professor Nishikawa, although in advance I was not very sure that he would agree to this new plan, in place of the old one which I had started such a short time ago, and was nevertheless planning to abandon. Therefore, I was rather surprised to find that my teacher not only willingly agreed to the new plan but immediately began to discuss details of the next apparatus to be constructed. In all probability, he had come to similar conclusions about the experiment himself. As he had a complete mastery of the techniques for producing cathode rays, a rough plan of the apparatus was drawn up on the spot. Construction of this new apparatus in the workshop took a couple of weeks.

In view of the fact that their first report sent to *Nature* was dated 14 May 1928, their switch to the new plan seems to have occurred in April. Therefore, it is probable that their idea was motivated by G. P. Thomson's success in observing Debye rings by electron diffraction from metallic films. A full account of this experiment had been published in the February issue of the *Proceedings of Royal Society* in 1928 (Thomson, 1928), but publications in Europe took about one month or so to reach Japan in those days. Similar to Thomson's experience with his preparation of thin metallic films, the success of Kikuchi's experiment depended ultimately upon the preparation of a sufficiently thin single crystal specimen. Indeed, mica was almost a unique choice as the single crystal to meet this purpose. Kikuchi writes:

> After having asked at several dealers of mineral samples in Tokyo, I could obtain mica samples of good quality from South America with cleavage faces as wide as the palm of the hand. After a few exercises in preparing thin crystal layers by cleaving, we soon became aware of the fact that extraordinarily thin flakes of mica, bearing no colour of interference on account of their extreme thinness, were occasionally found at peripheral portions of larger pieces, thick enough to exhibit interference colours. In our first observation of electron diffraction we used these thin flakes, thought to be of the order of 0·1 μm in thickness.

There was another point to be settled in advance about mica; at the time the crystal structure of mica (muscovite) had not yet been analysed, and even its cell dimensions were unknown. Nishikawa, thereupon, urged I. Sumoto, his expert assistant, to conduct an X-ray examination of mica. The lattice parameters $a = 5.17$ Å, $b = 8.96$ Å, $c = 20.05$ Å $\beta = 84°$ 10' for muscovite, and its space group

C_{2h}^6, that have been quoted in Kikuchi's paper as the crystal data, are those supplied by Sumoto.

Sumoto writes, in his essay in memory of Kikuchi (Sumoto, 1979), of an interesting episode:

Sitting at a table together, we three, Nishikawa, Kikuchi and myself began an exercise of cleaving thin layers from mica specimens, and eventually became skilled enough to make thin flakes, bearing no interference colours, of an area just sufficient to cover a hole of about 2 mm in diameter bored at the centre of a small circular metal disc used as the sample holder. When the first exposed plate was developed, we three together gazed keenly at the bottom of a developing tray in a small darkroom. In the meantime, our strained eyes seemed to have seen, in the dim light, something like a spotty image appearing around the central strong spot on the photographic plate still under development. Subsequently, impatient Kikuchi took out the plate from the fixing bath, though the fixing was hardly finished, and held it very close to the red lamp. We then became almost convinced that the spotty image was a real one. Curiously enough, however, at the very next moment the spots began to disappear! It was soon discerned that the heat from the red lamp was responsible for this curious effect; the heat softened the wet photographic emulsion and caused it to flow down the plate. Spots forming a hexagonal network on the plate were unmistakably confirmed in the next – thus more elaborate – observation.

Their excitement in the darkroom is indeed very similar to Friedrich's experience, also in a darkroom and described by himself as an episode at the time of the discovery of X-ray diffraction in 1912 (Friedrich, 1932).

The first report by Kikuchi was a communication by Terada at a monthly meeting of the Imperial Academy, Japan, on 12 June 1928. Kikuchi continued the same study to the end of that year with the best of his concentration and enthusiasm. He wrote six papers and one full account. Upon publishing these papers, Kikuchi considered that Nishikawa, to whome he was indebted so greatly and who had taken so important a role in the whole series of studies, should naturally be a co-author of some, if not all, of these papers, and more than once he solicited Nishikawa for his consent. But Nishikawa's response was always absolutely negative. To be a co-author of his pupil's paper for the simple reason that he was Kikuchi's boss would be the last thing Nishikawa's conscientious mind could ever conceive of. In the end, among the six papers, Nishikawa's name appeared only once in a paper sent to *Nature*.

As a short break after this exciting period, Kikuchi left Japan for Europe in January 1929, and stayed there for about two and a half years. He devoted this period mainly to the study of the new theory of quantum mechanics. First he stayed at Born's institute in Göttingen for about one year, and he wrote there a paper on the quantum statistical mechanics in collaboration with L. Nordheim. Then he moved to Heisenberg's institute in Leipzig, and published two papers relating to quantum electrodynamics, one dealing with the propagation of light and the other the Compton effect. He writes: "The last two papers had a personal meaning as a summary of my former concern about the dualism of light." These papers all appeared in *Zeitschrift für Physik*. He came back home in June 1931, and he was again engaged at IPCR in the study of electron diffraction by single crystals until the early spring of 1934, when he moved to Osaka University to assume the position of Professor of Physics.

For his achievement in electron diffraction, Kikuchi was awarded the

Mendenhall Prize from the Imperial Academy, Japan, in 1932. In the same year
he received his Doctorate of Science from the University of Tokyo.

4. KIKUCHI AS NUCLEAR PHYSICIST

Kikuchi's activity in Osaka University was devoted to nuclear physics almost
exclusively, and within only a few years he had energetically organized there a
very active centre of nuclear physics, which was powerfully equipped with a
Cockroft–Walton accelerator of 600 kV, a cyclotron of 60-cm diameter and also
a van de Graaff accelerator. Incidentally, another active centre of nuclear
physics in Japan at the time was the Nishina Laboratory in the IPCR, Tokyo.

The subject he took up in the new laboratory was the interaction of the
neutron, then a new particle, with various kinds of nuclei. With the aid of the
neutron beam generated from the Cockcroft–Walton accelerator by deuteron–
deuteron interaction, he studied the cross-sections of γ-ray excitation due to
slow-neutron capture for various elements, and also the cross-sections and the
directional distributions of fast neutron scattering. These studies continued from
1935 to 1943. Other main subjects were the study of the energy spectrum of
β-rays from artificial radioactive elements and the study of cosmic ray showers.

However, the happy and fruitful days of his nuclear laboratory did not last
very long due to Japan's entry into in the Second World War in December 1941.
Meanwhile, in 1943, when Japan had already been confronted with a serious
phase of the war, the Government appealed to Japanese scientists for their full
cooperation in the national defence effort. Although Kikuchi was far from being
pro-war, he responded to the appeal, and decided to move to the Japanese
Navy's radar research laboratory. In the autumn of 1943 he joined this
laboratory together with some of his colleagues.

With such a drastic change in environment, nuclear research in his own
laboratory in Osaka University was bound to run down. The situation did not
noticeably improve, even after the War ended in August 1945; for not only was
his cyclotron destroyed by the U.S. Army in October, but research in nuclear
physics was severely restricted by order of the Occupation Forces. Therefore, in
order to prepare for the future re-opening of his laboratory, he went to the
United States in February 1950. He stayed first at Cornell University, Ithaca,
where he studied the photonuclear reaction caused by high energy γ-rays using
an electron synchrotron. Subsequently, he moved to the Radiation Laboratory
in the University of California, Berkeley, in October 1951, and stayed there until
his return home in March 1952.

During his absence for two years, a change favourable to the recovery of
nuclear physics in Japan took place, and a new plan for the construction of a
second cyclotron of 110-cm diameter in Osaka University was waiting for him.
This cyclotron was completed in 1955. Also, a unanimous appeal of Japanese
nuclear physicists to have an inter-university centre for nuclear research was
accepted by the government, and their desire was finally realized with the

establishment of the Institute for Nuclear Study, University of Tokyo. This Institute started in 1955 and Kikuchi was made its first Director.

Almost in the same period atomic energy became a subject of national debate, and the Japan Atomic Energy Research Institute (JAERI) was established in 1956. This Institute also required Kikuchi, and he served as Director of the JAERI from 1959 to 1964.

Thereafter, he became President of the Tokyo Science University from 1966 to 1970, which was his last public position.

In November 1952, he was awarded an Order of Cultural Merits, one of the most honourable Orders in Japan. He was elected member of the Japan Academy (formerly the Imperial Academy) in 1957.

5. PERSONALITY AND THE LAST YEARS

Kikuchi was a man of unselfishness and self-sacrifice. He was respected by many people not only for his scientific ability but also for his sincere and noble character. He had a strong sense of justice and rejected unfairness and hypocrisy. At the same time he was broad-minded, and generous to others. His frank attitude gave the immediate impression of an affable gentleman, though he was in fact to be classified as taciturn. Having been unbiased, he could accept everything as it really was. He was also a man of action; once he set an aim, he used to concentrate all his energy to achieve it. On the contrary, he was utterly indifferent to a matter in which he was not interested. In short, he was a man of independence and of large calibre.

He had versatile tastes and various hobbies, such as playing golf, playing the game of *Go*, playing the piano, the cultivation of roses and so-on. By virtue of his inborn gifts as well as his concentration and enthusiasm, he showed great skill in most of them. He had taken lessons on the piano since his boyhood, so his pupils had occasionally to listen to his playing of pieces by Bach, Beethoven, Chopin, and so-on. Sometimes he also played duets with amateur violinists, among whom was the famous Terada. Kikuchi writes: "We enjoyed together some pieces by Mozart and Beethoven when I was a student in the University." As for sport, when he was young he enjoyed playing tennis, baseball, and was also good at field athletics. The sport on which he was very keen in later years was golf. He diligently visited a golf links every holiday, unless it snowed. As he was a tall man with a tough constitution, he held a powerful shot.

A little later after resigning from the post of President of Tokyo Science University, he became uneasy about his health, and in April 1972 the symptoms of pleurisy were discovered in him. In place of golf, which he had by then given up, the game of *Go* became his main hobby, and he played it at home with his old friends. In the same period he wrote an article entitled 'The Safety of the Atomic Generation of Electric Power and the Public Acceptance of It'. This suggestive article, which was published in April 1973, became his last publication. Both mentally and physically he looked quite active for his age, but his health never really returned. His disease was cancer of the lung. He died on 22 November 1974 in a hospital in Tokyo.

Kikuchi married Miss Taeko Kawada in December 1931. He was survived by his wife, one son and four daughters. Shiro Kikuchi, his son, is now a physicist on the research staff of the Japan Atomic Energy Research Institute.

6. KIKUCHI'S CONTRIBUTION TO ELECTRON DIFFRACTION

Kikuchi's investigation on *Diffraction of Cathode Rays by Mica* was motivated by the expectation that the study of electron diffraction using single crystals, not polycrystalline samples or aggregates of small crystals as was the case with G.P. Thomson's work, should be of most value. The results obtained were reported in the following six papers:

1A. S. Nishikawa and S. Kikuchi: *Nature*, **121**, 1019–1020 (1928).
1B. S. Kikuchi: *Proc. Imp. Acad. Japan*, **4**, 271–274 (1928).
1C. S. Kikuchi: *Proc. Imp. Acad. Japan* **4**, 275–278 (1928).
1D.* S. Kikuchi: *Proc. Imp. Acad. Japan* **4**, 354–356 (1928).
1E. S. Kikuchi: *Proc. Imp. Acad. Japan* **4**, 471–474 (1928).
1F. S. Kikuchi: *Jap. Jour. Phys.* **5**, 83–96 (1928).

Fig. 2. An early picture of the N-pattern from an extremely thin layer of mica (from paper 1F).

* In the original of this paper, the dates on which it was received and communicated have been printed, in error, as June 11 and June 12, respectively.

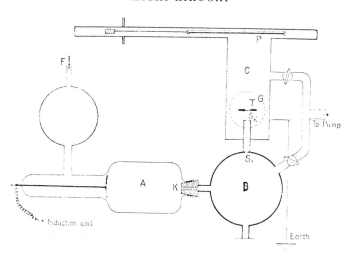

Fig. 3. Kikuchi's improved electron diffraction unit, with magnetic monochromator (from paper 1F). A: discharge tube, B: magnetic deflector, F: palladium tube, T: sample holder, P: photographic plate. The camera length (distance between T and P) is 14.9 cm.

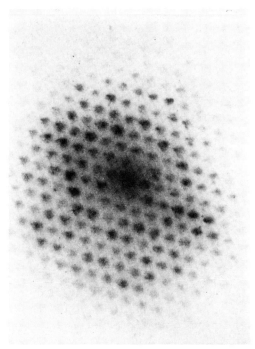

Fig. 4. An improved picture of the N-pattern, using electrons of an energy of about 34 keV (from paper 1F).

Among these, papers lA to lE are preliminary short reports, and lF is a full account.

Throughout his experiments he used the cathode rays generated from a gas tube which was operated by an induction coil with a mercury-jet interrupter. The applied potential was from 10 to 85 kV. The first observation of network-like electron diffraction patterns, obtained from a thin mica flake showing no interference colours was reported in paper 1A – in which Nishikawa is a co-author – and in paper 1B. Paper 1A, being a letter dated May 14, is thought to be the one first written in the above series of papers, while paper 1B, as well as paper 1C, are the reports which were published first. The network-like pattern was called an 'N-pattern' in paper 1F.

Nishikawa and Kikuchi originally suggested that the spot arrays in the N-pattern were the result of two-dimensional diffraction from a very thin sample, though as was later noted by W. L. Bragg (1929), it is more reasonable to ascribe them to three-dimensional diffraction from a distorted lattice, because the crystal flakes will have more than one layer, though they can hardly be free from distortion. Incidentally, the N-patterns obtained at the preliminary stage of their study were not yet aesthetically good, and generally showed elongated streaks rather than discrete spots, as shown in Figure 2. Elongation of the spots was due to the heterogeneity of the cathode rays as well as to crystal curvature. Later Kikuchi used a magnetic velocity selector for his cathode rays (Figure 3) and obtained N-patterns of a much better quality, as shown in Figure 4.

In paper 1C, characteristic patterns from thicker mica sheets – showing interference colours and estimated to be about 1 μm thick – were described.

Figures 5(a) and (b), taken from paper 1C, show an early example of a pattern of this type and a simplified sketch, respectively; the pattern consists of some intense spots, black (excess) and white (deficient) lines and a few concentric continuous rings. As to the intense spots, Kikuchi identified them as an analogue to the Laue spots in the X-ray case, thus as spots fulfilling the three-dimensional diffraction condition. He called them 'L-spots' in paper 1F. On the other hand, the pattern consisting of the black and white lines as mentioned above has now been called the 'Kikuchi pattern', and the lines 'Kikuchi lines'. However, the nature and origin of these lines were clarified first in the next paper, 1D, which was published one month later, although their resemblance to the γ-ray line pattern previously observed by Rutherford and Andrade (1914) in their study on the transmission of divergent γ-rays through a crystal plate had been duly quoted already in paper 1C.

As pointed out in 1D, the black and white lines usually appear as a pair, parallel to each other, the latter always nearer to the central spot than the former. Figure 6, taken from paper 1D, shows the geometry of a number of line pairs and the central spot. When the crystal was rotated, these lines moved as if they were fixed to the crystal. He postulated that these lines (Kikuchi lines) were formed by divergent electrons produced by inelastic scattering of the incident electrons in the crystal, by assuming that these scattered electrons possessed an anisotropic angular distribution and are at the same time subject to Bragg reflexions by lattice planes of various orientations with various spacings.

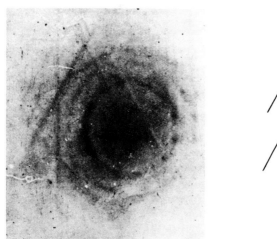
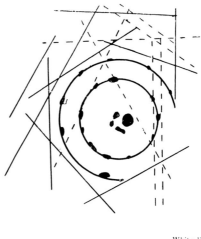

— — — — — White line.

Fig. 5. (a) An early picture of the electron diffraction pattern from a slightly thicker layer of mica, including L-spots and Kikuchi lines, (b) schematic sketch of (a) (from paper 1C).

Kikuchi himself called the Kikuchi pattern the 'P-pattern', as it consists of line pairs. Figure 7, taken from paper 1F, illustrates a pattern, bearing both Kikuchi lines and L-spots, which was obtained from a layer of intermediate thickness.

The continuous rings in Figure 5, Kikuchi originally considered (paper lC) were due to a one-dimensional diffraction effect of the lattice period in a direction nearly parallel to the incident electrons. However, this interpretation, though being correct as far as the geometry of the rings is concerned, was withdrawn in the final paper 1F, for he had by then realised that they were better described as – according to the later terminology – *Kikuchi envelopes.*

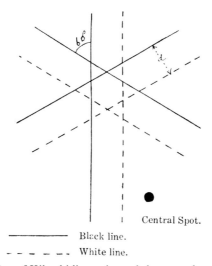

Central Spot.

———————— Black line.

— — — — — — White line.

Fig. 6. Configuration of Kikuchi line pairs and the central spot (from paper 1D).

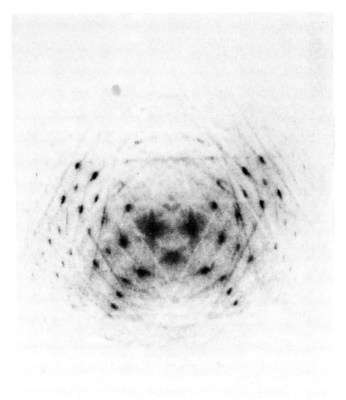

Fig. 7. An electron diffraction pattern from a mica sheet of moderate thickness, using electrons of an energy of about 66 keV (from paper 1F).

In the last paper, 1F, in which further details about the N-pattern, the L-spots and the P-pattern (Kikuchi pattern) are discussed, a brief mention is made about the so-called *Kikuchi band*. According to the explanation of the Kikuchi line-pair mentioned above, they should completely disappear if the lattice plane concerned happens to be parallel to the incident electron beam. In paper 1F, Kikuchi noted that under these conditions a band-like intensity distribution (or the Kikuchi band!) can still be observed, between the Kikuchi line positions. The origin of the Kikuchi band, however, was left unexplained until the advent of the theoretical treatments by von Laue (1935) and by Kainuma (1955).

In parallel with the study with mica specimens, Nishikawa and Kikuchi published jointly the following two papers on *Diffraction of Cathode Rays by Calcite*:

2A. S. Nishikawa and S. Kikuchi: *Nature*, **122**, 726 (1928) (in the Nov. 10 issue).
2B. S. Nishikawa and S. Kikuchi: *Proc. Imp. Acad. Japan* **4**, 475–477 (1928) (Rec. Oct. 1; Comm. Oct. 2).

where paper 2A is a letter dated September 17, but 2B was earlier in publication. It is worth noting that this study is the first example of the application of the reflexion high-energy electron diffraction method (RHEED). In this study diffraction patterns were obtained by allowing cathode rays to strike a crystal at grazing incidence. They used mainly calcite, though cleavage faces of mica, topaz, zinc blende (sphalerite) and a natural face of quartz were also examined. It is to be noted that the earliest description of the Kikuchi *band* is found in paper 2A rather than in 1F.

The last series of Kikuchi's papers on electron diffraction are as follows:

3A. S. Kikuchi and S. Nakagawa: Zur Reflexion der Kathodenstrahlen an der Einkristallober-fläche. *Sci. Pap. Inst. Phys. Chem. Res. Tokyo*, **21**, 80–91 (1933).
3B. S. Kikuchi and S. Nakagawa: Die Anomale Reflexion der Schnellen Electrone an die Kristalloberfläche. *Sci. Pap. Phys. Chem. Res. Tokyo*, **21**, 256–265 (1933).
3C. S. Kikuchi and S. Nakagawa: Zum Innere Potential des Kristall. *Z. Phys.* **88**, 754–762 (1934).
3D. S. Kikuchi: On the Theory of Refractive Index of Crystal for Cathode Rays and Breadth of Reflexion Lines. *Sci. Pap. Phys. Chem. Res. Tokyo*, **26**, 225–241 (1935).

These studies were those performed in the Institute of Physical and Chemical Research after his return in 1931 from Europe.

Papers 3A and 3B are especially important because a number of fundamental features of electron diffraction by single crystals are admirably described in them. From the observation of diffraction patterns from the cleavage face of zinc blende (sphalerite) with 100 keV electrons Kikuchi and Nakagawa found that the reflexion spectra depended characteristically on crystal azimuth, and that they very often showed intensity anomalies. They classified the observed anomalies into two kinds. 'The first kind of intensity anomaly' (to use their expression) was a splitting of the intensity peak. This phenomenon is caused by simultaneous Bragg reflexions and is, as pointed out by them, of the same origin as that responsible for the anomalous refractive index of electrons observed by Davisson and Germer (1928) in the slow electron diffraction by a single crystal of nickel.

On the other hand, 'the second kind of intensity anomaly' was a more striking new effect. This is characterized by the appearance of an intensity maximum not assignable to any Bragg maximum and also, in some cases, by appreciable intensity enhancement of ordinary Bragg spots. Figure 8 is an example of photometer curves for reflexion spectra of different orders from a cleavage face, including such extra peaks as indicated by the letters p, q and r. This remarkable phenomenon was left unexplained for about fifteen years following the

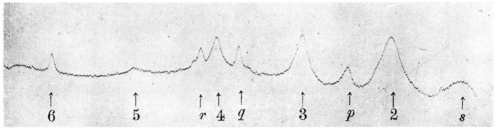

Fig. 8. Photometer curve for reflexion spectra of electrons from the cleavage face (110) of zinc blende, containing non-Bragg intensity maxima such as p, q and r (from paper 3B).

experiment, but was later interpreted as an effect which takes place under quite a special condition in which a strong Bragg reflexion is expected in a direction parallel or nearly parallel to the crystal surface (Miyake, Kohra and Takagi, 1954; Kohra, Molière, Nakano and Ariyama, 1962). More recently, in a study of low-energy electron diffraction (LEED), McRae and Caldwell (1964) found a characteristic diffraction effect due to the 'electronic surface resonance scattering'. It has been confirmed that this diffraction effect in LEED is essentially of the same origin as 'the second kind of intensity anomaly' in RHEED as found by Kikuchi and Nakagawa (Miyake and Hayakawa, 1970; McRae, 1979).

The last two papers, 3C and 3D, which are less important than the foregoing two, are concerned with the anomalous variation with the order of reflexion of the refractive index of electrons in a crystal, the effect found by Yamaguti (1934). Paper 3C experimentally confirms the Yamaguti effect for molybdenite, while paper 3D deals with the theory of this effect. This theory is one of the early examples of the many-beam treatment in the theory of electron diffraction. It is interesting that Kikuchi adopted in it a formulation somewhat different from the usual Bethe scheme.

By surveying Kikuchi's contribution as above, one will realize that indeed the majority of basic phenomena of high-energy electron diffraction by single crystals was unveiled so marvellously and at such an early stage by Seishi Kikuchi.

Emeritus Professor
Institute for Solid State Physics
Tokyo

REFERENCES

Bragg, W. L.: 1929, *Nature* **124**, 125.
Davisson, C. J. and Germer, L. H.: 1928, *Proc. Nat. Acad*, **14**, 317–322.
Friedrich, W.: 1932, *Naturwiss.* **10**, 363–366.
Kainuma, Y.: 1955, *Acta Cryst.* **8**, 247–257.
Kikuchi, S.: 1927, *Jap. Jour. Phys.* **4**, 143–158.
Kikuchi, S.: 1970, *J. Cryst. Soc. Japan* **12**, 170–173 (In Japanese).
Kohra, K., Molière, K., Nakano, S., and Ariyama, M.: 1962, *J. Phys. Soc. Jap.* **17**, Suppl. B-II, 82–85.
Laue, M. von: 1935, *Ann. Phys.* **23**, 705–746.
McRae, E. G. and Caldwell, C. W.: 1964, *Surf. Sci.* **2**, 509–515.
McRae, E. G.: 1979, *Rev. Mod. Phys.* **51**, 541–568.
Miyake, S., Kohra, K., and Takagi, M.: 1954, *Acta Cryst.* **7**, 393–401
Miake, S. and Hayakawa, K.: 1970, *Acta Cryst.* **A26**, 60–70.
Nitta, I.: 1962, *Fifty Years of X-ray Crystallography*, (ed. P.P.Ewald) pp; 328–334. Utrecht: Oosthoek.
Rutherford, E. and Andrade E. N. Da C.: 1914, *Phil. Mag.* **28**, 263–273.
Sumoto, I.: 1979, *Seishi Kikuchi in Memory and His Scientific Contribution*, pp. 223–226. (In Japanese. Private Edition).
Thomson, G. P.: 1928, *Proc. Roy. Soc.* **A117**, 600–609.
Yamaguti, T.: 1934, *Proc. Phys.-Math. Soc. Jap.* **16**, 95–105.

M. J. H. PONTE

I.6. Louis de Broglie
Académie des Sciences (Paris) Académie française (Paris)

1. ANCESTORS – STUDIES

Louis de Broglie (b. 15 August 1892) belongs to a family (Broglio) which came from the Piédmont to France in 1643. At that time, Italians and Savoyards under the rule of Mazarin, and Anne of Austria, participated actively in the government of France. The de Broglie family began dramatically by providing a Field-Marshall of France. Since then it has counted a good number of military men of high rank, diplomats, men of letters and ministers and under regimes of all types. The first manifestation of royal gratitude was the gift in 1716, of a Norman village baptised BROGLIE and made a Duchy by Louis XV in 1742. The castle still stands, having remained the property of the family. The title of Duke de Broglie came to Louis de Broglie in 1960.

Until 1900 the great family de Broglie had no men of science. The most one could say was that by coincidence, Fresnel, the father of pure wave theory of light was born in 1788 in the village of Broglie. Perhaps this factor helped to influence the career of Louis de Broglie. His brother Maurice said that 'the agitation of all these ancestors made the choice of his career an arbitrary one'.

2. ROLE OF MAURICE DE BROGLIE

However that may be, science first called Maurice de Broglie, brother of Louis and his elder by 17 years. The French role in the history of electron diffraction with Louis de Broglie in first place cannot be understood without knowing about these two personalities. They were very different and yet complementary. They defined themselves in their academic speeches and their writings. Maurice was initially a Naval officer, handsome, sociable and enjoyed meetings, functions, voyages, Congresses. Louis on the other hand was described by Maurice as 'an austere and fairly untamed scholar', who did not like leaving his own home. In fact he did not enjoy going to Stockholm in 1929 to receive his Nobel Prize.

According to Louis, Maurice was a little suspicious of theoretical generalisations and had never made a deep study of them.

Maurice reminded him of 'the educational value of the experimental sciences' and that 'the theoretical constructions of science have no value unless they are

55

supported by facts'. The mutual influence of these two intellectuals of such diverse age and experience working together in Paris has been a determining factor in the progress of work on electron diffraction.

3. 1900–1914

For the two brothers, moreover, family traditions presented obstacles to their entrance into a scientific career.

Maurice de Broglie, born in 1875, followed family tradition at first by passing through the Naval College and becoming an officer in the French Navy. This career showed him the military role of the Navy and the presence of France in the world. He stayed there for eleven years. but the young officer, already moulded scientifically by the Naval College and with a Bachelor's degree in Science, arrived in a period of rapid change in the Navy. Not only were the ships themselves being transformed but also attempts were being made to communicate between the ships. That was the time of Marconi's first trials, and the first submission by Maurice de Broglie in 1902, to the Academy of Sciences, of a paper concerning 'a thermal galvanometer for radioelectric waves'. Maurice was already determined to devote as much time as possible to personal research and to abandon the Navy.

His father, Albert de Broglie tried to put him on the right path at least for a time. "Science", he said, " is an old lady who does not fear mature men", but to no avail.

Maurice de Broglie, having become the head of the family, possessed a large house in the rue de Chateaubriand, Paris, in the Etoile quarter. He constructed, using his own finances, a laboratory devoted to his research on X-rays. In several years he obtained some fundamental results on the spectra of X-rays and their interpretation from atomic structure. These were communicated in numerous publications until 1914.

The creation of this family laboratory proved important for the development of Louis de Broglie's thinking. We shall see this later, but let us begin with his studies. His first orientation was traditional, and he studied for the classical diplomas of the times. In 1909, when he was seventeen, he gained a bachelor's degree in the history of the Middle Ages and ancient writings.

He was also studying law. In 1911, the mathematician and philosopher of science, Henri Poincaré, gave a lecture which determined the vocation of Louis de Broglie. He was attracted by the critical approach of Poincaré towards classical analytical mechanics, initiated by Planck's concepts on the quantum of energy and in the scientific method. Louis therefore enrolled in the Faculty of Science and in two years he obtained his degree in mathematics.

His discussions with his brother certainly had a great influence on him. The year 1911 was, in effect, also the year of an international meeting, the Solvay Conference, dedicated to quantum theory. Maurice, following his tastes and capacity, was secretary, and made a report of the papers and ensuing discussions. Louis therefore had at first hand the state of theories on atomic structure and matter and was able to exercise his critical faculties on them.

4. 1914–1919

The war of 1914–18 stopped the scientific work of the two brothers, since they were eligible to join the armed forces: they were 39 and 22 years of age, respectively. Maurice took up contact again with the Navy and conceived a system of communication between submarines by radioelectric waves.

It is probably due to him that General Ferrié, then in control of military radiotelegraphy, mobilized Louis de Broglie in the 'Corps des transmetteurs' (something like the Signal Corps). He was attached to the famous station of the Eiffel Tower under Commandant, later Colonel, Brenot, whom I knew very well and who well remembered his 'distinguished scientist'. Louis de Broglie was assigned to mechanical engineering work and to triodes, work which did not correspond to his tastes. He was described as lost in his thoughts in the midst of the infernal noise of arcs and switch gear which abounded in the station, which was set up underground, below the Champ de Mars. At least he acquired the conviction of the reality of electromagnetic waves and of the conditions of their propagation.

At the end of hostilities, in November 1918, he was demobilized with the grade of 'adjudant' (similar to sergeant-major).

It is then that a new period began – 1919 to 1929, when the association of the two brothers was the most fertile in its diversity.

5. 1919–1925

Maurice de Broglie developed a private laboratory already referred to, in a way that resembled the English laboratories, like those of Sir William Bragg at the Royal Institution or of Sir J. J. Thomson at Cambridge. The laboratory was still dedicated to X-Rays and to their spectral study, with their relationships to energy levels in the atom and the corresponding particle spectra. Therefore it welcomed both the French and foreigners of diverse nationalities: American, Spanish, Japanese. I had the good fortune to work there, in 1924, on the emission of X-Rays by mercury vapour and I well recall the principal laboratory, situated in a former living room. One entered it from a rear door of the main building situated on Rue Lord Byron and it had a little staircase and a glass porch which had an old-world charm. In the laboratory were held gatherings of scientists at which Maurice de Broglie was often present, and discussions on new developments were to the fore. From 1919, Louis de Broglie also occasionally went to the laboratory and he has recalled the teams of young physicists who surrounded his brother. It is certain that this group of scientists which worked on the relationships between photons, electrons, and particles must have motivated many of the reflections of the future founder of the theory of wave mechanics. I, who went there only in 1924, seldom saw Louis de Broglie at the laboratory at Rue Lord Byron; he didn't live there, having chosen Neuilly, 2 kilometers from there, and faithful to his image, he worked alone in his home. One saw him sometimes in the street, lost in his thoughts, and his lips concerned with an interior conversation. That was not the time to approach him; it is true that that was the period of the preparation and publication of his thesis (1924).

6. 1925–1929

I soon had an opportunity to observe how slowly new ideas of a purely theoretical nature, such as those of de Broglie, were disseminated. After my military service (October 1924 – September 1925), thanks to a Rockefeller Scholarship, I spent six months in Sir William Bragg's Davy-Faraday Laboratory at the Royal Institution, Albermarle Street. That was an extraordinary year with a team of research scientists of very diverse character and nationalities. The most original was Bernal, author of *Science and History*. Practically all the works were devoted to the determination of a wide variety of crystal structures by the rotating crystal method. The experimental training was unique and typically English at that time, where the scientists had to construct everything themselves, and lively discussions characterized the preparation of equipment. But the most fertile meetings were at afternoon tea, for Sir William came there in person and was familiar with the advances in research. It was like the atmosphere in Maurice de Broglie's Laboratory but on a larger scale.

Other memorable meetings were the conferences at the Royal Institution where English society – in dinner jacket and evening dress – came to listen to a celebrity of the day discussing a current scientific subject, which demanded particular skills on the part of the speaker, since the audience in the main was non-scientific. The most extraordinary talk that I heard was that of Sir J. J. Thomson, in 1926, in the same amphitheatre as was used for his famous lecture on the electron in 1897.

If I recall these memories, it is for the following reason. In 1925–26 in London, in that climate of discussions and exchanges which were quite open, concerning most often the field of electrons–radiation–matter, and among people who were well-informed, I do not remember that the thesis of de Broglie was mentioned and presented as work that was going to transform the physical sciences. It is perhaps because, as often happens in science, the need to modify the original quantum theory of Bohr, which had become evident in 1923, was approached almost simultaneously by Louis de Broglie, Schrödinger, Heisenberg, and Dirac; and the work of Louis de Broglie still appeared insufficient.

In fact, Louis de Broglie himself sought to refine his concepts and to create a more complete system for the new wave mechanics by building it up around what he called the double solution. . . . In that line of research, he did not have enough mathematical technique at hand nor mathematics students versed in his ideas; he still regretted this in 1927. It can also be asked if, in 1926, the single experimental verification of the existence of the associated wave would have appeared to him as being worthy of interest for its own sake.

It is the more astonishing that, at a short distance from there, a research worker from the Laboratories of Maurice de Broglie was not attracted to such a demonstration: Was it thought impossible or were the means unavailable? On returning from London, in October 1926, I obtained a job as laboratory assistant at the Physical Laboratories of the Ecole Normale Supérieure (E.N.S.), a sought-after position which readily permitted a thesis to be prepared.

I again made contact with the de Broglie teams and began my experiments at the E.N.S. in another direction. Strong in the experience required in the Laboratory of Sir William Bragg on electron beams and thin metallic foil, I began, by interesting myself in Lénard's rays, to study fluorescence phenomena and certain biological properties of electrons in air. A tube was constructed, fairly similar to an X-ray tube, but with an anticathode permeable to electrons. The high-tension was furnished by an old Sécheron machine used during the '14–18 war in the radioelectric station of the Eiffel Tower (again!). It generated a maximum of 10 000 volts and did not work too badly for its age. Experiments were started in January 1927 when the results of Davisson and Germer became known.

7. 1927–1930

For all of us, it was a revelation, and the writings of Sir George Paget Thomson show that we were not the only ones! Nevertheless it was necessary from all evidence to make another experimental proof, direct and capable of recording.

My idea was to apply to electrons the techniques acquired with X-rays. It appeared necessary to renounce the use of polycrystalline thin films which were difficult to prepare and likely to cause dispersion of the velocities of the electron beam. In order to generalize the demonstration it was necessary to use either an assemblage of microcrystals permeable to the beam, or to use a single rotating crystal. For the first method, zinc oxide quickly provided a solution: in effect, on melting a piece of zinc placed on a support in a Bunsen burner, the zinc caught fire in something like a small explosion and for a long time the air of the laboratory was sprinkled with little white flakes of zinc oxide. To collect one and place it on the hole of the diaphragm defining the cross section of the beam was relatively easy. A temporary tube, using the 'Lenard' tube as a starting point, was made with a movable glass capable of receiving a fluorescent powder or a photographic plate; the vacuum was maintained by a mercury diffusion pump. After numerous adjustments, tests and loss of time usual in such circumstances, the tube was ready at the end of 1927 with, as specimen, a film of zinc oxide. At the moment of the crucial experiment when the potential had reached 12 000 volts, the Sécheron generator gave up the ghost following a discharge in the tube.

It was necessary to start all over again using a new source, this one classical, with transformer and two valve-rectifiers followed by a condenser. The necessary delay was used to design and construct a special device for which the vacuum was created by a Hollweck mechanical pump, chosen because it did not give off any undesirable vapour. During the construction of all this apparatus, which took until September 1928, G. P. Thomson published his famous note in *Nature* in December 1927 and his articles in the *Proceedings*.

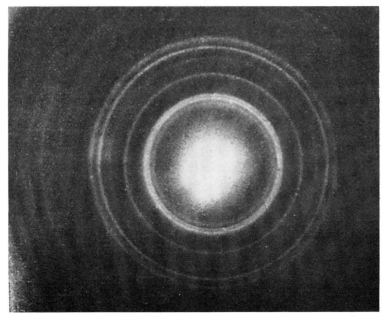

Fig. 1. ZnO diffraction pattern obtained with the new apparatus, using 12 500 volts acceleration. 17 rings are visible. (Ponte: Dr Thesis, University of Paris, 1930).

There is no doubt that the credit for having proved experimentally the existence of diffracted waves of de Broglie must go to Davisson and Germer and to Sir G. Paget Thomson. The experiments of M. Ponte were nevertheless conducted after September 1928 and were followed through to a conclusion defined by the goal set, namely that of generalising the demonstration with assorted substances and structures.

The first results with the new equipment were rapidly attained, in October 1928, and the first publication was made in January 1929 at the Academy of Science. The intensity and the resolution of the diffraction rings obtained with zinc oxide powder as shown in Figure 1 were surprising: for an electron velocity corresponding to 12 500 volts, 17 rings were present on a plate of 40×40 mm^2. Their measurement was easy with a photometric recorder, resulting in the length of the de Broglie wave being established to better than 1%; and the structure of zinc oxide, known from X-rays, was confirmed in all details. I went to announce these results to Louis de Broglie at his home in Neuilly. During 1929, the experiments were generalised and published in the *Comptes rendus* of the Academy of Sciences, at a Conference at the Collège de France, the whole assembled in a thesis presented verbally in April 1930 and published in *les Annales de Physique*. During 1929, Maurice de Broglie sent one of his students, Trillat, to familiarise himself with my work, which could possibly have had vast applications.

An amusing incident comes to mind. At the anouncement of the Nobel Prize for Louis de Broglie in 1929, my name was cited and I received at my laboratory, at the E.N.S., the unexpected visit of journalists who had understood that I had discovered new coloured rings, a new rainbow so to speak . . . ! They were very disappointed with the facts.

In reality, the scrutiny of my results assembled in my thesis since October 1928 leaves them a definite originality in the following areas which are more in the domain of applications but which should figure in the anniversary of diffraction itself:

The generality of the phenomenon: shown by tests on diverse metallic oxides, taken alone or on different supports,

The sensitivity of diffraction: detection of impurities like greases or vapours in vacuum, proved in the thesis.

The resolution of diagrams: obtained when the beam is well monochromatized. The electrons appeared from this fact like a new instrument for the detection of traces in minute quantities of matter and for the study of surface structures.

All these results led me to synthesize them under the name of *Electron Analysis* which appeared at the time of my publications at the beginning of 1929.

Académie des Sciences
Paris, France.

PART TWO

Subsequent Development at Various Centers of Research

Period I
1928–1935

Fig 2.1.a.

Equipment constructed in September 1928 by Ponte and used to obtain diffraction from ZnO in 1929. (See Article I.6.)

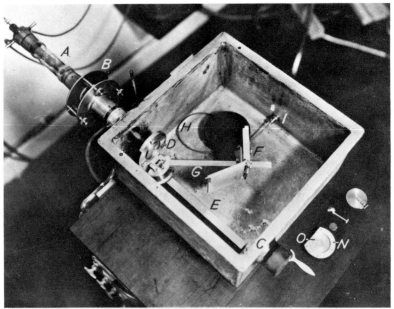

2.1.b.

Earliest camera constructed by Aminoff (during 1932–33). Arrangement inside metal box contains a plate-holder and specimen goniometer. (See Article II.9.)

2.1.c.

The first electron microscope constructed in France by Trillat and Fritz at Besançon (1933). (See Article II.3.)

2.1.d.

The De Lazlo gas-diffraction camera, constructed by Coslett (1932). (See Article II.7.)

CHESTER J. CALBICK

II.1. As I Saw It

Clinton J. Davisson was a perceptive and ingenious designer of experiments and an able mathematical physicist. In my early association with him and with Lester Germer, the emphasis was on experimentation. I remember Davisson saying "The results of an experiment must always be calculable using a proper theory." As Germer once said, we who were fortunate enough to work with C. J. Davisson – known as Davey to his associates – learned an enormous amount from him: how to do experiments; how to think about them; how to relate them to what other people had done; and the underlying importance of theory in physics. He was always a patient, wise counsellor, not only for us but for all of those who worked for him or who came to him for help in understanding physical problems.

My first experience at the newly-formed Bell Labs was with J. A. Becker, who had been a student of Richtmyer at Cornell University and a National Research fellow. He was engaged in research on thermionic emission from oxide-coated filaments, the same problem that had brought Davisson to the Engineering Department of Western Electric during World War I. Emission from alkaline-earth oxides is very complex, and even today is not fully understood. Becker turned to basic research, studying the thermionic emission from tungsten covered with layers of cesium. As a recent electrical engineering graduate, albeit with advanced courses in physics and chemistry, I quickly became fascinated with the intricacies of experimental procedures and the interpretation of data guided by physical theory based on a model. Becker devised a patch theory for less-than-monolayer films of Cs on W. At about that time, Langmuir was espousing a uniform-film theory. Each championed his case with considerable vigor, and it became evident to me that physical theories were in a constant state of evolution.

During my year with Becker, I became familiar with state-of-the-art vacuum techniques. Pyrex glass had only recently become available, and our talented mechanic-glassblower, George Reitter, had learned how to use it. Fore-pumps used ordinary oil, and mercury-vapor diffusion pumps trapped with liquid air were used to obtain pressure in the 10^{-6} mm Hg range while baking. Our ionization gauge tubes were ordinary triodes, whose limiting low-pressure reading was 10^{-7} mm Hg. Always there was a spurt of gas when the tube was sealed off. This gas could then be removed by using a barium 'getter' or, in the case of the Davisson–Germer tubes, by chilling, with liquid air, a side-tube filled with activated charcoal which had been torched during pumping to about

67

500 °C. By various adsorption-desorption techniques, the residual pressure could be measured roughly, certainly below 10^{-8} and perhaps as low as 10^{-9} mm Hg.

In the fall of 1926, following Davisson's return from Europe and his discussions of de Broglie's waves and the new wave-mechanics of Schrödinger with Born, Blackett, Hartree and others, I joined Davisson and Germer in their investigation of elastically scattered electrons reflected by a single crystal of nickel. During the spring, on a previous pumping of the tube, the results had been very disappointing. No sign of the sharp spikes – observed following the famous accident in which a polycrystalline target had been oxidized, following an implosion of a Dewar flask containing liquid air – had been found. The tube had been cut open to check its mechanical alignments and re-pumped. An immediate problem intruded. The resistance between the inner and outer boxes of the collector was too low. In order to measure the elastically scattered electrons, the inner box was near cathode potential, while the outer box was at anode potential, so a value of at least 10^{11} ohms was required. The low resistance was due, no doubt, to an organic film. Fortunately, application of a high voltage increased the resistance to an adequate value and the experiments could proceed!

There were other similar practical problems – the galvanometer had to be mounted on a suitable suspension to reduce vibration and shielded from air currents. Even so, it was necessary to alternate reflected current readings with zero readings – doubling the time of data collection. Filament current and anode voltage were supplied by batteries. Since the filament was grounded, care was necessary to avoid electrical shock. Finally we began to obtain reliable data!

In retrospect, the most remarkable thing about this period was Davisson's patience and persistence. He must have been keenly disappointed when the curves showed only slight maxima in the principal azimuths – those containing the (111), (100) and (110) crystallographic directions, and no peaks in the co-latitude curves. He had calculated, from the wavelength $\lambda = h/p = 12.25/V^{1/2}$ Å, that a beam should appear in the (100) azimuth at $V = 78$ volts. Then occurred an incident which was not recorded. Davisson and Germer had retreated into Davey's office to recalculate the beam location. We had just run a co-latitude curve at 78 volts, and no peak had appeared. So, left alone, I dropped the voltage to 70 volts, then to 65. Each curve required about 20 minutes to run. At 70 volts, there was a bump, and at 65 a definite peak which I plotted on polar paper and took in to Davisson. I remember asking "Is this what you are looking for?" They were, of course, very excited. We immediately returned to the tube, and ran a voltage curve. The date was January 6, 1927, as I recorded, and Germer wrote "First Appearance of Electron Beam." The co-latitude angle was 40°. The next day we ran voltage curves at four nearby co-latitude angles, and localized the peak at 45°, finally on January 8 running a co-latitude curve at 65 volts showing the maximum beam – curiously, just the voltage at which I had taken the unrecorded curve two days earlier.

Next we ran an azimuth curve showing that this beam maximized in the three (100) azimuths. Peaks also appeared in the azimuths half-way between [the (100) azimuths]. These were, of course, due to another set of reflecting planes.

The intensities of these peaks varied considerably and it was not immediately evident that they verified the three-fold symmetry expected. A period of intense activity followed during which we localized about 20 peaks.

In early March, Davisson sent an article to O. W. Richardson, his brother-in-law, with an accompanying note asking him to get in touch with the editor of *Nature* with the idea of securing early publication – this correspondence is recorded by Gehrenbeck, earlier in this volume. The article entitled, very conservatively, 'The Scattering of Electrons by a Single Crystal of Nickel', was published in *Nature* **119**, 558, 1927. The paper by G. P. Thomson and H. Reid reporting the diffraction of high voltage electrons passing through a thin metal layer appeared a month later, also in *Nature*. None of us were aware that this research was going on in Aberdeen. This almost simultaneous publication of the results of two research studies, each of which established the wave nature of electrons, resulted ten years later in the award of the Nobel Prize to Davisson and Thomson. I have often wondered why Germer and Reid were not included in the award but apparently in 1937 the Nobel Committee was unwilling to further divide the award. Also, Reid had unfortunately been killed in a motorcycle accident.

Why was the article in *Nature* so cautiously titled? Certainly we were convinced that we had verified de Broglie's theory. In March, Davisson and Germer gave an oral paper before the Physical Society at the Washington meeting, going somewhat beyond the *Nature* article. (During March I was called home by the death of my mother.) Upon my return we completed the localization of the diffracted beams appearing at voltages below 370 volts including the 'anomalous' beams ascribed to an adsorbed layer of gas. These were 'plane grating' beams ascribable to a monolayer on the surface with twice the spacing of the Ni atoms. They disappeared if the crystal was strongly heated, which also greatly strengthened the space-grating beams. But *none of the latter appeared in the right places in co-latitude angle and voltage*. The data were substantially completed by early summer, and the definitive paper, entitled 'Diffraction of Electrons by a Crystal of Nickel', was sent off in August, and was published in *Physical Review* **30**, 705, in December 1927. We were still unable to ascribe Miller indices to the space lattice beams. Davisson developed his own method of displaying the beams. If the index of refraction μ were unity, the Bragg equation $n\lambda = 2d \sin \theta$ requires that when λ is plotted against $\sin \theta$ the beams must fall on lines of slope $2d/n$. The actual beams did not fall on these lines. A correlation was made between each of the actual locations, and a theoretical location which it would occupy if $\mu = 1$. This location could be chosen in one of two different ways corresponding to $\mu > 1$ and one to $\mu < 1$. In the paper, the wrong correlation was made. This was in part because of a second possibility, namely, that the spacing between atomic layers parallel to the surface might be smaller very near the surface than in the bulk – and the electrons would penetrate only a few layers. The case $\mu > 1$ required that electrons be accelerated through a surface potential field of about 18 volts – far greater than the work function.*

* Bethe at the time was about to publish his theory of the inner potential, which gives a value of about 15 eV for nickel (*Naturwiss.* **15**, 787, 1928, *ibid.* **16**, 333, 1928).

Characteristically, Davisson and Germer decided to resolve this question by building a tube in which the angle of incidence of the beam in a selected azimuth could be varied, and the electrons could be reflected from the planes parallel to the surface. The indices and spacing of the reflecting planes would be known. Some preliminary data from this tube were published in April 1928 (*Proc. Nat. Acad. Sci.* **14**, 317, 1928) and a complete account was given in an article entitled 'Reflection and Refraction of Electrons by a Crystal of Nickel' (*ibid.* **14**, 619, 1928) in August. There was still a problem – the 1st order beams could not be observed because our electron gun was unable to produce usable beams at less than about 20 volts, where the wavelength was already too short to yield a first order beam at the maximum angle of incidence geometrically accessible. Bethe's papers were by this time available, and a smooth curve corresponding to an inner potential of 18 volts gave an index of refraction curve which seemed best to fit the data. But there were two regions of 'anomalous dispersion' which were very difficult to explain and remain so to this day. Philip Morse spent the summer of 1928 with Davisson, and observed that the 'anomalous dispersion' regions were near places where other sets of crystal planes would also diffract. Perhaps the splitting of the beam was a double-resonance phenomenon.

Were electrons waves polarized like electromagnetic waves? The reflected beams were quite strong, and this suggested a double reflection experiment in which a beam from one crystal would be reflected from a second crystal which could be rotated in azimuth. In analogy with the case of light, the final beam would be strongest when the two sets of planes were parallel, and weakest at $90°$ from these angles. While the tube was being built, I designed and constructed a d.c. amplifier. Only commercial vacuum tubes were available – electrometer tubes had not yet been invented – and the result of my efforts was a rather noisy amplifier, but still capable of measuring currents in the 10^{-12} amp range expected for the twice-reflected beam. Another problem was a variable magnetic field due to variable currents circulating in the steel frame of our building, which was in an area of New York City supplied by direct current. (No doubt this variation had also contributed to difficulties in reproducing data in earlier tubes.) At the low electron energy, variation in deflection of the order of half the diameter of the hole in the collector could be calculated to occur. Magnetic shielding – not very complete because of the necessity of mechanically moving the tube – was used, and best data obtained during times when the building was magnetically relatively quiet.

We reported (*Phys. Rev.* **33**, 760, 1929) no observed polarization, with some reservations about the 103 volt beam. Recently, however, Kuyatt (*Phys. Rev. B* **12**, 4581, 1975) has pointed out that the use of the polarization of light as a model for data analysis is incorrect. For spin polarization, the proper model for electrons, only one maximum and one minimum should be observed as the crystal is rotated in azimuth through $360°$, rather than two, as in the case of light. He re-analyzed our data, which did show a first-harmonic variation of 14.5% for the scattering of 120 volt electrons from the (111) face of nickel.

The difficulties with the polarization tube led Davisson and me directly into the research which culminated in our development of electrostatic electron optics, and to an early note entitled 'Electron Microscope'. It was clear that we did not know enough about the shape of the electron beam as it emerged from the electron gun, as it interacted with the crystal, nor as it entered the collector. We calculated the trajectories of electrons as they approached slits or holes in charged plates, which were shown to be, for electrons, cylindrical or spherical lenses. Experimentally, we built tubes in which the shape of the beam could be photographed using the light emitted by mercury atoms, present at a low pressure, and ionized by passage of the electrons. These photographs verified our formulae for the focal powers of cylindrical and spherical electron lenses. We then built a simple 'Electron Microscope' in which the object was two crossed pairs of wires, illuminated by electrons from a ribbon filament. Shadow images of this object were focused to verify the calculated magnification. A small area of the ribbon filament was also portrayed as an emission-type image.

Next we designed and built the Davisson–Calbick television receiving tube. Television had been demonstrated in 1927 at Bell Labs by a research group headed by H. E. Ives. An image of President Hoover addressing an audience in Washington, D.C., was transmitted to an audience in New York. They used an electromechanical scanning system, the image being reproduced by a Nipkow disk rotating at high speed. This was a steel disk on which were mounted 72 lenses in a one-turn spiral, each lens reproducing one line of the television image. Mechanical stresses in the disk limited the possible number of lines, and it was clear that a high-vacuum, high-voltage oscilloscope tube would permit reproduction of many more lines. Such a tube was not then available. Zworykin at RCA was developing one which, he called a kinescope, and a companion camera tube, the iconoscope. The first coaxial-cable had just been developed at Bell Labs, and it was important to test its transmission of TV signals. We were diverted into making a number of these tubes, rather than continuing into the more scientifically fascinating development of electron microscopy. Several patents were issued to us in the course of this development, of which I shall mention one, which was basic to ion pumping. To reduce the pressure after seal-off, we added a side-tube containing a triode with a tantalum filament. It was used as an ionization gauge but, more importantly, by operating the filament at a temperature such that it evaporated a few micrograms per minute, it was also an evaporated-film ion pump.

In 1937, Davisson was awarded the Nobel Prize and, curiously, this more or less coincided with the completion of the receiving-tube development.

After the report on polarization of electrons Germer began independently to develop an electron diffraction spectrometer. He and Davisson remained close friends. In later years he returned to low energy electron diffraction, developing the post-acceleration tube now almost universal in LEED studies.

I feel privileged to have been a colleague, especially of Davisson – the patient, persistent experimenter, always helpful to the numerous physicists with whom he was associated, and having a basic drive to relate experiment to physical theory.

Department of Physics
Washington State University

REFERENCES

Davisson, C. J. and Germer, L. H.: 1927, *Phys. Rev.* **30**, 705.
Davisson, C. J. and Germer, L. H.: 1928, *Proc. Nat. Acad. Sci.* **14**, 317.
Davisson, C. J. and Germer, L. H.: 1929, *Phys. Rev.* **33**, 760.
Davisson, C. J. and Kunsman, C. H.: 1927, *Nature*, **119**, 558.
Kuyatt, C. E.: 1975, *Phys. Rev.* **B12**, 4581.

H. A. BETHE

II.2. Reminiscences of the Early Days of Electron Diffraction

In the spring of 1926, I went to Munich to be a graduate student of Professor Arnold Sommerfeld. It was a most fortunate time. Schrödinger's papers were just coming out, one after the other, and Sommerfeld got advance copies of these from the *Annalen der Physik*. Each one of them was discussed in detail in Sommerfeld's theoretical seminar so that we all had a chance to learn wave mechanics thoroughly. Everybody had to report on one part of these papers, and it was my task to report on perturbation theory. Ever since that time I have been very fond of perturbation theory in all its forms.

Shortly after Schrödinger's publications there appeared the paper by C. J. Davisson and L. H. Germer on their observation of electron diffraction. This observation confirmed de Broglie's and Schrödinger's ideas beautifully, and made the entire subject much more real.

There was one problem with the Davisson-Germer experiment: The diffraction maxima did not occur at exactly the correct energy. Sommerfield suggested to me that I have a look at this. I soon found that all the maxima were at the correct place if I assumed that the electrons have a negative potential energy, of about 20 eV, in the Ni crystal which Davisson and Germer used for their experiments. This is what I have remembered for these fifty years. But when I now looked at my old paper, I saw that I proposed a *positive* potential energy in the crystal. Only later, between the first paper and my thesis, I changed the sign of the potential, and the index of the order of reflection, and it worked again. Anyway, the first paper was published in *Naturwissenschaften* in late 1927.

A little later I was rather dismayed that A. L. Patterson had found the same explanation; a great disappointment for a physicist who has just published his first paper. Only a few years ago I learned that the explanation had also been found by C. H. Eckart and F. Zwicky.

A few months later Sommerfeld got intcrested in the electron theory of metals, using the free electron model and the Pauli principle. Every phenomenon on which Drude's electron theory of metals in 1905 had been defeated could now be explained quite fully. One of Sommerfeld's results was that electrons in the metal have considerable Fermi kinetic energy – of the order of 10 eV for a dense metal like Ni. I then proposed to Sommerfeld that the observed work function of a metal is the difference between my negative potential energy and the positive kinetic Fermi energy. Of course, this is not accurate enough to calculate the work function, but it explains the phenomena qualitatively. After this suggestion, Sommerfeld considered me a great expert in solid state physics, which I was not. But his favorable estimate led to our joint article on the electron theory of metals in the *Handbuch der Physik*, Vol. XXIV/2, a report which gave me a lot of pleasure.

Returning to electron diffraction, Sommerfeld now proposed to me that I make a detailed theory of the diffraction in a crystal. He recommended to me to use as a model the theory by P. P. Ewald of the diffraction of X-rays, written in 1917. I studied this diligently, and found to my pleasure that electron diffraction was a great deal simpler. In the X-ray case one has to contend with a vector field which complicates things greatly. Having studied all these complications in Ewald's paper, I was only too happy to discard them, and to retain only his fundamental idea. About the most important quantitative part of Ewald's paper was the expansion of a spherical wave (i.e., the wave scattered by an atom) in terms of plane waves.

Having Ewald's theory as an example, it was easy to develop the theory of electron diffraction in first-order perturbation theory. In this work I found that the form factor for electron diffraction by an atom is

$$[Z - F(q)]/q^2 \tag{1}$$

where $F(q)$ is the well-known form factor for X-ray scattering. Formula (1) has been used many times since.

Being inexperienced and ambitious, I then proceeded to do a second-order perturbation theory. Nowadays I would have known better: If the first-order gives a non-vanishing result (as it did), and if the interaction is strong (as it is for electrons) the second-order perturbation generally becomes messy and not very revealing. So it turned out to be, but I spent many months on trying to develop it. When I first met Wolfgang Pauli, a year after my Ph.D., he greeted me with the words, "Mr Bethe, I have been very disappointed by your thesis." From my present point of view, I would agree with him.

In the meantime, experiments on electron diffraction continued at a great rate. The most important probably were by G. P. Thomson who used high energy electrons, of tens of keV, rather than the 100-eV electrons of Davisson. With the higher energy it was possible for Thomson to use the electron beam transmitted through a thin foil and study its diffraction pattern, much like von Laue's first experiments. Davisson and Germer had to observe the reflected electrons since there were no foils thin enough to transmit electrons of 100 eV. Moreover Thomson's electrons, being more energetic, had much less interaction with the atoms in the foil, so that now elementary interference theory (or first-order perturbation theory) was indeed sufficient.

Davisson and Germer themselves continued their work, and soon found very complicated diffraction patterns, as a function of the energy of the electrons. At certain energies, a given reflection might split into two, and then at a higher energy these two diffraction maxima might combine again. I was lucky that these experiments were not available when I wrote my thesis; they clearly defied any simple theoretical explanation. As far as I know, nobody has given a theory – although with modern methods this would probably be feasible.

A rather enigmatic figure in electron diffraction was E. Rupp who was working at the laboratory of the AEG, the German General Electric Company. He was apparently the first experimenter who observed diffraction of electrons by ordinary optical gratings. But thereafter many of his experimental results seemed very strange. For instance, he claimed to see effects of electron polarization at quite low electron energy which was contrary to theoretical prediction and could not be reproduced by any other experimenter. Moreover, he found diffraction maxima which seemed to require half-integral orders of reflection. This seemed absurd to me. While I did not publish any papers on this subject, I spoke in the discussion of a report on Rupp's experiments (I don't remember whether it was given by Rupp himself) at a meeting of the German equivalent of the AAAS, the Deutsche Naturforscher Gesellschaft. This was held in Danzig or in Königsberg, I don't remember which, and the year was about 1930. I was vigorously beaten down by another member of the AEG research team, Dr Ramsauer (an excellent physicist), who said, in essence, "How does such a young pipsqueak dare to attack the work of a member of our laboratory?". A few years later it was discovered that many of Rupp's experiments were faked. He withdrew them explicitly in a short note in *Zeitschrift für Physik* **95** (1935), complete with a note by his psychiatrist that "psychogenic dream-like states (had) entered his research."

A very useful application of electron diffraction was made by Mark and Wierl at the research laboratory of the BASF, the Badische Anilin und Soda Fabrik. They studied the diffraction of electrons by gases such as CCl_4, and found very beautiful results, giving the structure of the molecule. This led to the first industrial consultation in my life; Dr Mark asked me to come and help them interpret their results which I was very happy to do. Mark gave me a very friendly reception, and I even considered possibly joining their staff because academic positions in Germany were very rare in 1928–33.

On the basis of my thesis, I was invited by P. P. Ewald to give a talk at a small conference on diffraction which he was arranging in Stuttgart in 1928. Apparently my talk pleased him, because a year later he asked me to become his 'assistant', a title which in Germany corresponded approximately to a research associate in the U.S. I had a most enjoyable semester there, with a great deal of research, and close personal contact with Ewald and his family. Out of this, I got a wife: Ewald's daughter, then 12 years old, was already very attractive, but I did not dream of marrying her. Eight years later, I met her again, and in 1939 we got married. So I owe a great deal to electron diffraction.

Cornell University

REFERENCES

Bethe, H.: 1927, *Naturwiss.* **15**, 786.
Bethe, H.: 1928, *Naturwiss.* **16**, 333.

Bethe, H.: 1929, *Ann. der Phys.* **87**, 55.
Ewald, P. P.: 1917, *Ann. der Phys.* **54**, 519.
Rupp, E.: 1935, *Zeits f. Phys.* **95**, 801.
Schrödinger, E.: 1926, *Ann. der Phys.* **79**, 372; **79**, 734.

JEAN-JACQUES TRILLAT

II.3. The Start of Electron Diffraction in France: My Recollections

It was somewhat by chance that in 1924 I entered the X-ray laboratory, a private laboratory belonging to the great X-ray physicist Maurice de Broglie. That laboratory did not resemble a classic laboratory; devices, often built by Maurice de Broglie himself, were placed in the salons or other rooms of his mansion at 27 rue de Chateaubriand near the Place de L'Etoile.

Maurice de Broglie had several collaborators, some of whom became famous in subsequent years: A. Dauvillier, J. Thibaud, and later L. Leprince Ringuet, Magnan Cartan and others. It was a veritable honour to be admitted into that circle in which reigned an atmosphere both respectful of the Chief, and familiar and friendly for the researchers. But the principal collaborator of Maurice de Broglie was his own brother, Louis de Broglie who, at that time, began to reflect deeply on the relations between waves and matter.

I arrived there completely ignorant of X-rays, for I had a diploma of chemistry from the School of Physics and Chemistry of Paris. My aim was to utilise these rays in order to study the fine structure of certain crystals and molecules, like those which constitute fatty acids, and I wrote my thesis on this subject, discovering on the way new laws which permitted some applications to diverse phenomena such as lubrication, catalysis, etc. I was working then in a little attic room and cleaned this tiny laboratory myself. I installed my X-ray spectrometer constructed by the only mechanic available at the time who was none other than Maurice de Broglie's valet. I passed my thesis in 1927, but in between times I had begun to interest myself in wave mechanics and above all in its applications.

Indeed, the laboratory had grown and my little maid's room had been replaced by a vast cellar situated in rue Lord Byron, which saw daylight on the pavement by an air vent; a situation which was not very hygienic, but enthusiasm for research made me forget these little inconveniences.

What made the laboratory of M. de Broglie interesting was that each person enjoyed complete freedom in the choice of his subject of study; the office of M. de Broglie was always open to anyone who wished to discuss an experiment or to submit some ideas. But above all, each Wednesday, the personnel and the Chief met to keep up to date with the studies of each person and to discuss new articles, and submit ideas.

Louis de Broglie, who was not an experimenter but a theoretician, regularly attended these meetings and I had the privilege to see the appearance, little by little, of the theory of Wave Mechanics; a theory that was not really accepted until after the experiments of Davisson and Germer in 1927, and then that of G. P. Thomson, which demonstrated experimentally the phenomenon of electron diffraction predicted by Louis de Broglie. In France, M. Ponte was the first, in 1930, to make an apparatus clearly demonstrating this phenomenon. Needless to say that, after such proof, the Nobel Prize had to be given to Louis de Broglie, and I still remember the joy at the laboratory at the announcement of the great news.

The great discovery marked for me the point of departure for the major part of my research activities, opening for me new paths for research both theoretical and applied.

With the constant support and encouragement of the two de Broglie brothers, I threw myself right into this new domain starting from the idea that if the X-rays gave an image of a relatively significant thickness of matter, the diffraction of electrons allowed the scientist to obtain information on very superficial layers or on surface phenomena, due to their weak penetration and the extraordinary intensity of the observed diagrams. I can say that a large part of my scientific activity was concentrated on the combination of X-rays and electron diffraction, and later on electron microscopy.

I therefore constructed several instruments for electron diffraction which permitted the most varied experiments; certain of these still exist, and they were used for many years by my students who came to study these new techniques which were rapidly being brought to a peak of perfection.

In 1932 I left the de Broglie laboratory to take up the Chair of Physics of the Faculty of Science at Besançon. Naturally it was necessary to equip this provincial laboratory, which lacked everything, including alternating current. Thanks to the aid given by M. de Broglie, who made a gift of the equipment I was using, and thanks to the grants from the Faculty and Industry Research Funds, I succeeded, in less than two years, in creating a laboratory at Besançon. This laboratory was very well equipped for X-ray and electron diffraction and in 1933 I was able, with the aid of my assistant M. Fritz, to construct the first French electron microscope, an apparatus which, though rudimentary, aroused a keen interest among the physicists.

Numerous students came to work and to prepare Doctorate theses. Foreign students came also, and I am thinking particularly of our colleagues, Professor Takahashi and Professor Oketani, who carried back to their own country the fruits of their experiences.

The laboratory ceased functioning in 1939 at the beginning of the war. It never reopened for, in between time, I had been appointed to Paris and given the responsibility by the CNRS to construct at Bellevue a large laboratory of X-ray and electron diffraction and microscopy. But this leads us to more recent times which are out of the scope of this article; let us only say that, when I retired in 1971, there were about 50 research scientists and the laboratory was exceptionally well-equipped.

Louis de Broglie never ceased to be interested in my research and he often honoured me by coming to Bellevue to see new apparatus or to be present during new experiments.

Perhaps it is good, to finish these reminiscences, to recall the first experiments which seemed to be fundamental and which I carried out in the first years of electron diffraction.

I had, above all, the idea of using electron diffraction together with X-ray diffraction and the electron microscope to study the detailed structure of matter and of surface phenomena. This led me to study problems which were very varied and to create new apparatus or devices. I found myself thus to be, in France, a pioneer of electron diffraction and its applications, which attracted to me numerous students, particularly Japanese, and I could thus make a new contribution to the study of problems which were sometimes theoretical but above all practical. I was at all times guided by the desire to transpose the results into the domain of application. This led me frequently to follow my studies in collaboration with other research bodies or with industry.

I shall give several examples relative to this 'heroic period':

– Studies on light alloys and ferrous metals – process of cementation.
– Action of electrons on silver bromide – application of the mechanism of photography.
– Study of long chain molecules and applications to lubrication.
– Study of the structure and transformations of high polymers.
– Research on smokes and fogs.
– Combination of electron diffraction and ion bombardment.
– Study of the structure of semiconductors, etc.

Naturally this list is not exhaustive and corresponds to the first period of my research. Since that time, considerable progress has been accomplished as much in the theoretical as the practical areas, but that comes into recent studies, carried out in the last 20–25 years, and therefore I shall not discuss them here.

To conclude, it is in the country of Louis de Broglie that the first applications of his theory were conceived which in turn led to basic research on electron diffraction and microscopy.

Académie des Sciences (France)

II.4. Electron Diffraction at Imperial College Physics Department 1930–1939

G. P. Thomson moved from Aberdeen to Imperial College, London, in 1930, and the electron diffraction group became active as soon as a number of electron diffraction cameras had been constructed. My own connection with electron diffraction commenced when I joined the physics department in October 1937; because of this, my account is based partly on the information contained in the monograph by Thomson and Cochrane (1939) and partly on what was common knowledge in the group. The work carried out is discussed below under appropriate headings.

1. THE POLISH LAYER

The work on mechanical polish concentrated on the constitution of the polish layer. This 'Beilby layer' (Beilby 1921) was considered by its discoverer to have the properties of a viscous liquid. The electron diffraction reflection patterns showed two diffuse haloes (French 1933). The positions of these haloes were not inconsistent with a liquid (or amorphous solid) metal. There was, however, a good deal of disagreement on this point – Kirchner (1932), for instance, maintaining that the surface layer consisted of very small crystals. Support for this view was obtained by Cochrane (1938), who deposited a thin layer of gold on nickel, polished the gold layer and then stripped it. The transmission pattern showed three diffuse rings, instead of two in the reflection pattern from the original polished gold or nickel. Another investigation on the polished layer formed on the cleavage face of calcite (Hopkins 1936) showed that on careful etching the diffuse haloes gradually sharpened and the number of diffraction rings increased.

The work on polish ceased around 1935, but it should be noted that parallel work was carried out (and continued) in G. I. Finch's group in the Chemical Engineering department at Imperial College – Professor Finch was a staunch adherent of Beilby's point of view.

2. INNER POTENTIAL

The inner potential measurements were carried out at that stage, mainly on crystals for which good flat surfaces could be formed – by cleavage. This restricted group included zinc and antimony (Darbyshire 1933). In a few cases, diamond and haematite for instance, good surfaces were provided by natural crystals. The measurements were usually carried out with 30 keV electrons, though Tillman (1934) used 5 keV electrons as well.

The work of Beeching (1935) on diamond (111) surfaces was particularly thorough and advanced. He used a Faraday cage to measure the intensity of the 222, 333 and 444 reflections in sufficient detail to obtain the half widths. These were found to be much larger than predicted by the Bethe two beam theory (for the 333 and 444) as well as showing the 'forbidden' 222 reflection to be as intense as the 333. In addition Beeching used the photographic technique to find the intensity and shape of the horizontal Kikuchi lines. The inner potential of diamond was obtained from these Kikuchi lines, as well as by the standard rotation method.

3. THEORETICAL WORK

Beeching's measurements on the abnormal width of the Bragg reflections from diamond led to a theoretical investigation by Harding (1937) of the effect of an absorbed layer on the (111) surface, together with a spacing change, on the Bragg reflection. Harding applied the Darwin approach of 'reflecting' planes to the problem, and his solution led from the Schrödinger equation to a set of difference equations. He found again total reflection over a limited angular region in the Bragg case, with the additional consequence that the centre of the region of total reflection occurred at slightly higher angles than the Bragg angle. (This corresponds roughly to Bethe's third-order approximation bringing in higher-order Fourier coefficients).

In order to explain the enhanced width of the diamond reflections, Harding assumed that the first two spacings at the surface were 10% smaller than the others, as well as having a changed reflectivity. He was then able to obtain an approximation to Beeching's experimental width.

It was one of my first duties, as theoretical assistant, to provide material on Bethe's dynamical theory of electron diffraction (Bethe 1928), for the monograph which Thomson and Cochrane had in a fairly completed state. In particular I worked out the Laue case in the two beam theory, as Bethe had concentrated on the Bragg case. Another problem which then arose was the very small dynamical width in transmission as compared with that for reflection – a problem which led to my first paper on the subject (Blackman and Thomson 1939). A second question which presented itself was the 'dynamical' intensity of diffraction rings in transmission. This was raised by a paper by Orenstein, Brinkmann, Hauer and Tol (1938) on diffraction by polycrystalline copper, which suggested deviations from the Mott formula. I had fortunately roughed out the main outlines of the dynamical effect before Professor Thomson also brought up the question. He had himself published a paper on the intensity of diffraction rings from polycrystalline gold in the series from Aberdeen (G. P.

Thomson 1929) which appeared to fit in reasonably well with the Mott formula. Though the main conclusion – that the 'dynamical' intensity could differ considerably from that predicted by kinematic theory – was clear cut, the details proved rather troublesome. I was very concerned about the use of discrete Fourier coefficients when dealing with very small crystals, and also by the limitations of the two-ray theory (Blackman 1939). It took rather more time to sort out these problems than Professor Thomson was prepared to accept. This period I am referring to was March 1939, when the implications of the nuclear chain reaction had become apparent; work had begun by a nuclear physics group, and I was needed to carry out calculations on the diffusion of neutrons. Because of this, the electron diffraction paper was completed in rather a hurry.

I was completely unaware of the investigations of Pinsker and his co-workers, in which kinematic theory was used in the structural analysis of thin inorganic films. My paper (as well as the experimental paper by Orenstein *et al.*) apparently caused a good deal of perturbation judging by the remarkable attack on 'bourgeois science' which appeared in the preface of Pinsker's monograph (Pinsker 1953; the original Russian edition is dated 1948).

4. OTHER INVESTIGATIONS

Of the investigations which did not involve a number of studies I have picked out three of particular interest.

Murison (1934) carried out a comprehensive investigation of oils, waxes and greases. Of particular note is the pattern found in reflection from lard consisting of a few diffuse lines parallel to the shadow edge. This was the first indication of long chain molecules oriented normal to the base, the parallel lines being due to the spacing of carbon atoms along the chain.

An unusual experiment on epitaxy was performed by Cochrane (1936) on the electrolytic deposition of nickel and cobalt on the etched (110) face of a copper crystal. When the deposited film was very thin, the diffraction pattern was characteristic in spacing of copper only, though thick films showed the normal spacing of nickel and cobalt. Cochrane concluded that he had found a case of what Finch and Quarrel (1933) termed 'basal plane pseudomorphism', in which the lattice spacing of the deposit matched that of the substrate. The second feature of his 'thin film' patterns was a classic example of spikes in reciprocal space, which was interpreted as due to repeated twinning. It is a matter of interest that von Laue, in his monograph, *Materiewellen und ihre Interferenzen* (1948) devoted considerable attention to Cochrane's photographs, which he interpreted as being due to a shape effect.

Of the many studies of chemical attack, one of particular historical interest is the investigation of catalytic platinum which marked the first contact between G. I. Finch's group on catalysis in the Chemical Engineering group at Imperial College and the electron diffraction group under G. P. Thomson. Films which were non-catalytic were produced in Finch's laboratory and the electron diffraction investigation by Thomson and Murison showed diffuse platinum oxide rings (Finch, Murison, Stuart and Thomson 1933). This collaboration led to the formation of a separate electron diffraction unit under Finch, based on a newly designed, and much improved, diffraction camera.

5. REMINISCENCES

When I joined the electron diffraction group (in late 1937) the day to day responsibility for the experimental work rested on William Cochrane, also an assistant lecturer. He had worked on the intensity of diffraction rings from polycrystalline metals in Glasgow, and had come to Imperial College a year or so earlier. Professor Thomson carried out a general tour of the research groups in the morning and spent a short time with each research student; in our group the research students were expected to have the diffraction cameras in working order and the latest diffraction plates ready for inspection. As his time of appearance did vary, and in addition he moved very softly, it was not easy for anxious research students to achieve advance notice; fortunately he had a very penetrating (though not loud) voice which carried from a neighbouring floor. The students were required to provide a quick briefing, after which Professor Thomson took charge. He had a remarkably quick mind and a remarkable critical ability. He was able to concentrate completely on the problem on hand, switching his mind off when he was satisfied and disappearing very abruptly from the research room.

As I was acting as theoretical assistant, he was liable to appear at any time in my room, usually with a half sheet of paper on which he had written a few equations or calculations. If he was really concerned about a problem, a consultation at 5.30 p.m. one day would be followed by another appearance next morning to find out what I had done in the meantime. He had an excellent grasp of classical physics, and I learned a great deal from him, particularly the ability to make order of magnitude calculations. Curiously enough, he never felt at home in Bethe's dynamical theory of electron diffraction; this was fortunate as far as I was concerned, as I had to absorb as well as apply the Bethe calculations to deal with a number of problems, the answers to which were unclear at that time.

My other main contact was with Cochrane, an excellent experimental physicist with a thorough knowledge of diffraction theory. At that stage the use of the reciprocal lattice was novel and had not penetrated as far as textbooks. Even review articles were few and not particularly rewarding. I certainly learned mainly from Cochrane how to interpret reflection patterns from single crystals. He was an efficient supervisor of the research students, particularly adept at dealing with leaks in the diffraction cameras – a general bugbear for inexperienced research students. Our cameras were covered with black wax, and in addition plasticine was used to cover particularly suspect areas. Smoothing plasticine was an art at which Cochrane was particularly skilled. It was a great loss to the group when G. P. Thomson moved him to nuclear physics after the war, and a great loss to the department when he left shortly afterwards to take charge of medical physics at St Thomas's hospital.

My most vivid memory of the experimental apparatus, apart from leaking cameras, is of the high tension apparatus. This was essentially an induction coil (the Rhumkorff coil) with a mercury make-and-break filled with coal gas, a diode and fairly lethal microfarad condensers. The cameras had greased core joints which seized up in cold weather. It was normal to place electric radiators in front of the apparatus when arriving on a cold winter morning. It was a matter of some envy, too, that our cameras were markedly inferior to those used in G. I. Finch's group. It is a tribute to the quality of the experimental group that they not only produced work of high quality, but that the atmosphere in the group was one of which I have happy memories.

Blackett Laboratory
Imperial College
London, England

REFERENCES

Beeching, R.: 1935, *Phil. Mag.* **20**, 841.
Beilby, Sir George: 1921, *Aggregation and Flow of Solids*. London, Macmillan.
Bethe, W.: 1928, *Ann. Phys. Lpz.* **87**, 55.
Blackman, M.: 1939, *Proc. Roy. Soc. A.* **173**, 68.
Blackman, M. and Thomson, G. P.: 1939, *Proc. Phys. Soc.* **51**, 425.
Cochrane, W.: 1936, *Proc. Phys. Soc.* **48**, 723.
Cochrane, W.: 1938, *Proc. Roy. Soc. A.* **166**, 228.
Darbyshire, J. A.: 1933, *Phil. Mag.* **16**, 656.
Finch, G. I. and Quarrel, A. G.: 1933, *Proc. Roy. Soc. A.* **141**, 400.
Finch, G. I., Murison, C. A., Stuart, N., and Thomson, G. P.: 1933, *Proc. Roy. Soc. A.* **141**, 414.
French, R. C.: 1933, *Proc. Roy. Soc. A.* **140**, 637.
Harding, J. W.: 1937, *Phil. Mag.* **23**, 271.
Hopkins, H. G.: 1936, *Phil. Mag.* **21**, 820.
Kirchner, F.: 1932, *Nature* **129**, 545.
Murison, C. A.: 1934, *Phil. Mag.* **17**, 201.
Orenstein, J. S., Brinkman, H., Hauer, A., and Tol, T.: 1938, *Physica* **5**, 693.
Pinsker, Z. G.: 1953, *Electron Diffraction*. London, Butterworth.
Thomson, G. P.: 1929, *Proc. Roy. Soc. A.* **125**, 352.
Thomson, G. P. and Cochrane, W.: 1939, *Theory and Practice of Electron Diffraction*. London, Macmillan.
Tillman, J. R.: 1934, *Phil. Mag.* **18**, 656.

H. MARK

II.5. Early Work on Electron Diffraction from Gases

Sparked and sustained by significant progress in vacuum techniques and by improvements in calorimetry, a new branch of physics developed around the turn of the century, namely *cryogenics*, and with it physics and chemistry at very low temperatures. Two laboratories in Europe were leading in this discipline: that of Kamerlingh-Onnes in Leiden and that of Walter Nernst in Berlin. Both contributed enormously important experimental observations to the start of modern physics in its infancy; Kamerlingh-Onnes discovered superconductivity, a strange and incredible phenomenon which resisted a rational explanation for almost 50 years; Nernst (1914) on the other hand, established through careful and systematic experiments the anomalous behavior of ideal gases at very low temperatures, which also posed a startling enigma, because it could not be explained by the application of Boltzmann's statistics, which treat each molecule as a particle that maintains its individuality during its distribution over the cells in the phase space. A correct interpretation was only possible through a different statistical treatment which was first introduced by S. N. Bose (1924) and later expanded by Einstein (1924, 1925). In his article Einstein comments that in the degenerate state the gas molecules do not behave like permanently distinguishable particles but more like waves which interfere with each other and do not simply collide like solid particles. This was the first published inkling of the wave character of moving particles; it occurred in 1924, one year before de Broglie's fundamental paper (1925) which described in all details the particle-wave correlation by the equation

$$\lambda = \frac{h}{mv} \tag{1}$$

where m and v = mass and velocity of the moving particle; h = Planck's constant; and λ = wavelength coordinated to the particle.

The first experimental verification of this relation came for slow electrons from Davisson and Germer (1927) and for fast electrons from Thompson (1927). Immediately thereafter electron diffraction phenomena were used for two different research purposes: first to obtain information on the basic properties of the electron, particularly its spin, and second to use the electron waves to study the structure of matter in the same manner that had been in use with X-rays already for 15 years (Mark, 1926). Electron diffraction, however, offered new opportunities for structural studies in comparison with X-ray diffraction. X-rays are scattered by the electrons which surround the atoms of the scattering materials whereas electrons are primarily scattered by the electric field of the positively charged nuclei. The main consequence is a much stronger interaction of the electron beam with the irradiated material which clearly manifests itself in the much larger absorption of electrons (beta radiation) in comparison with X-rays (gamma radiation). Electron diffraction, therefore, provides an excellent means to study the *surface* structure of crystalline materials; it also makes it possible to localize hydrogen atoms – protons – in a lattice which cannot be 'seen' by X-rays because their only electron is used for bonding them to the structure (Mark and Wierl, 1930). If one proceeds from the scattering by a crystalline body to that by isolated molecules the strong interaction between the electron beam and the irradiated system gives much higher scattered intensities and permits a shortening of the exposure times by a factor of between 1000 and 10 000 in comparison with X-rays. Debye (1929) had already established that X-ray scattering from gases produces certain characteristic patterns but that the exposure times of several days rendered this approach difficult and impractical. It occurred to me that the scattering of fast electrons from gases should lead to much shorter exposure times, to a much better control of the experimental conditions and to the possibility of longer systematic studies of the interatomic distances. When I asked Dr. Raimund Wierl, one of the three high-level physicists in our laboratory, of his opinion he agreed that such experiments would be very interesting but certainly not easy. Wierl had received his PhD *summa cum laude* with Professor Willy Wien in Munich and had an excellent training in the physics of high vacuum and high voltage. The difficulties of the new experiment were rather formidable. The electron beam had to be *narrow* and well *collimated*, which is difficult to achieve because the negatively charged electrons repel each other as they travel together close to each other over a distance of a few centimeters. The beam has to be *monochromatic*; according to Equation (1) that means that all electrons should have the same velocity. In a strict sense this is impossible and there is always a certain velocity distribution which, for the purpose of this test, would have to be narrowed as much as possible. This electron beam would have to impinge perpendicularly on a jet stream of the gas which had to be as narrow and as dense as possible in order to give a maximum of interaction between the electron and the scattering molecules. All this had to happen in a vacuum camera in order to avoid any scattering by a gas that did not belong to the jet stream. Evidently the execution of this experiment required inventiveness in instrument construction and extreme care and skill in the execution. Fortunately Wierl had both to an admirable degree and already a few weeks after our first conversation he came with a beautiful photograph of carbon tetrachloride produced with 45 kV electrons in 1/10 of a second. What a tremendous difference between this test and the daylong exposure with X-rays with much more diffuse patterns of lower contrast! We had selected CCl_4 for these first experiments because the heavy highly charged chlorine nuclei would be responsible for most of the deviation of the electron beam and the pattern would simply reflect the Cl $\cdots$ Cl distance in the tetrahedral molecule; knowing this distance it is easy to calculate the C—Cl distance which was found to be 1.82 Å. This method, therefore, permits the direct experimental determination of the distance between atoms which are joined together by a covalent bond. In the late 1920s such 'interatomic distances' were of considerable interest for arriving at *quantitative* models of organic molecules. Some of them such as the C—C and C=O distance had been determined with some difficulty from X-ray crystal analysis and were known as the classical 'Bragg atomic radii', but they had to be somewhat laboriously separated from the much more pronounced effect of lattice scattering and the results of different independent sources did not at all agree in a satisfactory manner. Now electron diffraction of molecules in the gas phase offered the chance to measure these interesting and important quantities *directly* by the study of simple molecules in their natural state, namely in the gas phase. Thus the C—H distance was determined from CH_4, the C—F from CF_4, the C—C from ethane (C_2H_6) and propane (C_3H_8) and more complicated bonds like C=C from ethylene, C≡C from acetylene and so on. Wierl (Mark and Wierl, 1930) demonstrated the utility and reliability in a few typical cases. He found, for instance, for the aromatic C—C distance in benzene 1.41 Å, but for the aliphatic C—C distance in cyclohexane 1.58; the latter agrees very well with the 'Bragg value' for aliphatic chain molecules obtained from X-ray studies of aliphatic fatty acids. K. H. Meyer and I had proposed (Meyer & Mark, 1930) quantitative chain models for cellulose, silk, rubber, chitin, and starch with the use of the then available, approximate distance values for C—C, C=C, C—O and C—N; it was evident that better data for these fundamental bond distances would permit the construction of much more reliable models for these and other, even more complicated systems such as myosin, globulin and keratin. Obviously a large field for fundamental research had been opened by the new method and was inviting elaborate systematic work. But, our laboratory was part of an industrial organization and not part of a university. The members of our top management, Carl Bosch, K. H. Meyer and A. Mittasch, progressive and enlightened as they were, had viewed with satisfaction the development of the new method but certainly would disagree with extended studies of interatomic distances. In fact, Wierl already was aiming his electron beams on the surfaces of catalysts and of magnetic tapes.

Fortunately, at that time, Linus Pauling who had spent several months with Professor Sommerfeld in Munich visited our laboratory and was told and shown all we had established on the structure of polymers by X-ray and electron diffraction. His own interest was focussed on *quantum chemistry* with all its ramifications and he was, therefore, looking for all available evidence of a quantitative character. Electron diffraction was a very promising method to provide some of this background. Pauling immediately liked the new method and we gladly gave him everything we had: construction of the camera, operating conditions, precautions, possible sources of errors, etc.

It is well known from the literature how much Pauling improved our original procedure in his own laboratory in Pasadena, how he collected essential new data and how much all this improved and extended information ultimately contributed to the elucidation of the protein and nucleic acid structure.

Brooklyn Polytechnic Institute
New York

REFERENCES

Bose, S. N.: 1924, *Z. Physik*, **26**, 178.

Davisson, C. J. and Germer, L. H.: 1927, *Phys. Rev.* **30**, 705.

de Broglie, L.: 1925, *Ann. Phys. (Paris) Ser. 10*, **3**, 22.

Debye, P.: 1929, *Physik. Z.*, **30**, 84.

Einstein, A.: 1924, *Sitzungsber. Preuss Akad. Wiss., Phys.-Math. Kl.*, 261.

Einstein, A.: 1925, *Sitzungsber. Preuss Akad. Wiss., Phys.-Math. Kl.* 3.

Mark, H.: 1926, 'Die Verwendung der Röntgenstrahlen in der technischen Chemie' in *Ergebnisse der technische Röntgenkunde*, Leipzig, pp. 137–144. (J. A. Barth, ed.).

Mark, H. and Wierl, R.: 1930, *Naturwissenschaften*, **18**, 778.

Mark, H. and Wierl, R.: 1930, 'Die experimentellen und theoretischen Grundlagen der Elektronenbeugung', in *Fortschritte der Chemie, Physik, und Physikalischen Chemie*, vol. 21 (ed. A. Euken) Berlin: Goettingen Verlag von Bebrueder Berntrager.

Meyer, K. H. and Mark, H.: 1930, *Der Aufbau der hochpolymeren organische Naturstoffe, auf grund molekular-morphologischer Betrachtungen*. Leipzig: Akademische Verlagsgesellschaft.

Nernst, W.: 1914, *Z. Electrochem.* **20**, 357.

LAWRENCE O. BROCKWAY†

II.6. Diffraction by Gases

My own connection with electron diffraction began in the fall of 1930 when I appeared as a new graduate student at the California Institute of Technology. It was the custom there to start each new student on a research project very early in his first term, and when I expressed an interest in the structure of matter the department chairman, Professor A. A. Noyes, turned me over to Linus Pauling. Pauling's first suggestion was that I should embark upon the seas of crystal structure determination by X-ray diffraction. For some reason still unknown to me the various projects he suggested seemed unattractive and I kept refusing. Finally, in desperation he spoke of an experiment he had seen in the summer of 1930 while he was visiting the laboratories of the I.G. Farben Industrie carried out by Mark and Wierl. They had given Pauling some prints of diffraction

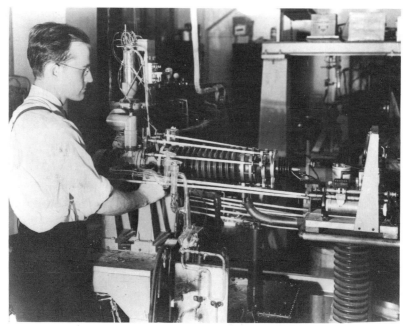

Lawrence Brockway with gas-phase diffraction apparatus, California Institute of Technology (about 1935).

† Unfortunately L. O. Brockway died in 1979, before this book could be published.

patterns obtained from carbon tetrachloride using a stream of electrons crossing a jet of the vapor.

Although there had been no publication describing either the equipment or the method of interpreting the recorded pattern, I felt I should agree to try the experiment before Pauling became completely disenchanted with his new graduate student. It was agreed that I should turn to Professor Richard M. Badger, an infrared spectroscopist, for guidance in the design and construction of a vacuum unit enclosing a source for an electron beam, a vapor jet and photographic plates for recording the scattered electrons. Our first try utilized a hydrogen discharge tube as a source of electrons with the hydrogen pressure adjusted by temperature control on a palladium filter. This was so unreliable that we next turned to a heated filament surrounded by an electron lens with adjustable bias potential. The vacuum pumps on the system were of the mercury-in-glass type necessarily constructed in our own shop but the pumping speed was too slow to cope adequately with the jet of vapor injected for each exposure. Nonetheless, after about three months we had obtained diffraction patterns from thin gold foils and also from carbon tetrachloride gas, showing barely detectable maxima and minima.

My contact with Pauling in the meantime had been very scant, and he was surprised and delighted when I was able to show him diffraction patterns made in our laboratory. After a series of improvements in the apparatus we began to get much clearer patterns and proceeded to apply the method to as many substances as we could readily obtain. It was early in 1931 when the first publication from Wierl's experiments appeared, and we used his simplified expression relating the intensity of scattered electrons at various angles to the interatomic distances in the gaseous molecules.

The calculation of theoretical intensity for various molecular models involved a double summation over terms of the type $(\sin sr_{ij})/sr_{ij}$. Our first calculations used a plot of $(\sin x)/x$, drawn on a very large scale, from which we tabulated values of the function with an appropriate adjustment of scale factor for each of the terms. The next improvement was a table of $(\sin x)/x$ constructed by Jack Sherman. After that Paul Cross assisted in the preparation of sets of strips containing $(\sin ax)/ax$ values with each strip having a fixed a value. To carry out the calculation for a given molecular model the procedure boiled down to selecting the set of strips with a values corresponding to the interatomic pairs in the model and then summing up at each value of the angular coordinate x the terms on the strips with an appropriate coefficient applied to each one. At first the strips were laid out on a drawing board with a straight edge to mark the particular x value being summed, but shortly we devised a cylinder around which the strips were wrapped, and later students and I spent many days with the cylinder-mounted strips adding up sums of products with the aid of the then standard desk calculator. Because of the extensive labor involved in the calculations, the selection of models had to be judged carefully or we found ourselves spending weeks on models which were unrelated to the structure at hand.

The comparison between theoretical curves and the experimental data was based at first on a comparison of the positions of the maxima and minima in the curves. The diffraction negatives were observed visually and the diameters of the apparent maxima and minima were measured using fine pointers. Measurements of the negatives on a recording microphotometer failed to show any maxima and minima but only fluctuations about a very rapidly falling background. For this reason our observations of the first hundred or so substances we analyzed were all reported in terms of the diameters of the maxima and minima which appeared in the visual measurements.

By 1933, when I received my doctoral degree, other laboratories had entered the field and major improvements in both the experimental technique and the methods of calculation were beginning to appear. I stayed on at the California Institute of Technology for four years, and then joined the faculty of the University of Michigan.

The last electron diffraction unit I designed and constructed at CalTech was supposed to be greatly improved and included a cylindrical camera which could record electrons scattered up to about 165°. I am not sure what enthusiastic dream led to this, but it is no surprise now that our molecular patterns never even reached the edge of our flat plate camera, and that the cylindrical camera was relegated, untested, to a museum.

Subsequent developments are well known. Experimentally the most important of these was the introduction of a rotating sector before the recording photographic emulsion; the sector opening could be cut in such a way as to compensate largely for the rapidly falling background, and real maxima and minima then appeared in the microphotometer records. The fit between calculated and observed modified intensity curves was considered over the whole continuous range of scattering angles. This was a great improvement over the primitive method of looking only for positions of maxima and minima. Major improvements in the calculating methods lay first in the use of better computing techniques, such as IBM punched cards, and later in the introduction of more complicated theoretical expressions, made feasible by the advent of high speed computers. The earliest refinement, which led to great improvement in the comparison between experiment and theory, was the use of proper atomic scattering amplitudes, expressed as functions of the scattering angles.

The developments after the late 1930s are fairly well known. My own recollections of the earliest days are centered around the sense of excitement and fun, and an appreciation of the opportunity to work in a major scientific development while still enrolled as a new graduate student.

L. E. SUTTON

II. 7. *The Earlier Studies in Great Britain of the Structure of Molecules in Gases and Vapours by Electron Diffraction, with an Epilogue*

1. THE WORK AT UNIVERSITY COLLEGE, LONDON

The first studies in Great Britain of the structures of free molecules by electron diffraction were made in the Sir William Ramsay Laboratory at University College, London. In the late 1920s or early 1930s the head of that laboratory, Professor F. G. Donnan, F.R.S., persuaded Imperial Chemical Industries Limited, newly formed (1926) by the merger of a number of companies, to support the formation of a group of senior research fellows to work freely on topics of their own choice. Donnan had become very influential in industrial circles because he had shown outstanding ability to apply science to practical problems during the 1914–18 war. He had worked closely with a former pupil, Dr F. A. Freeth, F.R.S., who was appointed a joint research manager of the new company in 1927. Dr Henry de Laszlo, son of an eminent portrait painter, became a member of the group and decided to build an electron-diffraction camera for gases and vapours. He must have done this in 1930 or 1931. Quite why he did we shall now never know for he was killed tragically in a road accident in the 1960s. What is certain is that he had a keen commercial sense – he later founded a very successful chemical manufacturing and supply company – so perhaps in addition to any scientific ambitions that he had he may have thought he saw an industrial future for this technique, as others did for infrared absorption spectroscopy. It was certainly a bold, pioneering venture. Herman Mark's first paper was published only in 1930: his famous work with Wierl, done at Ludwigshafen-am-Rhein, was a development of earlier work by F. Bewilogua and L. Ehrhardt, who were pupils of P. Debye at Leipzig, on the diffraction of X-rays by vapours. I happened to have seen this latter in progress (see below) without realizing at the time how it would later impinge on my interests. The original suggestion of using an electron beam was made by W. Bothe at a 'Röntgentagung' in Zürich after Debye had described the X-ray work (Mark and Wierl, 1930).

After the demonstration in 1927 of electron diffraction by solid materials, by Davisson and Germer in the United States of America and by G. P. Thomson in Scotland, the technique had been taken up by G. I. Finch at Imperial College,

London. He built a camera in the early 1930s and used it for the study of surfaces (Blackman, 1972). De Laszlo partly modelled his camera on Finch's. The cold cathode gun and the high-tension supply were much the same, but since it had no magnetic lens for focussing the beam it was somewhat less sophisticated than Finch's.

De Laszlo was the general designer. Some of the detailed design and all the assembly and commissioning were done, with little supervision, by Dr V. E. Cosslett (Cosslett, 1979) who had been offered a research assistantship with the University College group and who came there in September 1931 by which time many of the major components for the camera and the high-tension system had been made or bought: he had the whole system operating by May 1932. The camera was described by de Laszlo (1934a) who stated that over two hundred substances had been examined. Results were published for only nineteen (de Laszlo, 1934b, 1935): these were all organic halides. Cosslett left in August 1932 and later turned to electron microscopy in which he has done outstanding work: he was elected a Fellow of the Royal Society in 1972 and in 1979 he was awarded a Royal Medal. In June 1934 Donnan told Dr A. H. Gregg, who had come from Queen's University, Belfast, to work with the I.C.I. group in 1932, that he was to take over the camera because he was an expert on high vacuum technique and because de Laszlo's connection with the group would be ceasing. Gregg was not attracted by this change of activity because he knew nothing whatever about electron diffraction. For five months he was not allowed near the camera but eventually he took it over near the end of 1934. De Laszlo's technical assistant, P. L. F. Jones, stayed on to work with Gregg who improved the voltage measuring system and then took photographs of halides of phosphorus, arsenic and antimony (Gregg, 1979; Gregg, Hampson, Jenkins, Jones and Sutton, 1937).

2. THE WORK AT OXFORD: ITS ORIGINS AND DEVELOPMENT

In about 1932 I had become curious about the improbably large Cl—C—Cl valency angles in methylene dichloride and chloroform that were obtained from a very simple-minded treatment of the electric dipole moments of methyl chloride, methylene dichloride and chloroform. While it could be shown that the calculated angles were less if inductive effects were allowed for, this could not be done with any certainty so no satisfactory derivations of valency angles could be made from these data. During his first visit to the U.S.A. in 1931 my old tutor, Professor Nevil Vincent Sidgwick, F.R.S., with whom I had begun research in 1927, had met Professor Linus Pauling and they had immediately become great friends. Pauling told Sidgwick that electron diffraction studies should settle this problem and that one of his research pupils, Lawrence Brockway, was starting to use this technique in his laboratory. Because of this, and because the girl whom I married in 1932 – Catharine Virginia Stock – was American and I had not yet seen her native land, I decided that it would be a good thing if I could go to Pasadena to learn this new technique. I may say that I had already worked for six months in Leipzig (September 1928 to March 1929) with Professor P. J. W. Debye when I had learned the technique of measuring electric dipole moments so I was keen to import into Oxford another new physical technique for studying molecular structure. I was working in the organic chemistry laboratory at Oxford, the Dyson Perrins Laboratory, the head of which was Professor (Sir) Robert Robinson, F.R.S. Largely on his recommendation I was awarded a Rockefeller Foundation Fellowship which enabled me to go to CalTech in September 1933. I spent a year and one term there. It was a wonderfully stimulating experience to which I look back with unalloyed pleasure. Pauling and his group were making giant strides in bringing new, clarifying ideas into chemistry and it was very exciting to feel that one was taking part in this. Lawrence and I succeeded in determining the Cl—C—Cl valency angles that I was concerned about and found them to be only slightly larger than the tetrahedral angle. We also determined some oxygen angles. The people working with Pauling were delightful and extremely able. I have kept in touch with several of them ever since. George Wheland, alas, died in the early 1970s and Lawrence Brockway has died since I started drafting this article. They both had many friends who mourn deeply at their passing. Lawrence was the person I was closest to in that group, and I would like to take this opportunity to pay tribute to his power to stimulate, his generosity, his sound common sense and his integrity to say nothing of his dry, irreverent wit.

At the end of 1934 I returned to an England which seemed grey, cold, dank and very unlike California. Early the next year I started trying, through Sidgwick who also knew Freeth well, to get support from I.C.I. for building a camera similar to Brockway's second one, of which he had, characteristically, given me full working drawings. However, because de Laszlo's connection with the University College group and with electron diffraction had now ceased, I.C.I. said that they would instead arrange for the de Laszlo camera to be transferred to Oxford and this happened in late 1935. I cannot now remember why, but I know that we bought from de Laszlo parts of a second camera, some of which were later used. Gregg came to Oxford with the camera and he was of immense help in getting it re-started. We could not have managed without him. In Pasadena I had been introduced to electromechanical calculating machines. I had used a Munro. With a grant from the Royal Society I was able to buy a German Mercedes-Euklid machine which had a memory. Also I was able to buy a set of $(\sin ax)/ax$ strips which were painstakingly printed out by (Professor) Paul Cross whom I had met at CalTech where he was working with Professor R. M. Badger.

Gregg went to Manchester in 1937. Apart from what he did, all the experimental work was done by my research pupils, the earliest of them being G. C. Hampson, G. I. Jenkins, M. W. Lister, J. A. C. Hugill and H. A. Skinner. Hampson, Jenkins and Lister looked mainly at metallic halides, Hugill at organic halides and Skinner at organometallic compounds. The old high-tension generator began to give a deal of trouble so we had to replace it by a 50-cycle system bought from Metropolitan-Vickers Ltd. This included a generator of stabilized, 250 volt a.c. which consisted of a three-phase synchronous motor, with alternator and exciter dynamo on the same shaft. For our convenience this was installed in the room next door to the camera room, which was a men's toilet. The machine started up with a very loud and sudden whir. There were complaints.

In 1937 Lawrence Brockway, who was to assume a post at the University of Michigan at Ann Arbor, came to Oxford for a year on a fellowship and helped to design a camera which was intended to replace the one from University College, London. He also used the old one, with I. E. Coop, R. V. G. Ewens and M. W. Lister to examine some halogenoacetylenes, chlorsilanes and some metal hexacarbonyls. Parts of the new camera were built by 1939, although there were difficulties about this because the Dyson Perrins Laboratory had no workshop so the work had to be done by permission in the Inorganic Chemistry Laboratory workshop. The outbreak of war put an end to this. What had been completed was put into a box and left in a corner, gathering dust until 1947 when Henry Mackle and Peter Allen joined me in an effort to revive electron diffraction. (Professor) O. S. Heavens had previously spent a short time with me designing an electron gun as I remember but soon left to take a post elsewhere. The gun design was later modified by Peter Allen. The remaining parts were made in the workshop of the new Physical Chemistry Laboratory to which I had moved in about 1941 and Mackle and Allen, both of them very competent and energetic, had the whole camera assembled in 1948 (Mackle and Allen, 1951). They both have stories to tell of that exciting time. The head of the workshop was a very good mechanic and organizer but a fiercely independent character, with a remarkable flow of earthy language, who had to be handled with great care and respect when we wanted things made. Peter Allen says that Henry Mackle, whose father was a farmer in Ulster, found that the occasional pound of butter was a very useful lubricant of relations in those austere days. Henry remembers nothing about butter but asserts that Peter's brisk, quarter-deck manner (Downside and Christ Church) did not always work and modestly admits that his own Irish charm was more effective. Henry undoubtedly had a way with him, well illustrated by an incident, best told in his own words, concerning the setting up of the high-tension generator in its new home in the Physical Chemistry Laboratory which was presided over by that formidable genius Professor Sir Cyril Hinshelwood, later President of the Royal Society and a Nobel prize-winner.

You had pointed out to me that we really needed a stress distributor and had, by fair means or foul, managed to get me a couple of copper hemispheres about 8″ in diameter to use for this purpose. They needed, of course, to be mounted on an insulator. Fortunately, the new Botany Department was being built next door at the time so I went out to the builders and just asked the first young man I met for a 6″ i.d. ceramic drain pipe. I think I asked with enough authority to convince him that I was a foreman of some sort, because he gave it to me immediately. I then asked him for a bucket of cement grout, which he also got for me. Unfortunately, just as I was going in through the main P.C.L. door, I walked straight into the late Sir Cyril Hinshelwood, who stopped in his tracks, literally, but only momentarily. I did not stop at all and proceeded down the stairs to the basement, all too conscious that Hinshelwood was in fact silently following me. I went into the High Tension room, put the drain-pipe (which of course was flanged at the bottom end) in an upright position on the floor inside a wooden box 2′ × 1′ × 6″ deep, and poured the bucket of cement grout round it, and smoothed the whole thing as any professional plasterer would have done. It was only then that Hinshelwood asked me what, in fact, I was doing. I told him I was strengthening the foundations of E.D. studies in the P.C.L., and he shook his head, went away apparently either puzzled or satisfied and, no doubt, convinced that what he had frequently been told in Oxford was true, viz. all Irishmen are mad.

I think that Hinshelwood realized that for once he had met his match.

The new camera, somewhat modified over the years, is now in the Science Museum at South Kensington in London. It gave better results than the de Laszlo camera but it was disappointing and gave enormous trouble because of air leaks. Vincent Ewing eventually discovered, by machining off the outer skin, that the worst of these were in the plate box where heat used in welding two pieces of brass together had caused the zinc to be lost from some areas leaving the metal, in Vincent's words, 'like a well-matured cheese.' It must be remembered that in those days vacuum technology was still relatively primitive. There was not too much reliable information about what metals to use or how to use them. Wilson seals were known but O-rings were not yet on the market. A common practice was to cover the metal with a hard varnish. Later, Mr M. E. Haine of Metropolitan-Vickers gave us much helpful advice but could not save us from the effects of earlier errors. By 1947 the camera was, moreover, already obsolete.

Trendelenburg (1933) had described a simple system for reducing the fall-off of intensity with diffraction angle in electron diffraction photographs of powders: it was to use a rotating sector, with a simple rectangular slot between the original negative and a copying, positive plate. He suggested that it would be even better to have a rotating sector in front of the original plate, in the vacuum chamber; but he did not try this system which, he said, 'presents not insignificant difficulties', because the simpler one was adequate for his purposes. In 1937 Christian Finbak (1937) described a camera with a rotating sector, and in 1939 P. P. Debye (P. J. W. Debye's son) did so independently although in a later and fuller paper (P. P. Debye, 1939) he referred to Trendelenburg's suggestion and to Finbak's paper which he had found through the *Chemisches Zentralblatt*. During and immediately after the War, in Norway and the U.S.A., it was confirmed that the rotating sector was both effective and workable: it was a major advance, enabling a much more rigorous and quantitative examination of the diffraction pattern to be made. Professor Odd Hassel had invited me to Oslo in September of 1946 in order to re-establish contact, and I attended an electron-diffraction symposium in Cornell University in the summer of 1947; so I was well aware of recent developments. What was not clear was what I could do about them. I may say parenthetically that my visit to Oslo made a deep impression on me. Hassel was not long out of a German concentration camp and I heard much about the resistance movement from Otto Bastiansen. I remember that Hassel lived in a fragrant wooden house and that his hospitality was superb. He explained his ideas about what are now called axial and equatorial bonds in ring systems. It was then that I met Christian Finbak and saw his camera, and Henry Viervoll who had provided the theoretical basis for the analysis of photographs. My first visit to the U.S.A. after the War was also most memorable. Apart from the pleasure of seeing Lawrence Brockway again I met

Jerry and Isabella Karle and Simon Bauer. Probably, too, it was then that I first met Lawrence Bartell, Robert Livingston, Kenneth Hedberg and Verner Schomaker. All of them were pupils or grandpupils of Lawrence Brockway.

In spite of the defects and limitations of our new camera we used it for about ten years. In addition to Allen and Mackle, other co-workers in that time were A. D. Caunt, H. J. M. Bowen, C. J. Wilkins, Erwin Seibold, Andrew Gilchrist, David G. Jenkin and Vincent Ewing. Among the substances examined were acetyl halides, acraldehyde and crotonaldehyde, carbon suboxide, some silicon compounds, allyl and t-butyl halides, ferrocene, some inorganic fluorides and some trifluoromethyl compounds. Only the 'visual' method of examination could be used, but Erwin Seibold put a great deal of work into ingenious attempts to improve this, with some success. Henry Mackle remembers his work on acraldehyde very vividly, because he forgot to remove the liquid-air cooled trap from the camera so the stuff escaped to the workshop nearby. The remarks of the head of that department, made with tears in his eyes, about Henry's ancestry and future destiny, while not entirely original, were made with a conviction, a forcefulness and a richness of invective that Henry has never heard surpassed. Acetone was examined in collaboration with the Oslo laboratory: Vincent Ewing took some disilyl ether there, as 'deck cargo', and had a most enjoyable time solving its structure. We tried to improve the camera. I tried to attract enough money and manpower to enable us to design and build a newer one with a rotating sector and to buy a suitable microphotometer but these attempts, though not entirely fruitless, were quite inadequate. Despite the valiant efforts of Andrew Gilchrist, David Jenkin, Erwin Seibold and, especially, Vincent Ewing, no great improvements were made and no sector camera got built. I was busy with teaching, with the other chores that come from a College tutorship and a University lecturership, with some dielectric research and with a special task that I will enlarge on later. Moreover I was very conscious that the newer methods for analysing photographs required a degree of mathematical knowledge and expertise that I did not have and could not expect to acquire. Consequently, after Vincent Ewing decided to give up academic work and take an industrial post in 1960, I decided that I must abandon work on electron diffraction. Since then I have been an interested spectator.

3. THE TABLES OF INTERATOMIC DISTANCES

It is appropriate to say something about a related activity to which I have already referred obliquely. While waiting for pieces to be made in 1947, Peter Allen, always a very lively person, busied himself making a compilation of structure determinations for free molecules by electron diffraction. This was published in *Acta Crystallographica* in 1950 (Allen and Sutton, 1950), thanks to the good offices of Dr Dorothy Hodgkin, F.R.S.: it was of modest size, i.e. 27 pages. Thereafter various people urged that a revised version should be published. Humphrey Bowen tried for a while to keep the catalogue up to date. It became increasingly clear that there was too much to publish as a paper but not enough to make a saleable book. The addition of crystallographic results for molecules and molecular ions was obviously necessary to give a book of suitable size. Moreover it made no scientific sense to limit the compilation to gas-phase results or to structures obtained only by diffraction methods. Gradually, therefore, there grew the idea of a comprehensive book covering results from all possible methods for molecules and ions in any phase. I was very reluctant to become deeply involved, but it became clear that unless I assumed responsibility nothing would be done. From 1951 to 1957 I was an Honorary Secretary of the Chemical Society and I was able to persuade that Society to undertake the publication of this ambitious work. Promise of an advance purchase by the U.S. Army helped the Society to set an attractive price ($£2.10$ for non-Fellows and $£1.80$ for Fellows of the Society). Quite how the organization grew I cannot now remember; but a great deal of willing help came. (Professor) David Whiffen undertook responsibility for the spectroscopic entries and David Jenkin that for the electron-diffraction entries. At first Professor Jerry Donohue was responsible for the X-ray entries but later Mrs Olga Kennard took this over with Dr Peter Wheatley looking after the inorganic entries. Each had several helpers. I was enormously helped in the detailed organization by David Jenkin. By my standards the task was a huge one but by the enthusiasm and hard work of all who were concerned we finished it: the book was published in 1958 as *Tables of Interatomic Distances in Molecules and Ions* (Sutton, 1958). Writing the Introduction was a big task but I learnt some of the elementary ideas about symmetry and enjoyed this. Some other tasks were generally educative and interesting but there was a vast amount of administration, detailed checking of entries and drafting of diagrams for complicated molecules which I found tedious. A supplementary volume was published in 1965 (Sutton, 1965), this time with a rather different organization and more assistance with routine tasks, but after this I had had enough. The effort necessary to produce the compilation was such that I found no time to consider the meaning of the mass of data, although I must in honesty say that I doubt if I should have produced as good ideas as others have about the influences affecting interatomic distances. I was, too, very disappointed that, despite all our efforts, too many errors had sneaked through and that the task had taken so long to complete. It was clear to me that a professional organization was essential for the future. With the help of the Chemical Society, and particularly of Dr L. C. Cross who had by then become, I think, its Director of Publications, conversations were held with the Office for Scientific and Technical Information; but a major difficulty was that at that time no young person of adequate ability was willing to contemplate any career save one of research. No doubt I was showing the same attitude when I decided that I could not continue making compilation a major interest. I reckoned that my programme of publishing original papers had been set back by at least five years. So I left the field. Others have taken over (see Kennard and Watson, 1972; Landolt-Börnstein, 1974, 1976). It is ironic that what is probably the most useful task that I have performed was one that I did not want. In recent years I have returned to the field in a rather different way by acting as the senior reporter, responsible for electron diffraction, for one of the Chemical Society's Specialist

Periodical Reports *viz.* that for *Molecular Structure by Diffraction Methods* (Volumes 1–6). The long term aim has been to discuss all aspects – theory, methodology, apparatus, results and their interpretation. Most of the hard work has been done by a series of most admirable authors, and I feel that we can claim a good degree of success. There is only one trouble: sales have been much too low.

4. EPILOGUE

By way of epilogue I may tell how in the early 1960s I heard from Professor Durward Cruickshank that he had a grant, or hopes of one, which would enable him to buy one of the gas diffraction cameras which were then being built by the firm of Trüb, Täuber A.G. in Zürich (now owned by Balzers A.G. of Trübbach, also in Switzerland) to a design evolved between Professor W. Zeil and the firm (see Oberhammer, 1976). He wanted my advice. We went twice to Zürich, on the second occasion taking Dr Brian Beagley, who was to be in charge of the camera, and we suggested a number of modifications. Durward got his grant. The camera was installed at Glasgow and, after a few teething troubles, has been used successfully over a long period by a succession of workers. When Durward was later transferred to the University of Manchester Institute of Science and Technology Brian Beagley and the camera went with him.

Much more recently (*circa* 1977) Dr David W. J. Rankin at Edinburgh obtained a grant with which he purchased the latest camera that Simon Bauer had built at Cornell and which he no longer wanted because he was leaving the field. If it has not accepted political independence, Scotland has at least now got its own electron diffraction camera. In 1979 Dr Alan G. Robiette of Reading completed the building of a camera to his own design and with several new features. Altogether there is brisk activity today in this field of research in Great Britain and it is satisfying to be able to conclude on this hopeful note.

REFERENCES

Allen, P. W. and Sutton, L. E.: 1950, *Acta Cryst.* **3**, 46–72.
Allen, P. W., Mackle, H., and Sutton, L. E.: 1951, *J. Sci. Instrum.* **28**, 144–151.
Blackman, M.: 1972, *Biog. Memoirs Fellows Roy. Soc.* **18**, 223–239.
Cosslett, V. E.: 1979, personal communication.
Debye, P. P.: 1939, *Physikal. Z.* **40**, 66 and 404.
de Laszlo, H.: 1934a, *Proc. Roy. Soc.* **A146**, 672.
de Laszlo, H.: 1943b, *Comptes rend.* **198**, 2235.
de Laszlo, H.: 1935, *Nature* **135**, 474.
Finbak, Chr.: 1937, *Avh. norske Videnskap Akad.* M.-N., Kl. 1937 no. 13.
Gregg, A. H.: 1979, personal communication.
Gregg, A. H., Hampson, G. C., Jenkins, G. I., Jones, P. L. F., and Sutton, L. E.: 1937, *Trans. Faraday Soc.* **33**, 852–874.
Kennard, O. and Watson, D. G. (eds.): *Interatomic Distances* 1960–65; *Organic and Organometallic Crystal Structures*, Vol. A1, 1972.
Landolt–Börnstein: *Zahlenwerte und Funktionen aus Naturwissenschaften und Technik* Neue Serie, Gruppe II, Band 6 (1974); Band 7 (1976). Springer-Verlag, Berlin, Heidelberg, New York.

Mark, H. and Wierl, R.: 1930, *Naturwiss.* **18**, 205.

Oberhammer, H.: 1976, *Chem. Soc. Specialist Periodical Report No. 20, Molecular Structure by Diffraction Methods* Vol. 4, 25–31.

Sutton, L. E.: 1958 (as Scientific Editor), *Chem. Soc. Special Publication No. 11, Tables of Interatomic Distances in Molecules and Ions.*

Sutton, L. E.: 1965 *ditto, No. 18, Supplementary Volume,* 1956–59.

Trendelenburg, F.: 1933, *Naturwiss.* **21**, 173.

II.8. Electron Diffraction: The CalTech–Cornell Connection

1. THE EARLY YEARS AT CALTECH AND PENN STATE

I gladly accepted the invitation to write about my electron diffraction studies because it provided me with an incentive to assess half a century of growth of an important subdivision of molecular structure investigations.

My first contact with gas phase electron diffraction as a tool for determining molecular structures, was in 1931 when Professor T. R. Hogness (who was my thesis advisor at the University of Chicago) suggested that we undertake the construction of an apparatus to measure internuclear separations in gaseous sodium chloride. Such data would have augmented James Franck's studies of the UV spectrum of the alkali halide vapors. I proceeded to calculate the expected relative scattered intensities as a function of angle. For this I needed an extended $(\sin rs)/rs$ table which I computed and eventually published in 1933. Since we planned to record the diffraction pattern by collecting the scattered electrons in a Faraday cage, the small diffraction wiggles superposed on a $1/s^4$ background indicated that such measurements would be marginal at best. I chose instead to construct a mass spectrometer for the study of ionization processes in polyatomic molecules. Several years later I fully appreciated the wisdom of abandoning NaCl. When I visited Professor Rabi's laboratory in Columbia he spoke somewhat disparagingly of chemists who didn't know what to do with the very precise internuclear distances for the alkali halides which he was determining using the molecular beam–r.f. technique. It was evident then, as it is now, that electron diffraction should not be used for molecular structure analyses of molecules which could better be investigated by spectroscopic techniques. This does not exclude studies of which the primary purpose is calibration of distances or of mean amplitudes.

I returned to electron diffraction in 1935 as a post-doctoral fellow at CalTech. I worked in Professor Pauling's group with Lawrence Brockway, and jointly under the supervision of Professor R. M. Badger, recording molecular spectra in the near infrared. By then Mark and Wierl, as well as Pauling, had demonstrated that the human eye 'sees' diffraction rings, even though there were no maxima or minima in the photographic density; apparently our eyes record locally normalized derivatives of the density. I arrived in Pasadena about the time that the Mark II unit was set in operation with J. H. Sturdivant's help. One can get a feeling for the state-of-the-art from the following description of a 'run'. About two hours after the pumps were started the vacuum was checked, by

reading an untrapped McLeod gauge. If the readings were in the neighbourhood of 10^{-5} torr 'there were no leaks'. Otherwise various leak hunting procedures were initiated until we nursed the apparatus down to an operating
pressure. The electron beam was aligned to the approximate center of the
fluorescent screen by manipulating the electron gun. The beam was then
interrupted by a brass shutter. At that point two persons were required. One
monitored the galvanometer-potentiometer, which measured the accelerating
voltage. He watched the voltage fluctuate and when the galvanometer beam
slowly swung past zero (the set value) he shouted "GO!" The second operator,
properly positioned at a stopcock, with a twitch of his fingers allowed a spurt of
gas to be injected in the vicinity of the beam and, simultaneously with another
twitch of his foot, momentarily removed the shutter so that the beam passed into
the diffraction section. It took practice and a substantial amount of concentration on the part of the operator to synchronize these motions. For compounds
difficult to condense the high voltage cut off, the pumps gurgled, and one had to
wait several minutes before the pressure was low enough to initiate the next shot.
The diffraction patterns were carefully studied by several observers, who independently sketched their visual impressions until a consensus was reached on
'what was there'. The first crude radial distribution functions were obtained on
the basis of these sketches. Much depended on the background illumination, on
eye conditioning, and extensive self-calibration. Looking back it seemed that the
subjective aspects correlated with the inner conviction one had regarding the
electron diffraction technique. Since the rings one saw were not maxima or
minima in photographic density, some of us were somewhat skeptical of the
entire process, while others undertook self-calibration very seriously, correlating
visual impressions of densities recorded for well known compounds with computed patterns. Verner Schomaker stands out among the most successful practi-
tioners.

   The process of finding a combination of internuclear separations which fitted
the qualitative visual intensity patterns and quantitatively fitted the visually
measured positions of the maxima and minima was time consuming and boring.
The most advanced technique consisted of summing $(\sin rs)/rs$ functions, multi-
plied by coefficients printed on strips known as Shermann–Cross Tables and
using a motor-operated calculator. Nimble fingers and the acceptance of a
two-hour dull grind per model were required. Nonetheless, the search for a
model often proved exciting, but it is not surprising that as soon as one structure
was found which fitted fairly well, few others were tested, even though we knew
of Patterson's homometric structures for one-dimensional Fourier transforms.
At any rate, it was often stated, and always implied, that the reported structures
only represented models which best fitted the diffraction patterns within specified error limits of the data and the calculated patterns. That is all one could
claim unless he had exhaustively examined all conceivable models and demonstrated their unacceptability. There was always the clear implication that, were
another chemically acceptable model to be proposed which had not been tested,
but which fitted the recorded pattern more closely, the original author might be
surprised but not averse to accepting the new configuration. Unfortunately

there were many chemists to whom this concept was unacceptable, and regrettably polemical encounters ensued. No matter; every available substance which had sufficient vapor pressure was thus processed. The first routine procedure for optimizing a structure was suggested in 1936: use of a least-squares treatment of measured versus calculated peak and valley positions, described under the title 'An Analytic Method of Interpretation of Electron Diffraction Photographs'. When information is limited it is very tempting for imaginative researchers to supply the missing ingredients by speculation. This is troublesome only when the guesses become imbedded into positions which have to be defended.

With wisdom bestowed by hindsight one may pose the following question: since most of the early structure determinations were later repeated with much greater precision using the sector technique with the inclusion of atomic form-factors and molecular dynamics, were not the 1935–1950 investigations useless and inherently a waste of much effort? Possibly this position may be sustained by those who regard the determination of mean internuclear separations and interatomic root-mean-square amplitudes as goals unto themselves. But there was much more than bond distances and bond angles among the objectives of the Pauling school of structure analysis. First, it is doubtful that the extensive technical development which began about 1945 would have taken place in the absence of the experience derived from the operation of the earlier machines. Also several generations of research workers were trained, excited by the shapes of molecules but dissatisfied with the subjective procedures available during the first 15 years of electron diffraction investigations. Second, in many instances the primary qualitative results, such as determination of atom connectivities and stereochemical configurations, were inherently of interest to the chemical community. Determinations of internuclear distances to ± 0.03 Å (which amounted to a relative error of about 2%) proved useful chemical information for calculating rotational entropies, for resolving dipole moments, for estimating densities in condensed phases, etc. However, the most extensive use of these structures was the correlation of bond distances with bond type based on the valence-bond formulation of the electronic structure of molecules in their ground states.

The postulate of additivity of atomic radii for covalent bonds was a natural extension of the Bragg–Goldschmidt table of atomic and ionic radii, developed from the early crystal structure analyses. Thus, all negative deviations from the expected sums (corrected for electronegativity differences) were interpreted in terms of resonance contributions of Lewis-type electron pair structures. Many times I wondered why *distance* rather than energy was chosen as the experimental parameter. In either the variational or the perturbation calculation of molecular systems expressed in terms of linear combinations of idealized bond types, it is the *energy* which is minimized. One would have anticipated that Linus Pauling would have set off on a spree of determining heats of formation, to parallel more closely the computed results rather than to depend on an empirical correlation between the bond type and bond distance. There are several reasons for this diversion. Linus Pauling was first and foremost a structural chemist, who conceived molecules primarily in spatial terms. Second, as a practical matter, it is much easier to measure molecular structures, even with a crude electron diffraction apparatus, to a precision of several percent than to determine heats of formation to the required much greater precision. Finally, the interpretation of heats of formation proved to be far from straightforward, due to the importance of non-bonding interactions which are not incorporated in the valence bond theory (except for direct van der Waals overlap).

In 1938 Lawrence Brockway moved to the University of Michigan (after a fellowship year in England) and began construction of his Mark III electron diffraction unit; in 1936 John Beach, as an NRC Postdoctoral Fellow at Princeton, built a copy of the Mark II, and I went to Pennsylvania State University to develop a small device for testing the extent of swelling of coal when carbonized. On holidays and numerous weekends I traveled to Princeton to collaborate with John on electron-diffraction studies of samples which I corralled at Penn State. Those interludes saved me from the coal industry; all was well except for the summers when the high voltage unit malfunctioned because of electrical breakdown due to the high humidity at Princeton. We photographed various samples of fluorocarbons furnished by Professor Joe Simons. Analysis of these photographs proved frustrating. After several years of extended calculations to find satisfactory models, Professor Simons admitted that possibly his estimates of $3-5\%$ impurity levels were optimistic, and that the samples we had were less well fractionated. However, some results were obtained. These were classified for several years because of their relation to the Manhattan Project. By then I had moved to Cornell and begun the construction of our first e.d. unit.

2. THE CORNELL YEARS

In 1940 Professor P. J. W. Debye came to Cornell as the Baker Lecturer. In Berlin his son, Peter Paul, had built an electron-diffraction apparatus with a sector; only the basic unit was brought to the U.S.A. To complete his program for a Ph.D. degree in Physics, Peter Paul began to reassemble his apparatus, which required a power supply, pumps, and a microdensitometer. While conceptually our two instruments were similar, in execution they proved very different. However, both of us had to discover the difficulties of making quantitative photometric measurements of scattered electron intensities. Looking back I wonder why these proved so painful. We discovered the extent of extraneous scattering from the diffraction chamber walls, both of electrons and X-rays, of electrons scattered from the edges of the sector and particularly of multiple scattering from the excess gas injected into the chamber. All these generated a substantial non-uniform background. We realized the importance of placing the photographic plate very close to the sector, the need for good photographic material, and of a suitably designed microphotometer. Peter Paul abandoned his project even though he obtained several good diffraction patterns of ammonia which showed 'real' rings. He wrote his Ph.D. dissertation on light scattering from high polymer solutions, which proved to be much more interesting during the early stages of the war-time synthetic rubber program.

My first regulated 50 kV power supply occupied an area of about 10 m^2 and incorporated a large (discarded) medical X-ray transformer, two large keno-trons, a large condenser, 50 megohms of resistors encased in a tank of oil and an electronic feedback voltage regulator. Don McMillan began construction of this unit for his proposed studies of X-ray diffraction from liquids. I inherited the pieces when he left for Los Alamos. The supply and regulator worked quite well, but it was a lethal device, set up behind a grounded chicken-wire fence. The first 'Cornell structures' were obtained without the use of a sector (J. M. Hastings, 1942), but during the next 15 years many improvements were introduced. We looked into the proper design of nozzles and calculated the consequences of the extraneous gas (R. B. Harvey and F. A. Keidel). In 1974 we finally calculated the density and temperature distributions in jets from typical nozzles (K. L. Gallaher). We built several high-temperature injectors (T. Ino; J. V. Martinez). Photographic materials were greatly improved to meet the needs of the electron microscope community. We took our first halting steps to utilize the mechanical IBM computers then in use by the University Accounting Department (Judy Bregman). We began to look for ways to improve microdensitometry. Compari-son of scans by a microphotometer available at the Eastman Kodak Company (Rochester) with records of the same plates made with the Leeds-Northrup unit at Cornell showed that there was considerable room for improvement in the latter. All of this led to the design and construction of our second unit [K. Kimura, *J. Phys. Soc. Japan* **17**, (1962)] which incorporated a larger aperture lens for post-diffraction focusing, using a parallel incident beam. Over the years this unit served very well and is now in operation at the University of Edinburgh under the able supervision of David Rankin. The parallel-incidence feature did not prove particularly useful for structure analysis although Mark Cardillo did demonstrate that for low angles ($0.1 < s < 5$) the contrast in diffraction intensity was considerably improved over the conventional arrangement (being much closer to the theoretical). Indeed, this configuration would be most suitable for measuring atomic form-factors at very low angles. A Jarrell-Ash microdensit-ometer was purchased and modified for precision automated digitized oper-ation. A large array of interactive computing programs and CRT displays of radial-distribution and molecular-scattering functions (on a PDP-9) were deve-loped by several generations of dedicated postdoctoral and graduate students [R. K. Bohn, J. Carlos, C. H. Chang, J. F. Chiang, K. P. Coffin, W. Harsh-barger, J. L. Hencher, R. L. Hilderbrandt, R. R. Karl, K. Katada, R. E. LaVilla, H. Oberhammer, and Y. C. Wang]. Studies of the structures of extremely small crystals were begun by G. G. Libowitz. We developed a rational approach to the analysis of molecules which incorporated large amplitudes of motion [A. Andreassen; A. Yokozeki] and we analyzed the gas dynamics of jet expansion, so as to estimate the gas temperature and density at the point of diffraction.

3. SOME REMARKS

The phasing out of several electron-diffraction laboratories was the inevitable consequence of the decline in federal support for structure investigations which began in the early 1960's. The consequences are simple to state: there are now fewer than half-a-dozen gas-phase electron-diffraction laboratories in the United States and most of these are operating well below their capacity. By contrast, such investigations in the U.S.S.R., Hungary, Norway, England and Japan are continuing at a steady if not rising level. It seems that a number of factors contributed to the decline of gas phase e.d. activity.

For about two decades there was great excitement among chemists regarding the measurement of internuclear distances in gaseous molecules. Perhaps nowhere are the aesthetic aspects in chemistry as apparent as in the determination of molecular structures. The symmetry of molecules, the almost harmonic motions of the atoms, the involved but subtle intramolecular interactions are intriguing. This was particularly contagious when display graphics and computers became generally available. Several generations of graduate students undertook to construct and operate electron-diffraction units upon receiving their academic appointments. However, when competition for funding became tight, each applicant was under increasing pressure to demonstrate that his approach had some novel aspects and that his particular sequence of molecules was most interesting.

Perhaps electron diffractionists were a bit apologetic about their discipline. They could not match the complex structure analyses of X-ray crystallographers, who could play with 3-D Fourier transforms, nor could they match the precision of the microwave spectroscopists in their studies of relatively simple structures.

We still believe that there are many good reasons for determining molecular structures, at various levels of precision. Indeed for most chemical applications knowledge of structural trends in a family of compounds is far more informative than the precision determination of any one member of that group. Clearly, there is more than one answer to the question: why undertake a structure determination? – other than that we 'wish to know' the interatomic distances and bond angles with high precision. However, in an attempt to satisfy potential referee criticism some applicants were forced to propose questionable and in my view inappropriate novel applications. Perhaps now we can agree that a reasonable approach is to maintain a laboratory for structure analysis which serves many functions: to aid synthetically oriented organic and inorganic chemists in determining structures of their new compounds, to study thin films and layers adsorbed on well characterized surfaces, to measure with precision selected molecules, to integrate diffraction and spectroscopic investigations, to correlate observed structures with *ab initio* quantum mechanical calculations, to determine molecule dynamics parameters, etc. All this could be done in electron diffraction 'facilities'. Unfortunately that word was anathema to the NSF advisory panels in 1965; it appears to have become acceptable in 1979.

The phasing out of electron diffraction studies at Cornell occurred over a dozen or so years and was due (in part) to my conviction that our graduates should be more broadly trained to ensure their success in job hunting. We began to require that each graduate student do more for his dissertation than merely determine gas phase structures. Our interest in molecular structures was diluted with studies in chemical kinetics. Our last paper, on the structure of $(CF_3)_3CI$ [Yokozeki and Bauer], was published in 1976, exactly 40 years after the appearance of our first paper.

Department of Chemistry
Cornell University

REFERENCES

Bauer, S. H.: 1936, *J. Chem. Phys.* **4**, 406–412.
Bauer, S. H. and Andreassen, A. L.: 1972, *J. Phys. Chem.* **76**, 3099–3108.
Bauer, S. H. and Hastings, J. M.: 1942, *J. Am. Chem. Soc.* **64**, 2686–2691.
Bauer, S. H., Ino, T., and Porter, R. F.: 1960, *J. Chem. Phys.* **33**, 685–691.
Bauer, S. H. and Kimura, K.: 1962, *J. Phys. Soc. Japan*, **17**, 300–305.
Gallaher, K. L. and Bauer, S. H.: 1974, *J. Phys. Chem.* **78**, 2380–2389.
Harvey, R. B., Keidel, F. A., and Bauer, S. H.: 1950, *J. Appl. Phys.* **21**, 860–874.
Martinez, J. V. and Bauer, S. H.: 1965, Unpublished.
Oberhammer, H. and Bauer, S. H.: 1968, *Bull. Am. Phys. Soc.* **13**, 834.
Yokozeki, A. and Bauer, S. H.: 1975, *J. Phys. Chem.* **79**, 155–162.
Yokozeki, A. and Bauer, S. H.: 1976, *J. Phys. Chem.* **80**, 73–76.

II.9. The Development of Electron Diffraction in the Mineralogical Department of the Museum of Natural History, Stockholm

Electron diffraction applied to crystallographic problems was introduced in Sweden in 1932 by Gregori Aminoff of the Mineralogical Department of the Museum of Natural History in Stockholm. In much of this work Aminoff was assisted by his second wife Birgit Broomé (Aminoff and Broomé, 1933; 1943).

In the beginning Aminoff worked with an electron beam from a cold cathode. The high-tension was supplied by a transformer followed by a mechanical rectifier. As practically only the peak voltage was effective in this combination, fairly sharp interferences could be obtained. After 1934 a hot cathode was used as the source of electrons and the high-tension supply was improved by using a valve and a condenser for smoothing.

In his earliest camera, shown in Figure 2.1.b (page 64, bottom), Aminoff placed both specimen holder and plate holder in a large rectangular box. The arrangement inside the box was something like an optical bench with a goniometer, and very precise adjustments of specimen and plate could be made. Following the lettering of Figure 2.1.b we have: A – a gas discharge tube; B – the gun alignment; C – camera box; D – specimen table; E – optical bench rail; F and G – plate cassette, and plate-holder arm inscribed in cm (permitting a variable camera length), respectively; H – flexible cable drive operating camera shutter; N, O – clamp and specimen mount belonging to the goniometer (Aminoff and Broomé, 1935). In a later design, specimen and plate holder were housed in different chambers. The plate holder consisted of a rotatable octagonal drum permitting eight plates to be exposed successively without breaking the vacuum. In neither of the two cameras was magnetic focusing used. The wave length of the electrons was determined by diffractograms of crystals of known unit dimensions and was plotted against the primary voltage of the high-tension transformer. In the beginning the standard crystal was muscovite and later molybdenite. Wavelengths between 0.05 and 0.08 Å were used.

To begin with, Aminoff studied a great number of mineral crystals in order to examine the nature of the diffraction patterns and to see the possibilities of the method in crystal structure investigations. Most of these studies were performed in transmission. The specimens then consisted either of crystals which could be split into thin flakes or of wedge-shaped splinters in which the sharp edge was struck by the electrons. The results already determined by means of X-ray diffraction in 1921 (talc and graphite), were reported in greater detail. In the diffractograms of fairly thick flakes of graphite, Kikuchi lines appeared, and Aminoff discussed the conditions of their formation.

Later Aminoff devoted most of his electron diffraction work to studies of the oxidation of crystal surfaces, especially the formation of ZnO on the surfaces of single crystals of ZnS in the form of zinc blende (sphalerite). Crystals of zinc blende were heated in air until their surface (natural or cleavage surfaces) were covered by a thin film of ZnO, showing interference colours. The surfaces were then photographed in the electron diffraction camera with a very small angle between the incident beam and the surface and in different azimuths. Diffraction patterns of the transmission type were obtained, which showed that the film on a certain surface behaved as a single ZnO crystal or as an assemblage of parallel-oriented ZnO crystals. On certain ZnS surfaces two different orientations of the ZnO crystals appeared, both of these always being present at each point on the surface. For all observed ZnO orientations the c axis of ZnO was exactly or nearly parallel to a tetrahedron-normal of ZnS and the a axis of ZnO was exactly parallel to a tetrahedron edge of ZnS. The ZnO orientation on different ZnS surfaces follows the symmetry of the ZnS crystal.

Gregori Aminoff (1883–1947) in his laboratory.

Aminoff found that the geometry of the two structures could explain their relative orientation in the contact layer but not the absence of other orientations geometrically equivalent with the former. He concluded that this problem could only be solved by applying physicochemical considerations.

Aminoff was anxious to stress that investigations of surface reactions of the above-mentioned type could not have been carried out by means of X-ray diffraction. The thick oxide layers necessary for an X-ray study would not have been representative of the contact where the basic surface reaction takes place.

The investigations on the oxidation of crystal surfaces form the most prominent of Aminoff's contributions in the field of electron diffraction. They were completed in 1938 and from then until his death in 1947 Aminoff devoted his scientific work chiefly to geochemistry and, in this connection, chemical spectrography.

Emeritus Professor of Inorganic Chemistry
Uppsala University
Sweden

REFERENCES

Aminoff, G. and Broomé, B.: 1933, *Arkiv. Kem.* **1113**, 5.
Aminoff, G. and Broomé, B.: 1934: *Z. Krist.* **89**, 80.
Aminoff, G. and Broomé, B.: 1935. *Z. Krist.* **91**, 77.

Subsequent Development at Various Centers of Research

Period 2

1935–1945

The standard Finch electron diffraction camera with its characteristic sauterne bottle discharge tube gun (ca. 1936 design – See Article II.16).

First post-war microscope from Hitachi Company, used by Hibi for studies of the pointed filament source and operating at 50 keV. (See Article II.13.)

Simple gas-diffraction unit constructed in Nagoya by Uyeda and Morino. This unit had no rotating sector. (See Articles II.11 and II.12.)

SHIZUO MIYAKE

II.10. Echoes of Old Songs

1. BEAUTIFUL ELECTRON DIFFRACTION PATTERNS

I graduated from the University of Tokyo in March 1933, and in December of the same year I entered the laboratory headed by Iitaka in the Institute of Physical and Chemical Research (IPCR), Tokyo. Iitaka was a well-known metallurgist, and I was then to assist him in his new plan for extensive applications of electron diffraction, a novel method at the time, to the study of oxidation and corrosion of metal and alloy surfaces. Iitaka requested S. Kikuchi, a young colleague in the same institute, to join him as technical adviser. It was due to such a circumstance that I was accorded the privilege of receiving a full introductory guidance in electron diffraction practice from Kikuchi. He then gave me a small problem to study as well, concerning the behaviour of a certain kind of Kikuchi line in reflexion patterns of electron diffraction from a cleavage face of a crystal (Miyake, 1935).

My first study of electron diffraction was thus begun with observations of reflexion patterns from the cleavage face (110) of zinc blende (sphalerite), and I at once became captivated by the beautiful diffraction patterns seen on the fluorescent screen. It is true that electron diffraction patterns are generally beautiful, but reflexion patterns from crystal faces are particularly fascinating. It is really exciting to follow on a fluorescent screen a fantastic change in the reflexion pattern with rotation of the crystal, consisting of frequent flashes of sparkling stars due to Bragg reflexions and successive entry of various constellations formed by Kikuchi lines and bands.

From the beginning of my observation of reflexion patterns from zinc blende specimens, I had already acquired an impression that a number of very important diffraction effects of electrons, which needed to be theoretically elucidated, must be hidden behind these beautiful patterns. However, a fairly long time passed before I came to challenge such underlying problems. Actually, by the period around 1935 not only was my knowledge of the theory of electron diffraction still very poor, but the main activity in the field of electron diffraction was characterized by a strong inclination to practical applications, just as my main subject at the time was the electron diffraction study of surface oxidation of metals and alloys.

2. MY TREASURE BOX

In 1938 a private research institute on a small scale, called the Kobayashi

Institute of Physical Research, was established in a suburb of Tokyo as the result of a donation by U. Kobayashi, an industrialist engaged in tungsten mining, and I became a member of this institute by recommendation of Professor Nishikawa, my teacher. The building of this institute was finished in late 1940, and I made my own laboratory there, equipped with apparatus for X-ray diffraction as well as electron diffraction. I began there first an X-ray study on the ferroelectric domain structure of Rochelle salt and, at the same time, an electron diffraction study of epitaxial growth of evaporated metal films on crystal surfaces. The study with Rochelle salt brought about a fairly fruitful result, for which I was given the degree of Doctor of Science from the University of Tokyo in 1942.

As to the study of epitaxial growth of metal films, although I considerably enjoyed the experimentation involved in it, the results obtained were less conclusive. However, apart from its outcome, this study necessarily afforded me frequent occasion to observe reflexion patterns from fresh crystal faces used as substrates for evaporated metal films. During the course of this study, which I continued for several years, I became quite familiar with various reflexion patterns of electrons, particularly those from the cleavage face of zinc blende. And in the meantime, my knowledge thus accumulated of the patterns from zinc blende became my treasure-box, which was full of important diffraction effects and occasionally I could pick up valuable topics from it.

In fact this box continued to serve me for nearly thirty years as an almost inexhaustible source of subjects, and more than a dozen of my papers are more-or-less concerned with the reflexion patterns from zinc blende. Two such examples will be described in sections which follow, and the study on the dependence of the emission yield of X-rays from a single crystal of the diffraction condition of exciting electrons (Miyake, Hayakawa and Miida, 1967) and that on the nature of the horizontal Kikuchi lines (Kikuchi lines parallel to the shadow edge) in reflexion patterns (Miyake, Hayakawa, Kawamura and Ohtsuki, 1975) may also be considered as two further examples.

3. WHY NOT SYMMETRIC?

As soon as I began to engage myself in the study of evaporated metal films, I became aware of a curious symmetry feature in the diffraction patterns from zinc blende. The (110) cleavage face of this mineral contains in its plane the [001] zone axis which is a polar axis of the acentric structure of zinc blende. What I noticed was the fact that the diffraction patterns from zinc blende for incident electrons perpendicular to the [001] axis were definitely asymmetric with respect to the plane of incidence, namely (001): for instance, the intensities of the reflexions hhk and $hh\bar{k}$, with the structure amplitudes which are the complex conjugate of each other, were found to be generally not equal. Thus, such asymmetry allowed us to tell unmistakably the sense of polarity of the crystal.

This finding considerably confused me, as it meant a breakdown of Friedel's law, the law which eliminates the possibility of determining the sense of a polar axis in an acentric crystal from diffraction phenomena alone. Very soon, however, I became convinced that the observed phenomenon was a new and not trivial effect which was worth pursuing further. The experimental aspect of this effect was first reported at the meeting of the Physical Society of Japan held in July 1943.

Although Friedel's law had been known to fail in the X-ray case in the presence of the effect of anomalous dispersion for one of the constituent atoms in the crystal, I could not conceive of any possibility of a similar cause in the electron case. In the meantime R. Uyeda of Nagoya University was also much interested in this problem and we began to collaborate. We therefore together made various theoretical trials to elucidate the observed effect. However, as these trials all failed, we had to report our negative results more than once to the Society meetings in the period from 1946 to 1947. It may, however, be said that such a painstaking struggle as ours, with unsuccessful results, in those days still had its own merits, since it certainly awakened an interest in the dynamical theory of electron diffraction in some research workers of younger generation.

It was just at such a stage that K. Kohra, one of Shinohara's students in the graduate course in Kyushu University, came to stay in the Kobayashi Institute in order to work with me for a few months from late 1947 to early 1948. Fortunately, he had already had some training in theoretical calculation, and his participation in our research team brought about a definite progress in our approach to the solution of the problem. After a few preliminary trials, he dealt with a three-wave problem taking account of simultaneous Bragg reflexions hhk and $hh\bar{k}$, and succeeded in indicating the theoretical possibility of the failure of Friedel's law as experimentally observed. This result was orally reported to the meeting of the Physical Society held in April 1949. A short summary of our Friedel's law problem was published in 1950 (Miyake and Uyeda, 1950; Kohra, Uyeda and Miyake, 1950). In a paper subsequently published (Miyake and Uyeda, 1955), it was pointed out that the breakdown of Friedel's law may take place in electron diffraction even for a non-absorbing crystal, as a result of dynamical interference of electron waves involving a many-wave diffraction effect, where it is to be understood that 'three' is the smallest number of the waves concerned.

Nowadays, not only is the failure of Friedel's law known as quite a common effect in electron diffraction, but it seems that this effect plays some important role in the determination of crystal symmetry from electron diffraction patterns.

4. A PROLONGED RESONANCE

The surface resonance scattering of electrons is known as a characteristic phenomenon in electron diffraction by single crystals (cf. a recent review article by E. G. McRae, 1979). This phenomenon was initially discovered by Kikuchi and Nakagawa (1933) in reflexion patterns from a cleavage face of zinc blende; they found that the intensity of a bright spot due to the specular reflexion of electrons from the crystal surface was suddenly enhanced to a remarkable extent when this spot came to overlap a certain kind of Kikuchi line, even though the Bragg condition is then not always fulfilled for the specular reflexion. Naturally this visually attractive phenomenon also became my favourite as soon as I entered that branch of electron diffraction, and I used often to enjoy the observation of this dramatic intensity change on the fluorescent screen. As this phenomenon had not yet been theoretically explained, I gained from it, every time I observed it, an impression as if it were an apocalypse of a secret hidden behind the crystal surface.

Probably from around 1942, this effect became one of the targets of my challenge. I supposed then that the enhancement in intensity of the specular spot would be caused by the participation of a virtual Bragg reflexion which does not make an appearance on the fluorescent screen, being hidden beneath the shadow-edge due to the crystal surface. However, I could not so promptly develop this idea, because I could not by then readily identify the lattice plane responsible for the hidden Bragg reflexion.

After the end of the War, I renewed my interest in this problem. At this time I had the cooperation of M. Takagi (then Miss M. Kubo) and Kohra, and we came at last to recognize the fact that the hidden Bragg reflexion should be a surface wave, namely a unique kind of Bragg reflexion which is excited specifically in a direction parallel or nearly parallel to the crystal surface. At the same time we realized that this effect was a diffraction phenomenon relating characteristically to the presence of the crystal surface, and that it was this surface-wave that brought about a linkage between the incident wave and the specular reflexion so that the latter became enhanced in intensity.

The results of our study were orally reported to domestic society meetings from 1948 to 1950, and also to the Second Congress of the IUCr in 1951 held in Stockholm, which I attended, and finally summarized in a paper published in 1954 (Miyake, Kohra and Takagi, 1954). Later, Kohra and others carried out a more satisfactory calculation based on a four-wave treatment (Kohra, Molière, Nakano and Ariyama, 1962). They showed that the wave field of electrons in the crystal at the condition of the intensity enhancement of the specular reflexion is limited to a very shallow region beneath the crystal surface.

Meanwhile, McRae and Caldwell (1964) discovered the corresponding diffraction effect in low-energy electron diffraction (LEED), and named it 'electronic surface resonance'. Having been stimulated by their study, then, I subsequently performed some new observations of the same effect for electrons in low- and high-energy ranges (Miyake and Hayakawa, 1970; Hayakawa and Miyake, 1974).

Thus I have retained, over a period of nearly thirty years, a deep involvement in the problem of surface resonance in electron diffraction. Admittedly this series of studies is not exhaustive, but I believe that this important effect is well deserving of my long acquaintance with it. This effect is probably the only process in electron diffraction by crystals which produces a reflexion maximum not corresponding to a Bragg condition. At all events, I enjoyed this phenomenon throughout.

5. IN HAPPY CIRCLES; IN THE TOKYO INSTITUTE OF TECHNOLOGY

In 1949 I moved to the Tokyo Institute of Technology (TIT) from the Kobayashi Institute together with M. Takagi and G. Honjo, my collaborators at the time. Mrs Takagi, formerly one of Kikuchi's students in Osaka University, entered the Kobayashi Institute in late 1943, owing to the circumstance that Kikuchi by then was unable to continue research work in the University due to the war. She helped me with the study of the epitaxial growth of evaporated metal films. To my regret, however, some important part of the results thereby obtained was ultimately left unpublished, mainly due to the disordered circumstances of the post-war period. In this regard, it is indeed a relief to mention the fact that her well-known and unique contribution on the melting of small metal particles (Takagi, 1954; 1956), which she conducted in the TIT, was a work which was initiated on the basis of certain experimental evidence found in the course of our previous study on the epitaxial growth.

Honjo entered the Kobayashi Institute in February 1946, and began there a study of the fine structure of the electron diffraction Debye rings from smokes of metal oxides. He continued the same study in the TIT. Although in those days we had no high-resolution electron diffraction camera, he succeeded by his experimental skill in observing the fine structures of reflexions due to individual crystallites, each consisting of spot multiplets corresponding to the shape of the crystallite concerned. He gave a detailed analysis of these multiplets on the basis of the dynamical theory of electron diffraction, and at the same time found the first example of the Borrmann effect in the electron case, which he noted in a characteristic intensity distribution among each multiplet (Honjo, this volume).

In 1951 K. Kambe entered my laboratory in the TIT as a scholarship student of the graduate course. He engaged first, with Honjo, in a study of the equal-thickness fringes in cubic crystallites of MgO found in electron micro-scopic images that had been taken by Hibi in Tohoku University, and a little later, he became interested in the theoretical formulation of the three-wave problem of the dynamical theory of electron diffraction.

Possibly the latter problem was to some extent motivated by one of the impressions which I got at the Second Congress of the IUCr in 1951 held at Stockholm, namely an impression of the surfacing of the *phase problem* at this Congress as one of the major problems in X-ray crystallography at the time. I thought that the three-wave problem ought to be very important in this connexion, since an interference effect which is affected by the phases of crystal structure amplitudes would be found at a condition of simultaneous Bragg reflexions on two different kinds of lattice plane. Meanwhile, Kambe accom-plished an elegant formulation of the three-wave problem, and succeeded furthermore in confirming experimentally the expected phase effect in electron diffraction, by the use of Möllenstedt patterns from graphite specimens (Kambe and Miyake, 1954; Miyake and Kambe, 1954; Kambe, 1957).

Incidentally, in the TIT I was engaged also in X-ray studies, such as those on anharmonic thermal motions of atoms in the crystals having the zinc blende structure, with S. Hoshino, and on the structure of solid solutions of metallic halides, with K. Suzuki.

6. IN HAPPY CIRCLES; IN THE INSTITUTE FOR SOLID STATE PHYSICS

In 1957 I moved to the Institute for Solid State Physics (ISSP), which had just started as a new research institute in the University of Tokyo. However, because the construction of its building was considerably delayed, I could not engage in experimental work for a time. It was just during this period that I had the pleasure of carrying out a joint investigation with J. M. Cowley and A. F. Moodie in Australia. Initially I was interested in an important statement in their paper concerning the space-group extinction rule for transmission spot patterns of electron diffraction (Cowley and Moodie, 1959). I became, however, aware of the need for a slight amendment to their statement, and I worked it out in collaboration with S. Takagi and F. Fujimoto, then in the Faculty of General Education (Miyake, Takagi and Fujimoto, 1960). Subsequently, after exchanging comments several times between Australia and Japan, a short paper by the five authors describing their common understanding of the problem was published (Cowley, Moodie, Miyake, Takagi and Fujimoto, 1961). In those days, however, I did not envisage the possibility that this sort of study might be capable of finding an early application to a practical problem. Therefore, I was rather surprised at the work by Kimoto and Nishida (1966), in which they successfully applied our extinction rule to the determination of the space group for δ-Cr.

In the new laboratory in the ISSP, I planned to conduct a study of electron diffraction in a higher energy range, by the use of a 300 keV diffraction unit equipped with a small Van de Graaff generator. At that time K. Fujiwara, formerly one of Ogawa's students in the graduate course of Tohoku University, came to the ISSP to cooperate with me, and he immediately worked out a relativistic dynamical theory of electron diffraction (Fujiwara, 1961). In connexion with his work, then, we studied experimentally a number of such electron diffraction effects significantly manifesting the relativistic effect (Miyake, Fujiwara and Suzuki, 1962). The relativistic effect had been generally disregarded in most of the electron diffraction studies published until then. In this respect our work was quite timely since active studies of high-energy electron diffraction as well as electron-microscopy were just beginning in various countries.

Now, in another joint work performed with Fujiwara in the ISSP, let us refer to a computational study on the *lattice image*. I had often noticed the fact that the electron microscopic lattice images then published possessed such features that could only be interpreted in terms of many-wave treatments. We offered, therefore, the first example of the exact calculation of the lattice image on the basis of a many-wave treatment (Miyake, Fujiwara, Tokonami and Fujimoto,

1964), though we assumed in it a sinusoidal potential for which the exact analytic solution of the Schrödinger equation had been available, to ease the computational problems of that time.

I considered that this work was meaningful as an example of the reasonable treatment of a lattice image, but actually the complexity as well as the extraordinary practical importance of lattice imaging as it has now developed, went far beyond anything I could conceive of at the time.

From around 1963 I began with Fujiwara, and his successor K. Hayakawa, the work on low-energy electron diffraction (LEED). For this study we made a new type of diffraction apparatus which was so designed that it might facilitate the observation of diffraction patterns over an angular range wider than that feasible with the LEED apparatus of the conventional type (Fujiwara, Hayakawa and Miyake, 1966). The crystal sample used in the first trial in this series of study was again a cleavage face of zinc blende. Through these studies we became convinced that LEED was precisely a realm which was most strongly governed by the dynamical interference of electrons (Miyake and Hayakawa, 1966). A LEED study on spin-dependent exchange reflexions from nickel oxide, an anti-ferromagnetic crystal, which was performed with Hayakawa and K. Namikawa (Hayakawa, Namikawa and Miyake, 1971), became my last work in the University of Tokyo. In March 1971 I retired from the University.

REFERENCES

Cowley, J. M. and Moodie, A. F.: 1959, *Acta Cryst.* **12**, 360–366.
Cowley, J. M., Moodie, A. F., Miyake, S., Takagi, S., and Fujimoti, F.: 1961, *Acta Cryst.* **14**, 87–88.
Fujiwara, K.: 1961, *J. Phys. Soc. Japan* **16**, 2226–2238.
Fujiwara, K., Hayakawa, K., and Miyake, S.: 1966, *Jap. J. Appl. Phys.* **5**, 295–298.
Hayakawa, K. and Miyake, S.: 1974, *Acta Cryst.* A**30**, 374–380.
Hayakawa, K., Namikawa, K., and Miyake, S.: 1971, *J. Phys. Soc. Japan* **31**, 1048–1417.
Kambe, K.: 1957, *J. Phys. Soc. Japan* **12**, 13–25; 25–31.
Kambe, K. and Miyake, S.: 1954, *Acta Cryst.* **7**, 218–219.
Kikuchi, S. and Nakagawa, S.: 1933, *Sci. Pap. Phys. Chem. Res., Tokyo*, **21**, 256–265.
Kimoto, N. and Nishida, I.: 1966, *J. Phys. Soc. Japan* **22**, 744–756.
Kohra, K., Molière, K., Nakano, S., and Ariyama, M.: 1962, *J. Phys. Soc. Japan* **17**, Suppl. B-II, 82–85.
Kohra, K., Uyeda, R., and Miyake, S.: 1950, *Acta Cryst.* **3**, 479–480.
McRae, E. G.: 1979, *Rev. Mod. Phys.* **51**, 541–568.
McRae, E. G. and Caldwell, C. W.: 1964, *Surf. Sci.* **2**, 509–515.
Miyake, S.: 1935, *Sci. Pap. Phys. Chem. Res., Tokyo*, **26**, 216–224.
Miyake, S., Fujiwara, K., and Suzuki, K.: 1962, *J. Phys. Soc. Japan* **17**, Suppl. B-II, 124–128.
Miyake, S., Fujiwara, K., Tokonami, M., and Fujimoto, F.: 1964, *Jap. J. Appl. Phys.* **3**, 267–285.
Miyake, K. and Hayakawa, K.: 1966, *J. Phys. Soc. Japan* **21**, 363–378.
Miyake, K., and Hayakawa, K.: 1970, *Acta Cryst.* A**26**, 60–70.
Miyake, S., Hayakawa, K., Kawamura, T., and Ohtsuki, Y. H.: 1975, *Acta Cryst.* A**31**, 32–38.
Miyake, S., Hayakawa, K., and Miida, R.: 1967, *Acta Cryst.* A**24**, 182–191.
Miyake, S. and Kambe, K.: 1954, *Acta Cryst.* **7**, 220.
Miyake, S., Kohra, K., and Takagi, M.: 1954, *Acta Cryst.* **7**, 393–401.
Miyake, S., Takagi, S., and Fujimoto, F.: 1960, *Acta Cryst.* **13**, 360–361.
Miyake, S. and Uyeda, R.: 1950, *Acta Cryst.* **3**, 314.

Miyake, S. and Uyeda, R.: 1955, *Acta Cryst.* **8**, 335–342.
Takagi, M.: 1954, *J. Phys. Soc. Japan* **9**, 359–363.
Takagi, M.: 1956, *J. Phys. Soc. Japan* **11**, 396–405.

RYOZI UYEDA

II.11. Seven Stories in My Forty Years of Electron Diffraction

1. BEGINNING OF MY ELECTRON DIFFRACTION
(1934–1942)

I began my electron diffraction study in 1934 under the guidance of Professor Shoji Nishikawa (Figure 1:1.5) as soon as I graduated from Tokyo University. In the same year Kikuchi and his colleagues went into nuclear physics, probably

Fig. 1.   The author beside the 500 kV electron microscope of Nagoya University. Dr. Kamiya is at the controls.

because they thought that there were no more interesting problems in electron diffraction. Since I was also interested in nuclear physics, I felt a little disappointed when my teacher told me that I should carry out an experiment with an old electron diffraction camera constructed by Yamaguti. However, I took reflection spectra from graphite and molybdenite, and studied the so-called Yamaguti anomalous effect (Yamaguti, 1934), i.e. an apparent decrease of mean inner potential in low-order reflections. I learned Bethe's dynamical theory (Bethe, 1928) and applied it to the interpretation of this effect. This manner of study was not new in the Nishikawa school because Shinohara (1932) had applied Bethe's theory to the interpretation of Kikuchi envelopes. In any event, I was finally very pleased to obtain a definite result from my first research work (Uyeda, 1938).

Following that, I thought that the application of electron diffraction would become more important, and willingly worked on the surface oxidation of molybdenite and sphalerite. My doctor's thesis was on the epitaxial growth of evaporated metal films, an effect then found in Kirchner's laboratory in Germany (Lassen, 1934). I designed a new electron diffraction apparatus, the specimen chamber of which was equipped with a crucible for the evaporation of metal specimens, and a furnace for heating the substrate crystal. This led to: 'The Surface of the Thin Film Formed by the Deposition to be Examined *In Situ* During Formation' (Uyeda, 1942). I believe that this was the first *in situ* experiment in electron diffraction as well as electron microscopy. One of the most important conclusions drawn from this work was that "a uniform film is not formed initially, but isolated crystals are built here and there on the surface". This fact was found in Europe almost ten years later, according to Pashley (1965). I stopped working on epitaxy around 1950, because I was more interested in the dynamical effects and furthermore I did not foresee the importance of epitaxy in future electronics. At any rate, I am still interested in this topic and happy to learn about the remarkable results recently obtained in Japan by Ino (Ogawa, this volume) and Yagi (Honjo, this volume). Recollecting my old experiment I feel that my philosophy was exactly the same as that of the present workers, although it now seems ridiculous that I believed in the presence of clean surfaces while carrying out those experiments in dirty vacua.

2. DYNAMICAL PEST (1945–1955)

I moved to Nagoya University in 1942 and first worked there on the failure of Friedel's law, a remarkable effect then discovered by Miyake (this volume). The problem looked very difficult, but Miyake and I found we could finally solve it in collaboration with Kohra, then a graduate student of Professor Shinohara in Kyushu University. He was captivated by the dynamical theory, and visited us almost every month for discussions, in spite of the long distance from Kyushu to Nagoya and Tokyo. I remember how greatly excited he was at the paper of Lamla (1938) in which the concept of dispersion surface for many beams was introduced.

Fig. 2. My graduate students and technicians in 1947, taken by T. Ino in front of my laboratory. From left: H. Yoshioka, Y. Kainuma; (front row) the author, S. Ito and Y. Miura; (back row) Y. Sugiura, M. Nonoyama and N. Kato.

The success of our study aroused a whirlwind among our colleagues, which was jokingly called 'dynamical pest', a parody of 'Gruppenpest' by H. Weyle. Kohra was the first victim of this fever. It is to be mentioned here that we were studying the dynamical theory exactly during the worst time after World War II, in the period of acute shortages of food, clothing and shelter. Nevertheless, I was happy that I had some brilliant graduate students in my laboratory (Figure 2). This situation was possible because a very good law had come into force in Japan which exempted students of superior ability from military service so that they could devote themselves to research in universities.

The second victim of the dynamical pest was N. Kato, of my laboratory. He was at first carrying out a series of accurate measurements on the lattice parameters of evaporated alkali halides (Kato, 1951). He started his theoretical study from the refraction effect, which was necessary for the interpretation of his experimental results, and finally extended the dynamical theory, previously limited to parallel plates, to polyhedral crystals. Then I wrote an outline of his theory (Kato and Uyeda, 1951) and sent it to Professor von Laue. We were most happy when his answer arrived. The first few lines were: "Haben Sie herzlichen Dank für freundlichen Brief von 23.4.50 der Sonderdruck von Norio Kato und das Manuskript von Kato und Ihnen! Ich freue mich, dass die Erforschung der Elektronenbeugung auch in Japan betrieben wird, und dass insbesondere die dynamische Theorie dort gepflegt wird."

The third was Kainuma. He finished an experiment on epitaxy with a remarkable result (Kainuma, 1951) and then intended to work on the theory of the Kikuchi pattern. I told him that it was the most difficult problem passed on to me by Professor Nishikawa, and the application of the reciprocity theorem would be indispensable, as used by Laue (1935) for the interpretation of the Kossel pattern. Although I could do nothing more, he developed a theory of inelastic scattering of electrons by a crystal, following the orthodox quantum-mechanical formalism. After coping with a number of complicated equations, he arrived at an intensity formula for the Kikuchi pattern, which consisted of a symmetric term corresponding to a Kikuchi line and an asymmetric term corresponding to a Kikuchi band (Kainuma, 1955). In the classical interpretation by Kikuchi, the latter was missing because the direct wave and reflected wave were added in the form of intensity. In Kainuma's theory, they were automatically added in the form of amplitude, and for this reason the cross term appeared corresponding to a Kikuchi band. I thought that it was very strange that Professor Nishikawa had missed the simple idea of adding amplitudes. At any event, I credited Kainuma with having solved a problem so difficult that even my respected teacher could not find a clue.

The final victim in Nagoya was Yoshioka. He worked on the corrosion of iron and identified an unstable green rust as $2 FeO \cdot Fe_2O_3 \cdot H_2O$ (Yoshioka, 1949). Since he was a very reticent man, I did not know that he was developing a theory of such fundamental importance until he prepared a preprint in 1953. He had reformulated the dynamical theory, taking into account the absorption caused by inelastic scattering (Yoshioka, 1957). I have no intention of writing more on the importance of his theory, because it is well-known among specialists.

The dynamical pest spread also to Tokyo: Honjo, Mihama, S. & M. Takagi, Kambe, etc. were its victims. However, it gradually died down around 1955, although further developments were made by Fujimoto, Fujiwara, etc. One of the reasons for the flourishing of theoretical study after the War was the fact that it was easier for talented students to tackle theoretical problems than to carry out experiments with old equipment of pre-war type. However, I cannot forget that another reason was that we were encouraged very much by Professors von Laue and Ewald. I do not know how to thank them properly. When I sent Professor Ewald an invitation letter to the International Conference of Crystallography, Kyoto, 1972, with my compliments and thanks, he wrote to me:

I am touched by your statement that I was of help to the Japanese crystallographers – this help – if needed – would have come from anyone acquainted with the beautiful work on X-ray optics which was performed in Japan. I was perhaps early in recognizing this, owing to the talk you gave us at the Polytechnic Institute in 1950 or '51 (I don't remember which) which was the first time I met anybody who was fully familiar with the details of the dynamical theory (excepting, perhaps, Laue and his group in Dahlem), and who made the detailed confirmation of the theory his subject of research.

3. MOIRÉ PATTERNS (1950–1960)

In 1950 I constructed a new electron diffraction unit of high resolution. Since this equipment could not be completely home-built, I asked the electron microscope group of the Hitachi Central Laboratory, Tokyo for help. One day when I was there, Mitsuishi and Nagasaki, members of the group, showed me an electron micrograph of graphite in which a pattern looking like a fingerprint appeared, and asked me for an interpretation of it. It consisted of parallel lines with spacings slightly longer than 100 Å and was a picture of the highest resolution at that time. Although I had no experience in electron microscopy, I realized at a glance that the pattern was not produced by extinction contours. I believed that it was contrast of a new type, previously unreported. For this reason, I was captivated by this pattern.

In a train on the way back to Nagoya, various possibilities of interpretation circulated in my mind. All of them were in reciprocal space, probably because I was a diffractionist. Finally, the analogy with thickness fringes of magnesium oxide struck me. Since these were formed by the interference of two electron beams making a small angle to each other, I inferred that the fingerprint pattern must be formed in the same way. I calculated the angle between the two beams from the wavelength and the observed spacing, and tried to find a crystal arrangement which might result in two beams making an angle of the same order. Since it was of course impossible to produce this pattern with one crystal sheet, I had to assume two. Then I found that if they were tilted with respect to each other by a certain angle, the transmitted beam could make an angle of the required order with a beam produced by successive Bragg reflections in each sheet. A short paper describing the pattern was soon communicated by Professor Nishikawa to the Japan Academy (Mitsuishi, Nagasaki and Uyeda, 1951).

Even after the publication of this paper, I did not know the word 'moiré'. I learned it through a letter from Professor G. I. Finch, which arrived several months later. He also wrote to me how to demonstrate moiré patterns with macroscopic lattices, and pointed out that our pattern could also be interpreted as being caused by a rotation instead of by tilting. The second observation of a crystal moiré was made on sericite by Seki (1953) of the Taihei Mining Laboratory. In his case, the rotation angle was measured from the straight edges of the crystal and a complete interpretation was given.

A few years later, H. Hashimoto gave me a print of a beautiful moiré pattern of copper sulphide crystals. Provided with many examples, I was happy to talk on the subject of crystal moiré under the chairmanship of Sir George Thomson at the Electron Physics Conference, Washington D.C., 1956. After the conference, I was invited to Harvard University by Dr Lang and gave a talk on moiré patterns. I remember that the screen for the slides was very large, and when Hashimoto's picture was projected, I was very surprised to recognise, from the platform, the termination of a moiré line which I had never noticed before. After returning to Japan I tried to interpret this in real space, probably because I had by then become an electron microscopist. I found that it was due to a dislocation (Hashimoto and Uyeda, 1957), and demonstrated it with the moving slide, which readers can find in my article (Uyeda, 1974a). By the way, I recommended Kato to Lang on this occasion. The harmonious cooperation between these two is now well-known among crystallographers.

I visited Penn State University after Harvard, and there met Professor

Brindley. He showed me an electron micrograph of Yu-Yen Stone, a variety of antigorite, showing parallel, straight, and evenly spaced fringes about 100 Å apart, and asked me whether it was a moiré pattern. I studied it that evening and concluded that it was a lattice image with a long spacing. The next day, I convinced him that a slight modulation with a long spacing could produce an image of fairly high contrast. We were happy when we agreed to submit a paper produced by our international cooperation to the international periodical, Acta Crystallographica (Brindley, Comer, Uyeda and Zussman, 1958).

Since the 100 Å fringes of Yu-Yen Stone were the first lattice images I had ever seen, I was astonished at Menter's report of 12 å fringes of phthalocyanine (Menter, 1956), which I read just after returning from the United States. For this reason, I gave up the idea of proceeding with lattice images of small spacings, and instead studied the variation of contrast of lattice images with the degree of defocusing, using Yu-Yen Stone. We then found the reversal of contrast and the appearance of a half spacing (Uyeda, 1962), effects now well-known among electron microscopists.

At the same time I made some effort to improve the resolution available, and consequently we succeeded in resolving the 17 Å moiré fringes of an epitaxial silver film on molybdenite (Kamiya and Uyeda, 1962). This result stimulated G. A. Bassett to carry out his *in situ* study of epitaxial growth with the molybdenite sample supplied by me (Bassett, 1960). I was deeply impressed by his ciné-film which realized my dream of twenty years earlier.

4. ABSORPTION COEFFICIENT AT HIGH VOLTAGES (1961–1970)

I remember that one of the most important topics in the crystallography conference, Kyoto, 1961, was the problem of absorption of electrons in crystals. It was discussed on the basis of Yoshioka's theory (Yoshioka, 1957), and Fujiwara's relativistic diffraction theory (Fujiwara, 1962). It attracted the interest of experimental people, particularly metal physicists, because it was thought that the utility of a high-voltage electron microscope would depend on the decrease of absorption coefficient at high voltages. The news of Dupouy's 1 MeV microscope reached us at that time, but with no exact data. The only data available were those obtained by Hashimoto (1964) up to 300 kV.

After the conference I felt that the construction of a high-voltage electron microscope was urgent. I obtained funds from the Toray Science Foundation, Japan, and began the construction of a 500 kV machine early in 1962 with the cooperation of the Nagoya group (Uyeda, Sakaki, Maruse, Mihama and Kamiya) and the Hitachi group (Tadano, Kimura, Katagiri and Nishinaga). It was finished early in 1965 (see Figure 1) and I could present our preliminary result of absorption measurements in the electron diffraction conference, Melbourne, 1965 (Uyeda and Nonoyama, 1965).

I had started with measurements at 100 kV, adopting the method of Kohra and Watanabe (1961), viz. the intensity analysis of thickness fringes of MgO smoke particles. To obtain better results I later used a wedge of a macrocrystal, a technique originally developed by Lehmpfuhl and Molière (1962). Early in 1962 I wrote to Professor Dupouy to convince him of the importance of extending my measurements to very high voltages. At that time his grand microscope was in perfect operation, and he invited me to his laboratory after the Electron Microscopy Conference, Philadelphia, 1962. I flew over the Atlantic with some anxiety about my poor French. However, there was in fact no trouble at all, because I was warmly welcomed by young research workers, Ayroles and Mrs Mazel as well as Director Dupouy. Although my stay was extremely short, I finished my preliminary experiment with their help. Moreover, they continued my experiment for a few years, and the final result covering the range from 101 to 1204 kV was published late in 1965 (Dupouy, Perrier, Uyeda, Ayroles and Mazel, 1965). It showed that the absorption distance above 300 kV was much shorter than that predicted by a simple theory based upon the proportionality to $(v/c)^2$. The absorption of electrons in a crystal is indeed a complicated phenomenon. We have shown that the diffraction contrast is caused not only by the no-loss electrons, but also by the energy-loss electrons (Kamiya and Uyeda, 1962). For this reason the absorption coefficient in theory has quite a different implication from that in experiment. The interpretation of the absorption coefficient became more and more complicated and so difficult that I could hardly follow it when Ichimiya (1973) wrote his paper on this subject. In conclusion, I do not know now whether or not our final result in France has been made clear by theoreticians.

On the other hand, the utility of the high voltage electron microscope was gradually proved by practice (e.g. Dupouy and Perrier, 1964; Fujita, Kawasaki, Furubayashi, Kajiwara and Taoka, 1967), and early in the 1970s they were constructed here and there in spite of their high cost. In any event, the idea that the decrease of absorption would make objects visible was too primitive because contrast and resolution were not taken into consideration (Uyeda and Nonoyama, 1968).

5. DISCOVERY OF CRITICAL VOLTAGE EFFECT (1967–1970)

My hard work on the construction of a high voltage electron microscope was rewarded by the discovery of the effect of a vanishing of the second order reflection, which is now called more generally the 'critical voltage effect'. This was found in 1967 independently by the Hitachi and the Nagoya groups. In the former, Nagata, a young research worker, studied, at the suggestion of H. Watanabe, extinction contours of bent wedge-shaped aluminium crystals at various voltages. Nagata took pictures successively in bright field, in dark field using the first-order and then with the second reflexion. During the course of this study, he noticed that the second-order patterns became vanishingly weak at about 600 kV, and reappeared at higher voltages. He asked Fukuhara, an expert in dynamical theory, for his comments. Since Fukuhara was working on a many-beam problem, he at once gave the correct interpretation (Nagata and Fukuhara, 1967).

In Nagoya University I had meanwhile continued the measurement of absorption coefficients with the thickness fringes of MgO. I tried the measurement with second-order fringes as well as first-order ones. I was surprised to find that the spacing of the second-order fringes was much larger than the calculated value. Since I was familiar with the Bethe potential I realized that this was a relativistic effect in the many-beam case. Thus, I predicted that the effective structure factor would vanish at a certain accelerating voltage (Uyeda, 1968). Luckily enough, D. Watanabe of Tohoku University had come to my laboratory as a visiting research worker in 1967. I suggested to him that he should study this remarkable new effect. After some discussion he took Kikuchi patterns of an iron alloy at various voltages and observed the vanishing of the second-order lines at 140 kV (D. Watanabe, Uyeda and Kogiso, 1968). This was just before the 40th Anniversary of the Discovery of Electron Diffraction, London, 1967. I brought a series of pictures showing the vanishing effect to London, and showed them to a few of my friends in private discussion, although I was sorry that I had no time to arrange for the slides to be shown in the session.

The use of a Kikuchi pattern had the merit that the accelerating voltage could be calculated from the pattern without any other measurement (Uyeda, Nonoyama and Kogiso, 1965). Taking advantage of this, I proposed an accurate determination of the first-order structure factor by measuring the vanishing voltage. Watanabe immediately carried out experiments and obtained very accurate values for Fe, Ni and Al (D. Watanabe, Uyeda and Fukuhara, 1968). Even after the war, it was generally believed that, hampered by its dynamical effect, electron diffraction was inadequate for structure factor determination. At present, however, precisely because of this effect, it provides as accurate a determination as does X-ray diffraction. Although I have contributed no more to this field, I feel very satisfied to have initiated the first step.

6. CONSTRUCTION OF ELECTRON DIFFRACTION EQUIPMENT
(1935–1965)

I believe that my main contribution to Japanese crystallography was the construction of electron diffraction equipment rather than the study of dynamical effects. I designed three diffraction cameras while I was in Tokyo. The first one was for electrons of a few thousand kV, the MEED in the present terminology. In this equipment I tried to eliminate energy-loss electrons by an electron filter. However, I failed completely because I could not focus the energy filter properly. The second one was that already described in the first story. Since it was a special apparatus for *in situ* experiments, the specimen chamber and camera system were very complicated. For this reason, it was relatively large and difficult to manipulate.

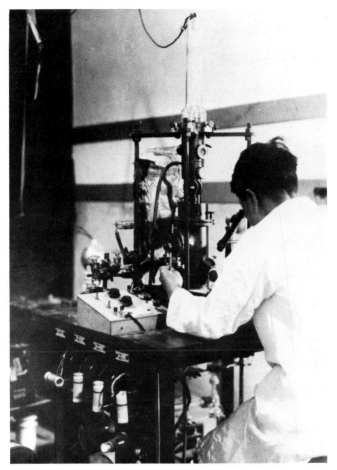

Fig. 3. The small electron diffraction camera and the author (1943).

After this experience, I constructed a third one for general purposes. In the design of collimator and crystal holder I followed Kikuchi. However, many improvements were made. My aim was to make the specimen-to-plate distance short, and to obtain patterns with the best possible resolution. Figure 3 shows that I was adjusting the crystal orientation looking at the viewing screen with the aid of a low magnification microscope. As shown in Figure 4, four or more pictures could be taken on a plate of about 4×16 cm, and if necessary, Debye rings of a standard specimen could be recorded for the measurement of wavelength. It is worth mentioning that by this small unit Kato (1951) measured the lattice spacing of KCl as accurately as $a = 6.2865 \pm 0.0007$ Å, and Honjo and Mihama (1954) studied the fine structure of diffraction spots.

The fourth one, the high resolution type, was almost an electron microscope as well as an electron diffraction unit, and the fifth one was the 500 kV electron microscope (Figure 1). Since these machines were beyond the capacity of our workshop, they were made in the Hitachi Co. However, I initiated these projects

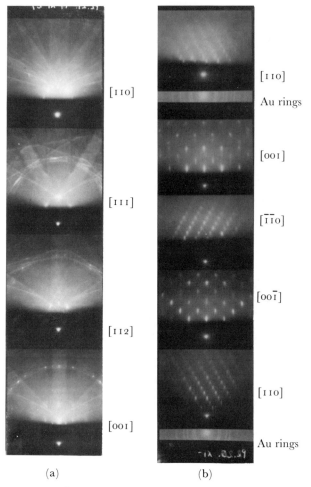

Fig. 4.   Diffraction patterns taken by the small camera (Fig. 3). (a) Cleavage surface (110) of sphalerite and (b) its oxidized surface.

and our group played an important part in the design, preliminary experiment, testing and improvement.

My Link with Gas Diffraction

Finally, to illustrate the technical collaboration with gas diffraction, which is described by Y. Morino in the next article, I shall mention my role in the design of their equipment. Because of my experience in electron diffraction instrumentation, I was able to help at the beginning of gas diffraction studies in Japan.

This story starts in 1943. When Professor Morino moved to Nagoya in that year he started his gas diffraction study with the equipment shown on the picture page preceding this section. Because of the war, and the lack of manpower and materials I tried to make the design as simple as possible. The next equipment with which I was associated was designed by my graduate student Tadashi Ino, and constructed by Mr S. Takahashi of our workshop. This was first used for the analysis of amorphous materials (Ino, 1953), and then for gases by Morino's group (Figure 1, Morino: this volume). Finally, we produced a more contemporary design (in 1962) shown in Figure 2 (Morino). These 3 units show the evolution of design over the years. Gas cameras of the 1962 design have been supplied to Professor Kimura at Hokaido University, and to Professor Kuchitsu of Tokyo University.

7. METAL SMOKE PARTICLES (1970–PRESENT)

The topic of my present study is fine metal particles made by the so-called gas-evaporation technique. When a metal is evaporated in a rarefied inert gas and the metal vapour is cooled in the gas, a metal smoke is produced. We study the smoke particles by electron microscopy as well as diffraction. There is a long story about the beginning of this study. During the War, the Japanese Army planned to develop a guided bomb, which could be aimed at a source of infrared radiation; for example, a funnel of a battleship. As the absorber of radiation in the detector they used zinc black, which was made by evaporating zinc in air at reduced pressure. They found that coatings made in this way, which were initially good, deteriorated after about a week. To make clear the cause of this deterioration I was asked to perform an examination by electron diffraction. Then Kimoto, in my laboratory, worked on this problem. He found that the zinc black was indeed a complicated material consisting of zinc metal, zinc oxide and n 'n tungsten oxides which came from the tungsten heater. To simplify the problem, he evaporated zinc in pure nitrogen, and obtained zinc black of pure zinc. He measured the particle size from the breadth of the Debye rings and found that it decreased below 100 Å with the decrease of nitrogen pressure. Meanwhile, the war came to an end, and Kimoto changed his research topic.

After about fifteen years, I was told by Professor R. Kubo, one of the leading theoretical physicists in Japan, that he had obtained an interesting result in his theory of fine metal particles (Kubo, 1962). Then, recollecting Kimoto's old result, I immediately began gas-evaporation experiments with the help of Nonoyama, my technician (see Figure 2). In the new experiment we used helium, argon, etc., instead of nitrogen, and the products were examined by electron microscopy as well as diffraction. We were excited when we found beautiful polyhedral crystallites of Mg, Cr, Fe, etc. At this stage I had to concentrate on the construction of the high voltage electron microscope. For this reason, I passed this study over to Kimoto, the initiator of the gas-evaporation experiment (Kimoto, Kamiya, Nonoyama and Uyeda, 1963). He developed the investigation with Nishida, and obtained many remarkable results; e.g. he found a new modification of chromium, δ-Cr, which crystallized in icosatetrahedra (Kimoto and Nishida, 1967).

I worked so hard on the high voltage business that I was almost exhausted in the late 1960s. I was happy to return to gas-evaporation, which I had begun almost as a hobby. With the collaboration of many graduate students, I have made a survey on the morphology and structure of almost all of the common metals (Uyeda, 1974b; 1978). We have found various beautiful polyhedra, plates and rods as reproduced in the articles cited. Although I retired from Nagoya University in 1975, I still continue this study in Meijo, a private university. The reason for this is that I love natural history as well as physics and intend to make Wulff polyhedra of various pure metals.

REFERENCES

[References marked * are reviewed in Uyeda (1974a).]

Bassett, G. A.: 1960, *Proc. Eur. Reg. Conf. Electron Micros. Delft*, 270–275.
*Bethe, H.: 1928, *Ann. Phys.* **87**, 55–129.
Brindley, G. W., Comer, J. J., Uyeda, R., and Zussman, J.: 1958, *Acta Cryst.* **11**, 99–102.
Dupouy, G. and Perrier, F.: 1964, *J. Micros.* **3**, 233–244.
*Dupouy, G., Perrier, F., Uyeda, R., Ayroles, R., and Mazel, A.: 1965, *J. Micros.* **4**, 429–450.
Fujita, H., Kawasaki, Y., Furubayashi, E., Kajiwara, S., and Taoka, T.: 1967, *Jap. J. Appl. Phys.* **6**, 214–230.
*Fujiwara, K.: 1962, *J. Phys. Soc. Japan* **16**, 2226–2238.
*Hashimoto, H.: 1964, *J. Appl. Phys.* **35**, 277–290.
*Hashimoto, H. and Uyeda, R.: 1957, *Acta Cryst.* **10**, 143.
*Honjo, G. and Mihama, K.: 1954, *J. Phys. Soc. Japan* **9**, 184–198.
Ichimiya, A.: 1973, *J. Phys. Soc. Japan* **35**, 213–223.
Ino, T.: 1953, *J. Phys. Soc. Japan* **8**, 92–98.
Kainuma, Y.: 1951, *J. Phys. Soc. Japan* **6**, 135–137.
*Kainuma, Y.: 1955, *Acta Cryst.* **8**, 247–257.
*Kamiya, Y. and Uyeda, R.: 1962, *J. Phys. Soc. Japan* **17**, suppl. B-II, 191–194.
Kato, N.: 1951, *J. Phys. Soc. Japan* **6**, 502–507.
*Kato, N. and Uyeda, R.: 1951, *Acta Cryst.* **4**, 227–231.
Kimoto, K., Kamiya, Y., Nonoyama, M., and Uyeda, R.: 1963, *Jap. J. Appl. Phys.* **2**, 702–713.
Kimoto, K., and Nishida, I.: 1967, *J. Phys. Soc. Japan* **22**, 744–756.
Kohra, K. and Watanabe, H.: 1961, *J. Phys. Soc. Japan* **16**, 580–581.
Kubo, R.: 1962, *J. Phys. Soc. Japan* **17**, 975–986.
Lamla, E.: 1938, *Ann. Phys.* **33**, 225–241.
Lassen, H.: 1934, *Phys. Zeitschrift* **35**, 172–178.
Laue, M.v.: 1935, *Ann. Phys.* **23**, 705–746.
Lehmpfuhl, G. and Molière, K.: 1962, *J. Phys. Soc. Japan* **17**, suppl. B-II, 130–134.
*Menter, J. W.: 1956, *Proc. Roy. Soc.* **A236**, 119–135.
*Mitsuishi, T., Nagasaki, H., and Uyeda, R.: 1951, *Proc. Jap. Acad.* **27**, 86–87.
*Nagata, F. and Fukuhara, A.: 1967, *Jap. J. Appl. Phys.* **6**, 1233–1235.
Pashley, D. W.: 1965, *Advances in Physics (Phil. Mag. suppl.)* **14**, 327–416.
Seki, Y.: 1953, *J. Phys. Soc. Japan* **8**, 149–151.
*Shinohara, K.: 1932, *Sc. Pap. Inst. Phys. Chem. Res. Japan* **20**, 39–51.
*Uyeda, R.: 1938, *Proc. Phys.-Math. Soc. Japan* **20**, 280–287.
*Uyeda, R.: 1942, *Proc. Phys.-Math. Soc. Japan* **24**, 809–817.
Uyeda, R.: 1962, *J. Phys. Soc. Japan* **17**, suppl. B-II, 155–161.
*Uyeda, R.: 1968, *Acta Cryst.* **A24**, 175–181.
Uyeda, R.: 1974a, *J. Appl. Cryst.* **7**, 1–18.
Uyeda, R.: 1974b, *J. Cryst. Growth* **24/25**, 69–75.
Uyeda, R.: 1978, *J. Cryst. Growth* **45**, 485–482.
Uyeda, R. and Nonoyama, M.: 1965, *Abstract, Inst. Conf., Melbourne*, I. M-3.
*Uyeda, R. and Nonoyama, M.: 1968, *Jap. J. Appl. Phys.* **7**, 200–208.

Uyeda, R., Nonoyama, M., and Kogiso, M.: 1965, *J. Electron Micros.* **14**, 298–300.
*Watanabe, D., Uyeda, R., and Kogiso, M.: 1968, *Acta Cryst.* **A24**, 249–250.
*Watanabe, D., Uyeda, R., and Fukuhara, A.: 1968, *Acta Cryst.* **A24**, 580–581.
*Yamaguti, T.: 1934, *Proc. Phys.-Math. Soc. Japan* **16**, 95–105.
Yoshioka, H.: 1949, *J. Phys. Soc. Japan* **4**, 270–275.
*Yoshioka, H.: 1957, *J. Phys. Soc. Japan* **12**, 618–628.

YONEZO MORINO

II.12. *Fifty Years in Tokyo, Nagoya, and Tokyo*

1. PREWAR PERIOD

When I look back on my research career, I feel myself fortunate in having started it early in the quantum mechanics era. I learned about the wave nature of an electron beam in my second-year class (1929) at the University of Tokyo. A few years later, I obtained a book entitled *Elektronen-Interferenzen* issued as one of the series of *Leipziger Vorträge,* in which Raimund Wierl had described his marvellous achievement on *Elektroneninterferenzen an freien Molekülen.* I was attracted to this novel technique: I was impressed with the straightforward way of deducing molecular structure from sets of interatomic distances. However, I was involved at that time in studies of the Raman effect under Sanichiro Mizushima.

Using this spectroscopic approach, the Raman effect, we had discovered two rotational isomers in 1,2-dichloroethane, $ClCH_2$—CH_2Cl. The *trans* form, one

Fig. 1.   Electron diffraction equipment with a rotating sector and T. Ino (1951).

136

of the coexisting conformers, was easy to identify because the Raman effect indicated the existence of the center of symmetry in this conformer. On the other hand, the structure of the other conformer, which was afterwards called 'gauche' by Mizushima, could not be determined because the relative position of the rotating —CH_2Cl groups could not be specified by the Raman effect. I thought that conclusive evidence for this relative position might be obtained by electron diffraction, since the $Cl \cdots Cl$ distance depends strongly on the dihedral angle of internal rotation. We followed this approach with the help of Shitego Yamaguchi, who had an electron-diffraction camera for his surface studies at the Institute of Physical and Chemical Research (Tokyo). He made a simple nozzle with glass, installed it in the camera, and after tremendous effort succeeded in getting a halo-photograph of the diffraction pattern of 1,2-dichloroethane. On this photograph we found a weak shoulder at a position corresponding to a $Cl \cdots Cl$ pair with a torsion angle of about 120° from the *trans* position. (Yamaguchi, Morino, Watanabe and Mizushima, 1943).

Encouraged by this success, we extended the study to hexachloro-, penta-chloro- and *asym*-tetrachloroethanes and demonstrated from the relative positions of the halos that these ethane derivatives were all in the staggered, rather than the eclipsed form (Morino, Yamaguchi and Mizushima, 1944). Such lucky experiences impelled me to get directly into the study of gas electron diffraction.

Fortune favored me again when I moved to Nagoya University in 1943. In the Physics Department was Ryozi Uyeda, a gifted pioneer of electron diffraction. Interested in gas electron diffraction, he made us aquainted with every detail of his techniques, useful for our study of molecular structure. With his generous aid we made a simple diffraction unit without a rotating sector as shown on the picture page preceding this section, and Masao Kimura started his studies of the structure of a series of halogenated methanes (Morino, Kimura and Hasegawa, 1946). Before long, however, we had to discontinue the work because of the war.

2. PURSUIT TO THE ACCURACY LIMIT

During wartime, we were absolutely isolated from the information of world scientific progress. After the war the startling news was brought us that P. P. Debye, and Chr. Finbak, Odd Hassel and B. Ottar (1941), had succeeded in bringing a halo-photograph into a more reliable form by using a mechanical device, the rotating sector. Inspired by this news, we decided to construct an apparatus based on this new principle. Among many efforts made by my colleagues, I must mention the outstanding achievement of Tadashi Ino in Uyeda's Laboratory, who designed our first sector apparatus shown in Figure 1 (see also the preceding article), and completed its construction in 1950, under circumstances in which we were ignorant of the design details of the foreign apparatus. I must also mention the remarkable contribution of Shigetoshi Takahashi in the machine shop of the University, who helped us in designing the apparatus, in leak-hunting during its construction, and by preparing excellent hand-made sector plates.

Fig. 2.   A new gas diffraction apparatus built in Nagoya University (1963).

In 1949 I returned to the University of Tokyo. After some delay caused by this moving, we constructed a pair of sector apparatuses, specially designed for our practical use for molecular structure studies, one for Nagoya University and the other for Tokyo University. These units were used actively in research for about ten years until they were replaced by a new model (in 1962) (shown in Figure 2).

During this period, Isabella L. Karle and Jerome Karle (1949) published their beautiful structure work on carbon dioxide and carbon tetrachloride by the sector method. Their results actually demonstrated that this new technique opened up a new road to a truly quantitative analysis of both interatomic distances and mean square amplitudes of vibration. Their subsequent report on 1,2-dichloroethane (Ainsworth and Karle, 1952) gave a more conclusive answer to our problem described in Section 1: the structure of the *gauche* conformer and its fractional population in the gas phase at room temperature were distinctly determined.

In order to obtain authentic diffraction curves, we critically examined all the techniques involved in our analysis. For instance, a finite spread of the diffraction center, which was assumed to be infinitesimal in the theory of diffraction, was one of the sources of significant errors, especially of those errors associated with the observed mean amplitudes. In order to measure the relative density of gas molecules along the path of the electron beam, we placed a circular aperture between the nozzle and the photographic plate and measured the blackness of the plate in the region where no electrons should be found if the center of diffraction were infinitesimal. Corrections for a finite size of the diffraction center were estimated on the basis of this distribution curve (Morino and Murata, 1965).

Here I recollect the pleasant discussion on the theoretical background of the diffraction experiment with Russell A. Bonham, Indiana University, who stayed with us in Tokyo in 1964. He gave us lots of helpful and timely remarks at the discussion (Morino, Kuchitsu and Bonham, 1965).

Improvement of the computational techniques for our analysis of diffraction data was another problem to be solved. At present one can use a computer fast enough to finish a complicated structure analysis within a few minutes, but in the old days computation of even one radial distribution curve by use of an abacus required a week of hard work. After reading a paper by Philip A. Shaffer, Verner Schomaker and Linus Pauling (1946) on the use of punched-card machines, we prepared several thousands of punched-cards and carried out the calculations by borrowing a Remington Rand adding-machine installed in a governmental office of population statistics. Even today I remember the happy moment when we first obtained a radial distribution curve in *three hours*.

When digital computers became available, the advantage we enjoyed was not only the saving of time but also the improvement of the accuracy of our experimental results by the use of the least-squares treatment (Bastiansen, Hedberg and Hedberg, 1957). A puzzling question arose in connection with the actual process of the least-squares calculation. Our observations were made as a continuous curve of a microphotometer trace, whereas our least-squares analysis was carried out on a finite number of discrete intensity points read from the curve. If one increases the number of observed points, one can decrease the apparent standard deviation, but it is impossible to make the standard deviation infinitesimal because of the increase of correlation among the neighbouring observations. Given a continuous curve, what would be the optimal interval? We examined the mutual correlation of such observations by the use of a non-diagonal weight matrix and estimated the minimum value of the standard deviation. From this we obtained the optimal intervals between the observed points with which one could assume that the points were all independent (Murata and Morino, 1966). We also estimated the random and systematic errors of our observations by taking into account all possible sources of experimental errors. Thus finally we arrived at a conclusion that the limit of accuracy of our structure was ± 0.002–0.005 Å for an interatomic distance (Morino, Kuchitsu and Murata, 1965), in comparison with ± 0.02 Å for that obtained by the old visual method.

Most of our earlier studies with the use of the sector apparatus were directly related to *internal rotation* in which I had kept up a deep interest for many years. Kozo Kuchitsu carried out a study on *n*-propylchloride CH_3—CH_2—CH_2Cl (Morino and Kuchitsu, 1958) and Machio Iwasaki on the halogeno-ethanes

F_2ClC—$\dot{C}ClF_2$ (Iwasaki, Nagase and Kojima, 1957), FCl_2C—CCl_2F (Iwasaki, 1958) and F_2ClC—CCl_2F (Iwasaki, 1959). They confirmed the *gauche* and *trans* isomerism for these molecules. The torsion angles of the *gauche* conformers were found in all cases to be about 120°. The challenging problem of the determination of the height of the barrier hindering internal rotation was tackled by Eizi Hirota for a series of molecules. The barrier height was found to decrease in the order of Cl_2C—CCl_3, Cl_3C—$SiCl_3$ and Cl_3Si—$SiCl_3$, with increase in the $Cl \cdots Cl$ distance between the two rotating groups (Morino and Hirota, 1958).

3. FROM STATIC TO DYNAMICAL MODEL

During my stay in the USA in 1955, I heard serious concern about the future of gas electron diffraction: as the precision of our structure determination was increased by the sector method, the results were found to be significantly different from the corresponding structures determined by microwave spectroscopy. Some people thought that the structure determined by electron diffraction was systematically biased and inaccurate. I disagreed with this opinion and thought that the systematic discrepancy would be solved if the physical significance of the observed structures was clearly understood by taking into account the fact that a molecule is a dynamical system, vibrating about its equilibrium structure. With electron diffraction we measure the mean of interatomic distances in various vibrational states in thermal equilibrium, whereas with microwave spectroscopy we measure the rotational constants which are related to interatomic distances in an entirely different way. They are both mean values, but there exists a definite difference between them, caused by the different way of taking vibrational averages. We saw the necessity of introducing new definitions of interatomic distances, r_g and r_z,* which are simpler and more fundamental with regard to molecular dynamics, than the directly-observed *effective* interatomic distances. Our studies in this respect were closely related to our microwave spectroscopic studies in the 1960s for determining the equilibrium structures (r_e) of simple polyatomic molecules.

According to the theory of molecular vibrations, the mean square amplitudes of thermal vibration are directly connected with the intermolecular force field. It was easy to derive the general formula using Wilson's FG matrix formalism, and the calculated values showed good agreement with the observed values (Morino, Kuchitsu and Shimanouchi, 1952). Conversely, we used the observed mean square amplitudes as additional information for the determination of harmonic force constants. Examples are those of tetrahedral molecules (CCl_4 and $GeCl_4$; Morino, Nakamura and Iijima, 1960) and those of planar triangular four-atomic molecules (BF_3 and BCl_3; Konaka, Murata, Kuchitsu and Morino, 1966).

---

* The distance r_g denotes the thermal average of internuclear distance (Morino, Nakamura and Iijima, 1960), which was first introduced by Lawrence S. Bartell (1955) as r_g (o), while r_z denotes the distance between the average nuclear positions in the ground vibrational state (Oka, 1960; Morino, Kuchitsu and Oka, 1962).

In 1959 I visited Trondheim and had a chance to talk with Otto Bastiansen about the future prospects of gas electron diffraction. My presentation at a colloquium there on the method of calculation of mean square amplitudes apparently stimulated Sven S. Cyvin, who later became an expert in this field (Cyvin, 1968).

Bastiansen took me to a ski-house on the top of the Holmenkollen Olympic Stadium near Oslo, where he told me about his new discovery that distance between two non-bonded atoms in a linear molecule, such as allene or dimethylacetylene, deviates appreciably from the sum of their bond lengths (Almenningen, Bastiansen and Munthe-Kaas, 1956; Almenningen, Bastiansen and Traetteberg, 1959). An idea came to me that this systematic difference would be attributable to the bending vibrations of linear molecules. Upon my return to Tokyo, we made a calculation and found that the observed magnitudes of the 'shrinkages' were what we expected from the force constants obtained by spectroscopic measurements (Morino, Nakamura and Moore, 1962).

4. AIMING AT UNIFIED MOLECULAR STRUCTURE

Thus it became evident that gas electron diffraction and spectroscopy provide us with slightly different information about the molecular structure. Then, as a natural consequence, it followed that if the two methods were properly combined, we might gain more advanced knowledge of molecular structure. For instance, the symmetry of a molecule determined by spectroscopy could be transmitted into the analysis of electron diffraction data, or the rotational constants could be used to calibrate the scale factor of the diffraction experiment. The idea was applied by Kuchitsu and Tsutomu Fukuyama (1968, 1969) to acrolein and related molecules and Takao Iijima and his collaborators to acetaldehyde (Iijima and Kimura, 1969; Iijima and Tsuchiya, 1972) and the acetyl halides (Tsuchiya and Iijima, 1972; Tsuchiya, 1974).

It was true that some criticisms had been raised against the idea of combining the results from different techniques. Was our use of data obtained by these two methods a mere compromise of conflicting results? As we succeeded in rearranging the basic principles on a unified basis of molecular dynamics, it is evident that all our data really form non-conflicting sources of information for the final molecular structure.

Probably the strongest evidence for the success of the collaboration between electron diffraction and spectroscopy may be found in the prosperity of the Austin Symposium on Gas-Phase Molecular Structure organized by James E. Boggs. The central theme of the Symposium is the discussion of the fundamental problems in molecular structural studies by electron diffraction, spectroscopy and other techniques including *ab initio* calculations. The Symposium has been held every other year since 1966, with increasing numbers of participants and reports.

Finally I would like to say a few words on the compilation of structural data, though it does not belong to my own work. Determination of the structure of an individual kind of molecule might be a tiny contribution, but, when the structural data were collected with a systematic treatment, they have proved to be a powerful source of information on the construction of our Nature. Indeed, Leslie E. Sutton's *Tables of Interatomic Distances and Conformations in Molecules and Ions*, published by the Chemical Society (London) in 1958 and supplemented in 1965, is an excellent example which has given us immense benefit for many years. Another recent publication *Structural Data for Free Polyatomic Molecules* in the New Series of *Landolt-Börnstein* was featured by a collection of molecular structures *evaluated* on the complete lists of published data of electron diffraction and microwave spectroscopy. The evaluation work was done by J. H. Callomon, E. Hirota, K. Kuchitsu, W. J. Lafferty, A. G. Maki and C. S. Pote, and the collection of the basic data for the evaluation was done by Barbara Starck and her collaborators. It is my great joy to find there our definition of interatomic distances having been used in the presentation of the evaluated results.

In concluding my recollections over the past fifty happy years, I wish to express my gratitude to all my friends, colleagues and collaborators for their kindness extended to me. I also thank Professor K. Kuchitsu for his help in preparing this manuscript.

Sagami Chemical Research Center
Sagamihara
Japan

REFERENCES

Ainsworth, J. and Karle, J.: 1952, *J. Chem. Phys.* **20**, 425–427.
Almenningen, A., Bastiansen, O., and Munthe-Kaas, T.: 1956, *Acta Chem. Scand.* **10**, 261–264.
Almenningen, A., Bastiansen, O., and Traetteberg, M.: 1959, *Acta Chem. Scand.* **13**, 1699–1702.
Bastiansen, O., Hedberg, L., and Hedberg, K.: 1957, *J. Chem. Phys.* **27**, 1311–1317.
Cyvin, S. J.: 1968, *Molecular Vibrations and Mean Square Amplitudes*. Universitetsforlaget, Oslo, and Elsevier, Amsterdam.
Finbak, C., Hassel, O., and Ottar, B.: 1941, *Arch. Math. Naturvidenskab.* **44**, No. 13, 8–11.
Iijima, T. and Kimura, M.: 1969, *Bull. Chem. Soc. Japan* **42**, 2159–2164.
Iijima, T. and Tsuchiya, S.: 1972, *J. Mol. Spectrosc.* **44**, 88–107.
Iwasaki, M.: 1958, *Bull. Chem. Soc. Japan* **31**, 1072–1080.
Iwasaki, M.: 1959, *Bull. Chem. Soc. Japan* **32**, 194–200.
Iwasaki, M., Nagase, S., and Kojima, R.: 1957, *Bull. Chem. Soc. Japan* **30**, 230–236.
Karle, I. L. and Karle, J.: 1949, *J. Chem. Phys.* **17**, 1052–1058.
Konaka, S., Murata, Y., Kuchitsu, K., and Morino, Y.: 1966, *Bull. Chem. Soc. Japan* **39**, 1134–1146.
Kuchitsu, K., Fukuyama, T., and Morino, Y.: 1968, *J. Mol. Struct.* **1**, 463–479.
Kuchitsu, K., Fukuyama, T., and Morino, Y.: 1969, *J. Mol. Struct.* **4**, 41–50.
Morino, Y. and Hirota, E.: 1958, *J. Chem. Phys.* **28**, 185–197.
Morino, Y., Kimura, M., and Hasegawa, M.: 1946, *J. Chem. Soc. Japan.* **67**, 93–94.
Morino, Y. and Kuchitsu, K.: 1958, *J. Chem. Phys.* **28**, 175–184.
Morino, Y., Kuchitsu, K., and Bonham, R. A.: 1965, *Bull. Chem. Soc. Japan.* **38**, 1796–1797.
Morino, Y., Kuchitsu, K., and Murata, Y.: 1965, *Acta Cryst.* **18**, 549–557.
Morino, Y., Kuchitsu, K., and Oka, T.: 1962, *J. Chem. Phys.* **36**, 1108–1109.
Morino, Y., Kuchitsu, K., and Shimanouchi, T.: 1952, *J. Chem. Phys.* **20**, 726–733.

Morino, Y. and Murata, Y.: 1965, *Bull. Chem. Soc. Japan* **38**, 114–119.

Morino, Y., Nakamura, J., and Moore, P. W.: 1962, *J. Chem. Phys.* **36**, 1050–1056.

Morino, Y., Nakamura, Y., and Iijima, T.: 1960, *J. Chem. Phys.* **32**, 643–652.

Morino, Y., Yamaguchi, S., and Mizushima, S.: 1944, *Sci. Papers Inst. Phys. Chem. Research (Tokyo)* **42**, 5–9.

Murata, Y. and Morino, Y.: 1966, *Acta Cryst.* **20**, 606–609.

Oka, T.: 1960, *J. Phys. Soc. Japan* **15**, 2274–2279.

Shaffer, P. A. Jr., Schomaker, V., and Pauling, L.: 1946, *J. Chem. Phys.* **14**, 648–664.

Tsuchiya, S.: 1974, *J. Mol. Struct.* **22**, 77–95.

Tsuchiya, S. and Iijima, T.: 1972, *J. Mol. Struct.* **13**, 327–338.

Yamaguchi, S., Morino, Y., Watanabe, I., and Mizushima, S.: 1943, *Sci. Papers Inst. Phys. Chem. Research (Tokyo)*. **40**, 417–424.

II.13. The Pointed Filament Source:
A Personal Account

1. MY CONTACTS WITH ELECTRON DIFFRACTIONISTS IN JAPAN

I have never regarded myself as an electron diffractionist, but I have been closely associated with the subject through the work of others. One of my early contacts was with N. Kato who visited me in Sendai around 1951, although I do not remember the exact date. We discussed our problems late at night, because during the day I was working experimentally on thickness fringes of MgO, and he was developing the dynamical theory for polyhedral crystals. A few years later, Honjo and Kambe worked out a detailed analysis of my results on thickness fringes (Hibi, Honjo and Kambe, 1955). Around 1952, I took many pictures of mica, one of which is reproduced in figure 2. I gave a print of this picture to R. Uyeda and asked him for an interpretation of the linear patterns. After a few years, he reproduced my picture in his paper and wrote "The set of V-shape contours is due to imperfections in the crystal." (Uyeda, 1955). If he had written dislocations instead of imperfections, I would have been one of the discoverers of dislocation images.

Fig. 1.   The author – Tadatosi Hibi.

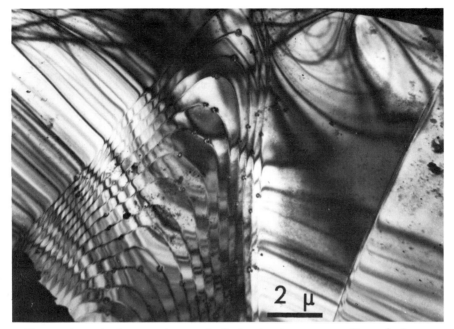

Fig. 2.   An early electron micrograph of mica showing contours of imperfections.

2. POINTED FILAMENTS AND THE RESOLUTION OF THE ELECTRON MICROSCOPE

In 1936 I constructed a magnetic lens microscope of 3 kV as shown in Figure 3 and studied with this the emission mechanism of oxide cathodes. After the 2nd World War (around 1949) I bought a magnetic lens electron microscope of 50

Fig. 3.   Magnetic-lens microscope constructed in 1936.

kV made by Hitachi as shown on the picture page preceding this section and looked for an interesting project which could be studied with this apparatus. I was happy to find an interesting paper entitled 'A Trial Holography Using an Electron Beam' by Haine and Mulvey (1952). Here, a defocused electron image of MgO having Fresnel fringes was taken as a hologram, and then the hologram was reconstructed by a light beam and a pinhole aperture, to obtain a high resolution electron image. This trial was an attempt to confirm Gabor's idea. However, it was not successful because the laser source was not known at that time. I tried to repeat their experiment by using a fine slit of 5 μm, which was adjustable from outside the apparatus. I asked the head of our mechanical workshop whether he could make such a fine slit. He replied "I suppose we can make it, but a lot of money and time would be needed". Needless to say I gave up the idea of making such a slit in our workshop. Then I realised that by placing a large slit in front of a pointed filament I could achieve the same result as one could expect from a fine slit and a normal filament. The idea of using a point filament was not mine originally: Müller had used it in his field emission microscope. In his case field emission took place in a vacuum of $10^{-8} \sim 10^{-9}$ torr, while in my case thermally-assisted field emission was used in a vacuum of $10^{-4} \sim 10^{-5}$ torr. The number of Fresnel fringes of MgO which Haine and Mulvey could observe was only two or three, while to my surprise twenty fringes were obtained in my experiment. The electron beam was very bright and coherent. However, I could not obtain a good electron reconstructed-image at that time because this was still prior to the invention of the laser. Later on, however, Tomomura and his group could obtain a very beautiful reconstructed image of MgO, by using a reference electron beam and a field emission source for obtaining an electron hologram, and also could use a laser light beam for image reconstruction. At any rate, many people in Europe and the U.S.A. were interested in our holograph showing many Fresnel fringes. By using a pointed filament and also an objective lens of short focal length, we could obtain better resolution (ca. 15 Å) than that of a commercial two-stage electron microscope (ca. 30 Å) (Hibi, 1954). Recently we have obtained an electron image of around 1 Å lattice image by using a three-stage electron microscope and a pointed filament (Yada and Hibi, 1969).

3. COHERENCE OF AN ELECTRON BEAM AND CONTRAST OF AN ELECTRON IMAGE

To understand the origins of good coherence from a pointed filament, we constructed an electron interferometer having a Möllenstedt-type electron biprism (Hibi and Takahashi, 1965). By using this apparatus we studied the relationship between electron beam coherence and electron image contrast. In this case, γ_{12} was used as the parameter of degree-of-coherence, as in the optical case. We found that the contrast of electron images of specimens, including biological ones, increased markedly with γ_{12} (Hibi and Takahashi, 1965). However, I was very sorry, generally speaking, that we could not get many researchers to accept our conclusions. Nevertheless we continued our research on electron beam coherence, and we next constructed a high resolution interference microscope by inserting a special electron biprism into a commercial electron microscope (Yada, Shibata, Takahashi and Hibi, 1973). By using this, the inner potential of a small specimen of several elements was measured fairly accurately. I have maintained my interest in improving the quality of electron sources over a very long period. However, I am very sorry to say that our work concerning electron beam coherence was very largely overlooked by contemporary electron diffraction and electron microscope workers, and a neglected topic at international conferences. At the present time, however, in order to increase the contrast for high resolution electron microscopy, a pointed filament is commonly used as a field emission source or as a thermally-assisted field emission source. I have very mixed feelings in contemplating this situation.

REFERENCES

Haine, M. E. And Mulvey, T.: 1952, *J. Opt. Soc. Amer.* **42**, 763.
Hibi, T.: 1954, *Proc. Intern. Conf. Electron Microscopy*, London, 460.
Hibi, T., Kambe, K., and Honjo, G.: 1955, *J. Phys. Soc. Japan* **10**, 35.
Hibi, T. and Takahashi, S.: 1965, *Proc. Intern. Conf. Electron Diff. and Crystal Defects*. Melbourne, I O–4.
Uyeda, R.: 1955, *J. Phys. Soc. Japan* **10**, 256.
Yada, K. and Hibi, T.: 1969, *J. Electron Microscopy* **18**, 266.
Yada, K., Shibata, K., Takahashi, S., and Hibi, T.: 1973, *J. Electron microscopy* **22**, 223.

S. OGAWA

II.14. Electron Diffraction Studies in Sendai

1. ORIGIN OF THE SENDAI SCHOOL

I graduated from the Department of Physics, Tohoku Imperial University, in March 1934 and got a position in the Research Institute for Iron, Steel and Other Metals – RIISOM – belonging to the University. My first project was to confirm a diffraction phenomenon of a hydrogen atom structure on a lithium fluoride surface found by Johnson (1930). Soon after beginning this study I changed my project at the specific request of Professor K. Honda, a founder of the RIISOM. Professor Honda had studied in Professor G. Tammann's laboratory in Göttingen and subsequently became a pioneer of metal physics, or physical metallurgy, in Japan. He looked on electron diffraction as a new tool for the study of metals and alloys and had invited Mr M. Miwa, a graduate from Professor S. Nishikawa's department, to carry out electron diffraction studies in the RIISOM. Mr Miwa was, however, about to move to the National Cancer Center Research Institute after having completed a two-year study in Sendai on amorphous carbon (Miwa, 1934–1935), and on the polished surfaces of metals (Miwa, 1935). Thus I began electron diffraction work as his successor.

My first study was made on nickel films formed by sputtering in hydrogen or nitrogen (Ogawa, 1937). The apparatus was left by Mr Miwa, who kindly instructed me in the elementary technique, and helped me through the initial stages of the experiment. An anomalous hcp structure of sputtered nickel film which was first found by Professor G. P. Thomson (1929) was confirmed, and some other new anomalous structures were found. These nickel films were not ferromagnetic, so I also studied their paramagnetic properties. My electron diffraction study was, then, temporarily interrupted by some studies of magnetism, but my concern and interest finally returned to electron diffraction, and I completed a new apparatus of horizontal type, with a camera length of 50 cm, in 1944. Unfortunately, nothing was done with this apparatus, because my laboratory was entirely destroyed by the air raid on 10th July, 1945 and the apparatus was reduced to a charred and smoldering waste. It could have been saved if only it had been kept in a cellar under the laboratory.

Early in the post-war period, in 1946, I started the electron diffraction work again. An electron diffraction camera of the horizontal type with a cold cathode was bought from the Shimazu Company. A high voltage transformer was given to us by Professor I. Obinata, but a paper condenser for 50 keV could not be obtained. I got about twenty condensers of 2 μF rated at 3 to 5 keV at some suburb of Sendai, where goods having belonged to the old army were concealed.

These condensers were connected in series. As the emission of the cold cathode was very unstable, this was substituted by a hot cathode of usual filament type.

In April 1948 a student belonging to the Department of Physics, Mr D. Watanabe, started at my laboratory. He graduated at Tohoku University in 1949, immediately got his position in the RIISOM and was my coworker for a long time before becoming a professor in the Department of Physics in 1967. Professor S. Shirai temporarily joined my laboratory at this point bringing his electron diffraction apparatus with him, as the National Second High School to which he belonged had been burnt out. He had studied the orientation relationships between metal films epitaxially grown on alkali halide crystals, and these substrates, for many years. We learned the technique of epitaxial growth from him during his stay in our laboratory.

While we were engaged in studies of identification of surface products, resulting from oxidation and corrosion of metals and alloys, by reflection electron diffraction, another paper on a similar topic came to my notice. Miyake and Kubo (1947) had observed by reflection electron diffraction that patterns of metals such as copper and silver, evaporated onto galena heated at, e.g., 400°C, disappeared after, say, ten minutes. They interpreted this phenomenon as a diffusion of metal atoms into the galena, and estimated diffusion constants as well as activation energies. Stimulated by this study, I undertook the determination of diffusion constants, not by reflection but by transmission electron diffraction. We formed a double layer from two different metals by successive evaporation and heating at various temperatures, and estimated the change in composition of the two films with the progress of diffusion, by measurements of changing lattice constants. But the attempted application to a double layer of copper and gold failed owing to the unexpected formation of a Cu_3Au superlattice at comparatively low temperatures. However, this failure led to a study of superlattices using thin films, as rapid diffusion was expected in films. Existence of the $CuAu_3$ superlattice had been predicted by Cowley (1950), but had not been experimentally confirmed. We found this new superlattice by electron diffraction (Ogawa and Watanabe, 1951, 1952), but the first discoverer was not us but Mr M. Hirabayashi, a postgraduate student of Tokyo Institute of Technology, who observed the superlattice lines by X-ray diffraction using a bulk alloy a little earlier than the presentation of our paper (Hirabayashi, 1951). He soon joined our group. Nevertheless our electron diffraction work was very useful, because it soon became clear that structure of superlattices could easily be studied in thin evaporated films, as we had expected, owing to a marked shortening of the ordering period. In our study, the size of ordered domains formed in the course of ordering was estimated from widths of superlattice diffraction rings.

2. LONG-PERIOD ORDERED ALLOYS AND THE MONTREAL CONGRESS

Long-period ordered alloys such as CuAuII have regular arrays of antiphase domains of definite size, causing superlattice reflections to split into several components. Electron diffraction patterns of single crystal films represent the intensity distribution in reciprocal space with little distortion. Thus, the use of evaporated single crystal alloy films contributes much to the structure analysis of the long-period ordered alloys, in distortionless observation of the splitting of superlattice reflections as well as in easy and rapid preparation of specimen films. The two advantages which thin films have over bulk samples are the relative ease of single-crystal preparation, and the relatively short exposure times needed for the electron – as opposed to the X-ray diffraction study. We have thus analysed structures of the long-period ordered alloys with success by transmission electron diffraction using single crystal alloy films epitaxially grown on rocksalt face by evaporation (Ogawa, 1962, 1973). Some of these alloys exhibited very complicated splittings of superlattice reflections owing to superposition of several orientations, but we carried out the analyses, utilizing a little tetragonal or rhombic distortion accompanying the ordering as a clue. Most of these analyses were confirmed by X-ray diffraction using bulk single crystals. There was scarcely any difference between the results from thin films and those from bulk crystals, and this fact convinced me that the combination of evaporated single crystal films and electron diffraction would prove to be an excellent method of structure analysis in alloys. That is to say, alloy structures formed in thin films several hundred Å thick are not different from those in a bulk state. Structures of $Cu_3Pd(\alpha'')$ and Ag_3Mg had never been analysed by X-ray diffraction. We made the analyses directly by electron diffraction using single crystal films (Watanabe and Ogawa, 1956; Fujiwara, Hirabayashi, Watanabe and Ogawa, 1958), and it was clear that what made the X-ray analyses difficult was the non-integrality of the periods, and the two-dimensional domain structure in $Cu_3Pd(\alpha'')$. These properties were clearly elucidated from the electron diffraction patterns.

I presented results obtained from CuAuII, Cu_3Pt, Ag_3Mg, Cu_3Pd and Au_3Mn in the symposium on electron diffraction held at the Montreal Congress in 1957, laying emphasis on the gold-manganese alloy. I believe that our structure analyses made a great impression on the audience. Spontaneous applause came unexpectedly from them when our diffraction patterns showing many split superlattice reflections were shown. The Montreal Congress was the first international meeting in which I had participated, and the speech in English was also my first. Owing to my unpracticed English speaking, my talk exceeded the time limit by ten minutes in spite of repeated prompting by the chairman, Professor M. Blackman. My participation in the Montreal Congress was a memorable event for me, and I shall never forget the success of my lecture, symbolized by the unexpected applause. Studies of the long-period ordered alloys by electron diffraction were carried out during the period 1952 to 1964 in our laboratory.

Industries in Japan began to recover from about 1950, and it became possible for us to use commercial electron diffraction equipment. Moreover, development of electron microscopes in Japan enabled us to obtain better results by combining electron diffraction with electron microscopy. The resolution of Japanese electron microscopes was becoming higher and higher and Dr H. Watanabe, of the Hitachi Central Laboratory, wanted to demonstrate the resolution of Hitachi electron microscopes, applying it to problems on metal physics. Since we were wanting at that time to make direct observations of regular antiphase boundaries, having spacings of about 20 Å, in CuAuII, a cooperative study was made. The objective aperture was chosen for a bright field image to extend (symmetrically placed around the central beam) up to the neighborhood of [100] superlattice reflections and to include the satellite reflections flanking the direct spot which resulted from the regular antiphase structure. Thus we succeeded in revealing for the first time the periodic antiphase boundaries in CuAuII as parallel lines with intervals of about 20 Å on electron micrographs. The parallel lines often included irregularities which might originate from imperfections of the lattice. I presented this result at the Montreal Congress informally, as an additional short paper. Though this electron-microscopic study was not included in the abstract owing to its late completion, immediately before my start for Montreal, a short abstract was distributed in the lecture hall. Following a delayed publication of this direct observation by electron microscopy (Ogawa, Watanabe, Watanabe and Komoda, 1958), a study made on the same topic by Dr D. W. Pashley and his coworker was published (Glossop and Pashley, 1959). They utilized split components of a (110) superlattice reflection in dark field image formation and obtained more beautiful electron micrographs showing regular arrays of antiphase domains.

We made further studies of the direct observation of antiphase domains by electron microscopy and succeeded in revealing antiphase domains of about 8 Å in size in an alloy film of copper–gold–zinc (Ogawa, Watanabe, Watanabe and Komoda, 1959), as well as the two-dimensional antiphase domain structure in a Cu₃Pd(α″) film (Ogawa and Watanabe, 1961). This two-dimensional structure which had already been deduced from the analysis by electron diffraction was vividly imaged in electron micrographs, the existence of six orientations assumed in the electron diffraction analysis being clearly demonstrated.

3. CLEAVAGE IN VACUUM AND MULTIPLY-TWINNED PARTICLES

A number of studies have been made on epitaxy since the 1930s, but most of them were on the orientational relationship between deposit and substrate, while no conclusive explanation seemed to be given as to the origin of the epitaxy. I visited laboratories in the United States on my way back from the Cambridge Congress in 1960 and was much astonished to see the marked development of techniques of ultrahigh vacuum. This factor encouraged us to study the effect of high vacuum on epitaxy. We obtained ultrahigh vacuum by an ion pump and used valves of molten indium, so chosen because of the very low melting point of this metal and the very low vapor pressure of its molten state ($\sim 10^{-12}$ mm Hg). In order to obtain a clean surface, substrate rocksalt was cleaved in vacuum by forced rotation of an iron piece to which the substrate was attached, by an electromagnet placed outside the vacuum. This device was designed by my student, Mr S. Ino. Cleavage of the substrate in ordinary high vacuum of $\sim 10^{-5}$ mm Hg markedly lowered the epitaxial temperatures of gold, silver, copper, nickel and aluminum. Cleavage of the substrate in ultrahigh vacuum ranging from 10^{-9} to 10^{-8} mm Hg, on the other hand, much deteriorated the orientations of gold, silver and copper (Ino, Watanabe and Ogawa, 1964). This curious result was against our expectation and considered to be connected with the origin of the epitaxy. According to later studies, this deterioration of the epitaxy might be caused by lack of water vapor.

It is noteworthy that, in the 1960s, fine particles with abnormal structure, formed in an early stage of film formation by vacuum evaporation of fcc metals, were discovered by electron diffraction and electron microscopy. Mihama and Yasuda (1964) observed anomalous (111) diffraction spots, which were not explained by usual orientations, from a gold film epitaxially grown on rocksalt face, and found the spots to arise from some minute particles with some compound structure. They called these 'compound particles'. The anomalous (111) spots were also observed by Allpress and Sanders (1964) from a gold film grown on a (001) single crystal film of silver. As they did not determine the structure of the particles in question, I advised Mr. Ino to carry out an analysis. He skillfully applied the dark field technique in electron microscopy and revealed the particle's structure (Ino, 1966). Gold particles grown on a rocksalt face and giving rise to the anomalous (111) spots consisted of many tetrahedral twins, and therefore, we named them 'multiply-twinned particles'. They are of three types: an icosahedron consisting of twenty tetrahedra, a decahedron consisting of five tetrahedra and a polyhedron with a structure similar to the decahedron. The first kind appears to be hexagonal in an electron micrograph, the second to be pentagonal and the third to be rhombic. All of them contain inherent strain. The observed anomalous (111) spots were well explained by the above structures. Dr T. Komoda (1968) observed lattice images of pentagonal and hexagonal particles of gold and proved Ino's analysis to be right. Allpress and Sanders (1967) reported observations similar to those of Ino on the particles of gold, palladium and nickel grown on mica and agreed with Ino in explaining them as multiply-twinned particles.

The multiply-twinned particles were known to exist in an early stage of film formation of various fcc metals, and they were also found in metal fine particles prepared in argon (Kimoto and Nishida, 1967). Ino (1969) calculated internal energies of the multiply-twinned particles and estimated sizes above which they became unstable. It is very interesting that a greater part of extra reflections occurring from evaporated films and having never been elucidated by any reasonable origin since the 1930s could be just explained in 1966 as arising from successive diffraction of different twinned parts of the multiply-twinned particles (Ino, 1966).

4. AMORPHOUS FILMS PREPARED BY LOW TEMPERATURE CONDENSATION

Thin metallic films formed by vacuum condensation onto substrates kept at very low temperature often have amorphous structures. I advised Mr S. Fujime in our group to apply the radial distribution analysis to electron diffraction patterns of thin films of various metals prepared by low temperature condensation, in order to confirm their amorphous nature and to investigate the origin of the amorphous state. I should say that his studies (Fujime, 1966) were carried out before, and independent of, recent fashionable studies on amorphous materials. According to him (Fujime, 1967), metals with the close-packed structure, other than transition metals, do not take any amorphous state, while those with structures deviating considerably from the close-packed structure are apt to become amorphous. Transition metals with the fcc structure adopt, however, the amorphous state. He discussed this behavior by considering the effects of abrupt condensation upon an intrinsic crystal structure, and the property of each metal. Mr T. Ichikawa in our group succeeded Mr Fujime and obtained information on local arrangement in amorphous metallic films from the radial distribution analysis. It seems very interesting to me that the local atomic arrangement is characteristic of each metal (Ichikawa and Ogawa, 1974).

5. PRESENT STATE OF THE SENDAI SCHOOL

Since I moved from Sendai to Tokyo upon my retirement in 1975, active electron diffraction studies in the Sendai School have mainly been, and will be, carried out by Professor D. Watanabe, Dr M. Tanaka and Dr S. Ino. Determination of structure factors by measurement of the critical voltage of the disappearance of the second-order reflection of Kikuchi pattern (Watanabe, Uyeda and Kogiso, 1968; Watanabe, Uyeda and Fukuhara, 1969) and observation of short-range order scattering in several alloys (Ohshima and Watanabe, 1973), made by Professor Watanabe and his coworkers, and a study by convergent beam electron diffraction made by Dr M. Tanaka produced excellent results.* Dr Ino (1977) obtained beautiful reflection patterns of HEED from silicon surfaces in ultrahigh vacuum, and his study is opening up a new field in surface structure analysis, with an extraordinary clarity and resolution of pattern which is not markedly lowered even at high temperature, as well as by a new technique introducing LEED-like patterns.

Shibaura Institute of Technology
Tokyo, Japan

REFERENCES

Allpress, J. G. and Sanders, J. V.: 1964, *Phil. Mag.* **9**, 645–658.
Allpress, J. G. and Sanders, J. V.: 1967, *Surf. Sci.* **7**, 1–25.
Cowley, J. M.: 1950, *Phys. Rev.* **77**, 669–675.
Fujime, S.: 1966, *Jap. J. Appl. Phys.* **5**, 778–787, 1029–1035.
Fujime, S.: 1967, *Jap. J. Appl. Phys.* **6**, 305–310.
Fujiwara, K., Hirabayashi, M., Watanabe, D., and Ogawa, S.: 1958, *J. Phys. Soc. Japan* **13**, 167–174.

* See article III.6 of this volume

Glossop, A. B. and Pashley, D. W.: 1959, *Proc. Roy. Soc.* A**250**, 132–146.

Hirabayashi, M.: 1951, *J. Phys. Soc. Japan* **6**, 129–130.

Ichikawa, T and Ogawa, S.: 1974, *J. de Phys. Colloque C4* suppl. No. 5, **35** (Mai), C4-27-29.

Ino, S., Watanabe, D., and Ogawa, S.: 1964, *J. Phys. Soc. Japan* **19**, 881–891.

Ino, S.: 1966, *J. Phys. Soc. Japan* **21**, 346–362.

Ino, S.: 1969, *J. Phys. Soc. Japan* **27**, 941–953.

Ino, S.: 1977, *Jap. J. Appl. Phys.* **16**, 891–908.

Johnson, T. H.: 1930, *J. Franklin Inst.* **210**, 135–152.

Kimoto, K. and Nishida, I.: 1967, *J. Phys. Soc. Japan* **22**, 940.

Komoda, T.: 1968, *Jap. J. Appl. Phys.* **7**, 27–30.

Mihama, K. and Yasuda, K.: 1964, Lecture at the meeting of Phys. Soc. Japan held in October.

Miwa, M.: 1934–1935, *Sci. Rep. Tohoku Imp. Univ.* **23**, 242–258.

Miwa, M.: 1935, *Sci. Rep. Tohoku Imp. Univ.* **24**, 222–239.

Miyake, S. and Kubo, M.: 1947, *J. Phys. Soc. Japan* **2**, 20–24.

Ogawa, S.: 1937, *Sci. Rep. Tohoku Imp. Univ.* **26**, 94–105.

Ogawa, S. and Watanabe, D.: 1951, *J. Appl. Phys.* **22**, 1502.

Ogawa, S. and Watanabe, D.: 1952, *J. Phys. Soc. Japan* **7**, 36–40.

Ogawa, S.: 1962, *J. Phys. Soc. Japan* **17**, Suppl. B-II, 253–262.

Ogawa, S.: 1973, *Proc. International Symposium on Order–Disorder Transformation in Alloys* 240–264.

Ogawa, S., Watanabe, D., Watanabe, H., and Komoda, T.: 1958, *Acta Cryst.* **11**, 872–875.

Ogawa, S., Watanabe, D., Watanabe, H., and Komoda, T.: 1959, *J. Phys. Soc. Japan* **14**, 936–941.

Ogawa, S. and Watanabe, D.: 1961, *Report of Technical Conference held in St. Louis, Missouri, March 1–2*, 523–542.

Ohshima, K. and Watanabe, D.: 1973, *Acta Cryst.* A**29**, 520–526.

Thomson, G. P.: 1929, *Nature* **123**, 912.

Watanabe, D. and Ogawa, S.: 1956, *J. Phys. Soc. Japan* **11**, 226–239.

Watanabe, D., Uyeda, R., and Kogiso, M.: 1968, *Acta Cryst.* A**24**, 249.

Watanabe, D., Uyeda, R., and Fukuhara, A.: 1969, *Acta Cryst.* A**25**, 138–140.

Z. G. PINSKER

II.15. Development of Electron Diffraction Structure Analysis in the USSR

1. INTRODUCTION

In the USSR, high energy electron diffraction studies were started immediately after Thomson's early works: these were Tartakovsky's and Lashkarev's experiments (Tartakovsky, 1928, 1929, 1932; Lashkarev, 1933, 1935) carried out at the Ioffe Institute (its present name) in Leningrad.

Essential results were obtained during investigation of NaCl and CdI_2 polycrystalline films by Pinsker (1935) and then by Pinsker and Tatarinova (1936) at Vernadsky's laboratory and, since 1943, at the Institute of Crystallography of the USSR Academy of Science. Two types of pattern have been obtained, namely, regular spot patterns and arc reflection patterns. On the basis of a thorough analysis of experimental conditions it was considered that the patterns represented scattering from great numbers of crystals in reflecting positions. Electron diffraction patterns were interpreted as formed by the intersection of the Ewald sphere with the reciprocal lattice spikes perpendicular to the incident beam. Spot patterns correspond to a mosaic single crystal, and permit us to determine the symmetry, unit cell parameters and extinction law for certain zones of reflections. The second type of electron diffraction pictures, oblique texture patterns, are from thin film specimens in which all the crystals are oriented parallel to the substrate, with some azimuthal disorientation. These patterns have received detailed geometrical interpretation. It was shown that, even in the case of an unknown structure, oblique texture patterns permit us to obtain unit-cell parameters, index all the reflections and determine crystal symmetry. The formula required to calculate the parameter c has been obtained for the particular case of orthogonal textures (Pinsker, 1941). The developed geometric theory and the estimation of reflection intensities, on the basis of kinematical theory, were successfully applied to the structure determination of CdI_2, $CdCl_2$ and PbI_2. The results and the literature data have been generalized, thus constituting one of the early works on polytypism in layer crystals (Pinsker, 1939, 1941; Pinsker and Tatarinova, 1941; Pinsker, Tatarinova and Novikova, 1944). Pinsker's *Electron Diffraction* appeared in 1949 and dealt not only with the physics of electron diffraction, but also stated the general scheme of electron diffraction structure analysis and provided many examples of its use (Pinsker, 1949).

In the late 'forties and early 'fifties intensive development of the theory and experimental technique of HEED was initiated. The kinematical theory of the structure analysis of polycrystals received further development (Vainshtein, 1949; Vainshtein and Pinsker, 1949); the law of atomic scattering was investigated (Yamzin and Pinsker, 1949) and the investigations of clay mineral structures were also started (Zvyagin and Pinsker, 1949; Zvyagin, 1952).* Electron diffraction has also been applied to the study of polymers (Distler and Pinsker, 1949, 1950) and semiconductors (Semiletov and Pinsker, 1955).† An essential contribution to electron diffraction development has been made by Vainshtein in his book *Structure Analysis by Electron Diffraction* (Vainshtein, 1956). He has worked out the complete geometric theory of the formation of various electron diffraction patterns, obtained formulae to calculate unit cell parameters, and expressions for calculating reflection intensities (local and integral).

Passing to the second stage of structure analysis, Vainshtein has analyzed in detail the Fourier expansion of potential, the scaling of experimental amplitudes Φ_{obs} and the value of potential in the centre of an atom $\phi(o)$, and derived formulae for the calculation of errors in coordinate and potential determinations. These relationships have been presented, in a convenient form, as a function of the widely used R-factor.

It is well known that the subsequent development of electron diffraction structure analysis has proved the practical importance of structure determination within the frame of kinematical theory, as developed by Vainshtein. New theoretical and experimental investigations carried out in this period should also be mentioned, such as diffraction from single crystals and experimental and computational methods of electron diffraction structure analysis.

2. KINEMATICAL THEORY

First of all, let us consider the applications of the kinematical theory and possible numerical estimates. The corresponding considerations made by Blackman (1939) were elaborated by Pinsker (1949) and, in more detail, by Vainshtein (1955, 1956). Kinematical relationships were treated as the limiting case of Bethe's two-beam theory in the case of thin crystals and weak reflections. Of significance is the expression for the mean value of structure amplitude for a structure consisting of n identical atoms, given by Vainshtein:

$$\Phi = f_{el} \sqrt{n} \tag{1}$$

where f_{el} is the atomic scattering factor, which leads to the following estimate of the limiting thickness of the kinematically scattering crystal:

$$t \approx 200 \sqrt{n}/f_{el} \tag{2}$$

From here, one can obtain the known numerical estimates, coinciding with those obtained from extinction distance τ_o, in accordance with Bethe's theory

---

* See also (Zvyagin *et al.*, 1979).
† See also an article by Semiletov & Imamov in this volume.

$$\tau_0 = \pi \Big/ \left\{ \lambda \frac{\Phi}{\Omega} \right\} \tag{3}$$

where Φ is the modulus of the structure factor, Ω the volume of the unit cell, and λ is the electron wavelength.

For structures of various complexity, composed of either heavy (or light) atoms, these criteria give the following thicknesses: 100–200 Å and about 1000 Å respectively.

And yet the above mentioned criteria and estimates are not quite correct, as has been proved by the progress of the many-beam dynamic theory developed by Cowley and Moodie (1957), Fujiwara (1959) and Fujimoto (1959) and by the intensive development of computational methods of many-beam diffraction, and the experimental study of diffraction patterns obtained from individual very-thin crystals.

Of fundamental importance is the fact that, while kinematical scattering can be investigated on polycrystalline objects, many-beam diffraction is used, for the time being, only in the case of unlimited, thin, single crystals. Thus it follows from the above that rigorous physical substantiation of a widely used kinematical approximation has not yet been given. At the same time, the criterion (2) is still of importance. It is widely accepted that the weak reflections can be treated as kinematical. It should, however, be mentioned that Kreutle and Meyer-Ehmsen (1971) and Ando, Ichimiya and Uyeda (1974), using only 3–18 beams, have derived the potential values for several strong reflections with high accuracy.

3. SURVEY OF RESULTS FROM POLYCRYSTALLINE THIN FILMS

Below, the main stages of the electron diffraction structure investigation of polycrystalline thin films will be considered. The most reliable experimental data, for the control and investigation of phase composition and structure, are given by two types of electron diffraction pictures: spot patterns from mosaic crystals and oblique texture patterns. These patterns, the latter especially, provide the main experimental data for the first stage of the study of unknown structures. After indexing all the reflections it is possible to measure, more or less precisely, their intensities. The most accurate results for intensity measurement, and the smallest R-factors, have been obtained for electron diffraction patterns from non-oriented polycrystalline films (Debye–Scherrer patterns). The use of oblique texture for structure determination of clay minerals is also considered in this volume (see Zvyagin's article).

In electron diffraction structure determinations, the main part is played by Fourier potential syntheses. The reliability of these syntheses, i.e. parameter accuracy, the absence of spurious maxima and their sensitivity to the localization of light atoms, depend, to some extent, on the series termination and, to a lesser degree, on the accuracy of structure amplitude determination. The advantage of oblique texture patterns for this purpose lies in the fact that they permit us to register the maximum possible number of reflections (up to maximum values of $\sin\theta/\lambda$, which are often indistinguishable in Debye–Scherrer patterns, even though the latter yield the most accurate intensities (I_{hkl}) and Φ_{hkl} values.

In such measurements two fundamental difficulties are encountered. In some cases strong reflections and their higher orders reveal the presence of dynamic scattering. The possible conditions promoting the appearance of these effects cannot be established with sufficient accuracy for the time being; these are: relative simplicity of a structure, a large atomic number, Z, for atoms contributing to corresponding scattering amplitudes and, lastly, the thickness of the scattering crystals. Also, during the registration of many medium-weak and weak reflections a significant error is due to the build-up of a continuous diffuse line in the background.

To take into account extinction effects (two-beam scattering), Blackman's method has been used first (Dvoryankina and Pinsker, 1958), and then Bethe's dynamic potentials in the form corresponding to the so-called systematic extinctions (Udalova and Pinsker, 1972). In the majority of cases it permitted not only a decrease in the final R-value but, in some cases, also improved the reliability and accuracy of light-atom parameters, as well as of some other important parameters of the Fourier synthesis of potential.

One of the most important results of our long experience is the establishment of the fact that Fourier-synthesis has a low sensitivity to errors in the observed structure amplitudes. It is essential to include in the calculations the amplitudes corresponding to the maximum possible $\sin\theta/\lambda$. To obtain reliable, and sufficiently accurate, coordinates and R values for light atoms in the presence of one or several heavy ones, one should use differential synthesis, thus excluding the influence of termination waves due to the heavy atoms. One should also include in the Fourier series values for amplitudes whose reflections are within the 'cone of inaccessibility', using oblique texture patterns only.

Our experience shows that errors in coordinates for heavy atoms are of the order of 0.01 Å and even less (at $R \approx 20\%$), while for light atoms they amount to 0.03–0.05 Å (Man, 1970).

For comparatively simple structures, and especially in the case of Debye–Scherrer patterns, the reliability factor does not exceed 5–8%. It is noteworthy that such low R-values are obtained for structures composed of atoms greatly varying in atomic number Z. Evaluation of data may be made by direct comparison of observed and calculated potential peak height values, $\phi(o)$, in real space. Alternatively, the root mean-square error (of a determination) may be estimated from the relationship obtained by Vainshtein (1956):

$$\sqrt{\overline{\Delta\phi^2}} = R\frac{1}{\Omega}\sqrt{\sum|\Phi|^2} \qquad (4)$$

The absolute value of the mean-square error is usually about 2 to 5 V, which means 1% to 2% for medium-weight atoms, and up to 10% for light atoms. These values, in the majority of cases, are in good agreement with the experimental values for known stoichiometric composition.

A discrepancy arises for the structures whose atoms are statistically distributed among several positions of the corresponding space group. Such is the situation with hydrogen atoms in NH_4Cl, NH_4 Br and NH_4I.

The significant disagreement between calculated and observed values of the potential maxima heights $\phi(\mathbf{o})$ for W-atoms in one of the positions was observed for WN (Khitrova and Pinsker, 1959). Later, many structures have been observed with 'defect' filling by atoms of their positions. The comparison of $\phi_{obs}(\mathbf{o})$ with $\phi_{cal}(\mathbf{o})$ not only proved the 'defectiveness' but also permitted us to refine the composition (chemical formula) of the nonstoichiometric phases under consideration.

4. LIGHT-ATOM STRUCTURES AND HYDROGEN-ATOM LOCATION

A typical problem of electron diffraction structure analysis, the localization of light atoms in the presence of heavy ones, was solved in many works. Examples are: the localization of hydrogen atoms in organic structures $(Z_H/Z_O = 1/8)$, in ammonium chloride, bromide and iodide, in hydrides of Ni and Pd; the localization of nitrogen in nitrides of Ti, Cr, Fe, Ni and Mo, and lastly that of oxygen in Nb, Ta and Bi-oxides $(Z_O/Z_{Nb, Ta, Bi} = \frac{1}{41}, \frac{1}{73}, \frac{1}{83})$. The localization of light atoms was also complicated by the fact that in many phases they occupy statistically more than one position in the corresponding space group. In these cases the effective structure amplitude of such a 'partial' atom diminishes, and it is rather difficult to determine.

A sufficiently small R-factor was made possible in the first place by the use of Debye–Scherrer patterns. In all cases the R-factor decreases if one takes into account the contribution of light atoms. The following two conditions are desirable in order to achieve low R-values: a structure should be simple or of moderate complexity, and its electron diffraction pattern should contain a sufficient number of reflections including those with high value of $\sin \theta/\lambda$. Furthermore, for an unknown structure, the symmetry should be sufficiently low to prevent the coincidence of non-equivalent reflections for the given point group (Man and Imamov, 1972), and the pattern should not have a background indicative of dislocations, or of structural disorder or variability.

A distinctive feature of our investigations is the small discrepancy achieved between derived atomic coordinates and those obtained by the X-ray method. The same is true for the heights of potential maxima $\phi(\mathbf{o})$. It should be underlined that the localization of light atoms has been made both for structures previously investigated by X-rays, and for unknown structures.

One of the early works here was that on diketopiperazin (Vainstein, 1954; 1955) where the author's goal was the study of hydrogen bonds. Oblique texture patterns permitted him to estimate the intensities with an error of about 30%, and the amplitudes to about 15%. And yet, using various Fourier-syntheses of potential (including three-dimensional), C—N,C—C and C—O distances were determined, their values differing only slightly (0.8–3.5%) from those obtained in X-ray work.

Very indicative also are root mean square errors in atomic coordinates $x_{C,N,O} = 0.008$ Å and $x_H = 0.028$ Å.

In the hexagonal carbide WC the positions of the C-atoms (Butorina, 1960) were determined, the final value of the reliability factor being $R = 8.1\%$; Blackman's corrections were introduced into several strong observed amplitudes. The section in the plane $(11\bar{2}0)$ of three dimensional potential series revealed maxima due to W and C-atoms, their heights being in good agreement with theoretical predictions. The results obtained have also been confirmed by neutron diffraction (Leciejewisz, 1961). In the course of investigation of polycrystalline films of cubic WN (Khitrova and Pinsker, 1959) the heights of potential maxima had the following values $(R = 7.75\%)$: for W 2865 V (complete value of $\phi_{cal}(o)$ 2865 V with the accuracy $\sim 0.6\%$) and for N 200 V $(0.67 \times 300$ V with the accuracy $\sim 3\%)$ which corresponds to the composition $WN_{0.67}$. This result is in good agreement with the data obtained by Hägg (1930). All the calculations were carried out in kinematical approximation. Most precise data on the localization of light atoms have been obtained by Khodyrev, Baranova, Imamov and Semiletov (1978). Using electron diffraction patterns from polycrystalline film of palladium hydride, they determined the structure and hydrogen composition of the cubic hydride. Hydrogen atoms occupy octahedral sites of the fcc lattice of Pd-atoms. Using experimental intensities, the authors determined the composition from the analysis of the heights of potential maxima, and refined it by the least squares method. For the composition $PdH_{0.64}$ the final R-value was 6.5%.

Localization of hydrogen atoms in NH_4Cl (Avilov, Imamov, Karakhanyan and Pinsker, 1973) and NH_4Br (Karakhanyan, Udalova, Imamov and Pinsker, 1974) was carried out on the basis of polycrystalline patterns. Halogen and nitrogen atoms are located according to a CsCl-structure. For NH_4Cl the final value was $R = 11\%$ (48 reflections, 100 kV, without the filter for inelastically scattered electrons), whereas it decreases to 7.2% for 18 reflections (50 kV, using the filter); $r_{N-H} = 1.4 \pm 0.02$ Å. These latter improvements being introduced, the two-dimensional section of a three-dimensional Fourier-series revealed not only maxima from Br and N but also from H, in the case of NH_4Br; the exact value $\phi_{obs}(o)$ for H/2 was obtained* from differential synthesis, and equals 17 ± 3 V; the accuracy of the localization of Br and N was 1.5% and 0.3% respectively.

A similar investigation into NH_4I was carried out with the use of oblique texture patterns. It included the choice between two different positions for H known from the literature (Karakhanyan, Udalova, Imamov and Pinsker, 1974). Not considering details, we mention only that dynamic correction was introduced into 11 of the total of 31 amplitudes, and that to exclude the maximum due to termination waves from the I-atom, the authors used differential synthesis. The reliability factor was about 13%, the heights of maxima being determined with an accuracy of ± 4 V.

---

* Since four hydrogen atoms take a statistically eight-fold position, these peaks correspond to half the potential of the hydrogen atom (see Vainshtein, 1956, page 372).

5. METAL STRUCTURES

The technique of thin film preparation of W, Mo and Ni-nitrides have been developed at the Physical-Technical Institute of the city of Gorky (Pinsker and Kaverin, 1957). Some new phases were obtained and their structures determined from oblique texture patterns. Most of these phases are defective as some metal and nitrogen sites are only partly filled. The atomic coordinates and the degree of occupancy of the corresponding sites (obtained from comparison of $\phi_{obs}(o)$ and $\phi_{cal}(o)$) provide full and reliable information on these important compounds at $R = 12$–20% (Khitrova and Pinsker, 1961). Work should also be mentioned wherein the polycrystalline film of TiN with the NaCl structure was used to check the data obtained for defective phases (Troitskaya, 1965). In the course of an investigation of Bi oxides, a series of phases was revealed which, during oxidation, transformed into a tetragonal phase $Bi_2O_{2.7-2.8}$. Thermodynamically, this phase is probably the most stable as there are no indications in the literature as to the existence of Bi_2O_3. At $R \approx 19$–21% the difference between $\phi_{obs}(o)$ and $\phi_{cal}(o)$ is about 1% for Bi and about 5% for O, which confirms the reliability of the investigation (Zavyalova, Imamov and Pinsker, 1964; Zavyalova and Imamov, 1968, 1969, 1971).

Many phases of the systems Nb–O and Ta–O have also been determined. Usually these structures are nonstoichiometric and very sensitive to the conditions of formation. The structure of cubic NbO and TaO is typical (Klechkovskaya, Troitskaya and Pinsker, 1965; Khitrova, 1966). Here, only metal atoms in the center of a properly chosen cubic unit cell have 100% filling of their sites. The rest of the metal sites have an occupancy of from 15% to 90%. O-atoms occupy four 24-fold positions with an occupancy of about 90%. The composition of the phase was determined as $NbO_{1.66}$ at a density $3.05\,g/cm^3$ and $R = 20\%$. Again, Fourier-syntheses of the potential played a key-role in the analysis of such phases, i.e. determination of their composition, and the relationships during the structure transformations under various physico-chemical conditions.

Of importance also are the experimental curves of atomic scattering, obtained from systems of equations connecting specific values of f_e with potentials at the centers of the atoms. Not considering other similar works of this series (Khitrova and Pinsker, 1970), we only note that the results obtained can be useful in studies of the structural mechanism of oxidation kinetics of these metals, so important for modern technology. Polycrystalline films of Al, Ni, Pd and Au were studied by Udalova and Pinsker (1972) and Imamov and Udalova (1976). Using Blackman's corrections and Bethe's dynamic potentials, they obtained the following values for the R-factor: 4.8, 8.0, 12.8 and 12.7% for Al, Ni, Pd and Au, respectively. The discrepancy between $\phi_{obs}(o)$ and $\phi_{cal}(o)$ is only about 3%.

The investigations of phases having new structure types for which no preliminary data had been known can serve as examples: α-Ag_2Se (Pinsker, Chou Tsin-lian, Imamov and Lapidus, 1965), Ag_7Te_4 and Tl_5Te_3 (see further: Semiletov and Imamov, this volume). In the case of α-Ag_2Se, oblique texture patterns were used, and also patterns were obtained from mosaic crystals with additional symmetry due to partial ordering of mosaic blocks. The total number of reflections amounted to 175. The structure possesses no center of symmetry and contains 8 Ag and 4 Se-atoms in the unit cell. The structure is of interest from the point of view of correlation between structure and properties. Due to the diamond-like frame of Ag-atoms it can be related to semiconductors, a fact which was confirmed by studies of its physical properties.

The investigation of Ag_7Te_4 was also carried out using oblique texture patterns (intensities of 450 reflections were used). The structure was determined by projections and three-dimensional Patterson functions, and refined by Fourier syntheses. It contained 55 atoms per unit cell and had 11 positional parameters. For both structures the final R was 22–23%.

Progress in the development of the experimental technique of electron diffraction structure analysis has been achieved by the study of the scattering law from thin films of mosaic single crystals (Avilov, 1979). Rather accurate methods to measure reflection intensities have been developed. In the case of LiF, a good agreement of intensities experimentally measured with those calculated on the basis of kinematic theory was shown. The final value was $R = 5.2$, for 37 reflections. Blackman's corrections were introduced into the 6 most intense reflections at the calculated crystal thickness 315 ± 15 Å.

It should be noted, in conclusion, that this article does not deal with many works whose authors used kinematical approximation to study thin films and surfaces in connection with various problems of solid state physics and modern technology.

Institute of Crystallography
USSR Academy of Sciences
Moscow, USSR

REFERENCES

Ando, Y., Ichimiya, A., and Uyeda, R.: 1974, *Acta Cryst.* **A30**, 600.
Avilov, A. S.: 1979, *Kristallografiya* **24**, 178.
Avilov, A. S., Imamov, R. M., Karakhanyan R. K., and Pinsker, Z. G.: 1973, *Kristallografiya* **18**, 49.
Blackman, M.: 1939, *Proc. Roy. Soc.* **173**, 68.
Butorina, L. N.: 1960, *Kristallografiya* **5**, 233.
Cowley, J. M. and Moodie, A. F.: 1957, *Acta Cryst.* **10**, 609.
Distler, G. I. and Pinsker, Z. G.: 1949, *Sov. J. Phys. Chem.* **23**, 1281; Distler, G. I. and Pinsker, Z. G.: 1950, *Sov. J. Phys. Chem.* **24**, 1152.
Dvoryankina, G. G. and Pinsker, Z. G.: 1958, *Kristallografiya* **3**, 438.
Fujimoto, F.: 1959, *J. Phys. Soc. Japan* **14**, 1558.
Fujiwara, K.: 1959, *J. Phys. Soc. Japan* **14**, 1513.
Hägg, G.: 1930, *Z. phys. Chem.* **B7**, 339.
Imamov, R. M. and Udalova, V. V.: 1976, *Kristallografiya* **21**, 907.
Karakhanyan, R. K., Udalova, V. V., Imamov, R. M., and Pinsker, Z. G.: 1974, *Kristallografiya* **19**, 946.
Khitrova, V. I. and Pinsker, Z. G.: 1959, *Kristallografiya* **4**, 545.
Khitrova, V. I. and Pinsker, Z. G.: 1961, *Kristallografiya* **6**, 882.

iya **11,** 204.

.: 1970, *Kristallografiya* **15,** 540.

'., Imamov, R. M., and Semiletov, S. A.: 1978, *Kristallografiya*

, N. V., and Pinsker, G. G.: 1965, *Kristallografiya* **10,** 37.

.: 1971, *Phys. stat. Sol.* **A8,** 111.

T.P. (J. Exp. Ther. Phys.) **3,** 510. *Z. Phys.* **86,** 797. *Diffraktsia*
ed. G.T.T.I.; 1935, *Trans. Far. Soc.* **31,** 1081.

, 200.

}, 471.

972, *Kristallografiya* **17,** 4, 757.

5, 520.

s. Rep. **10,** 10; 1941, *Sov. J. Phys. Chem.* **15,** 559; 1944, *C. R.*
95.

ektronov (Electron Diffraction) ed. AN USSR; 1953, *Electron*

mov, R. M., and Lapidus, E. L.: 1965, *Kristallografiya* **10,** 275.
1954, *C. R. (Doklady) Acad. Sci. USSR* **95,** 4.
1957, *Kristallografiya* **2,** 3.
I.: 1936, *Zhur. Exptl. Teoret. Fiz.* **6,** 13.
: 1941, *Zhur, Fiz. Khim.* **15,** 1006; *Acta Phys.-Chim. U.R.S.S.* **30,**

nd Novikova, V. A.: *Zhur. Fiz. Khim.* **16,** 148.
1955, *C. R. (Doklady) Acad. Sci. USSR,* **100,** 1079.
t. Sci. USSR, **A14;** 1929, *Z. Phys.* **56,** 416; 1932, *Experimental*
Matter (in Russian) ed. G.T.T.I., Moscow-Leningrad.
fiya **10,** 284.
a **20,** 640.
1972, *Kristallografiya* **17,** 90.
st. Cryst. (Trudy Inst. Krist.) **5,** 7, 113; 1950, **6,** 193.
Krist. (Trans. Inst. Cryst.) **10,** 115; *C.R. (Doklady) Acad. Sci.*

ady) Acad. Sci. USSR **104,** 537; *Sov. J. Phys. Chem.* **29,** 327.
a Electronographya, ed. Acad. Sci. USSR; *Structure Analysis by*
mon Press, Oxford.
.: 1949, *C.R. (Doklady) Acad. Sci. USSR* **64,** 49.
9, *C.R. (Doklady) Acad. Sci. USSR* **65,** 645; *Trudy Inst. Kristal.*

M.: 1968, *Kristallografiya* **13,** 49; 1969, *Kristallografiya* **14,** 369;

nd Pinsker, Z. G.: 1964, *Kristallografiya* **9,** 857.
Acad. Sci. USSR **86,** 149; **95,** 1305.
949, *C.R. (Doklady) Acad. Sci. USSR* **68,** 505.
Zhukhlistov, A. P., Sidorenko, O. V., Soboleva, S. V., and
Electron Diffraction Analysis of Clay Minerals (in Russian), ed.

II.16. *Electron Diffraction by the Finch School*

An excellent account of the life (1880–1970) and work of Professor George Ingle Finch, M.B.E., D.Sc., F.R.S. has been given by Blackman (1972). This touches well on his very practical and enterprising nature, his early work as a chemist at the Woolwich Arsenal, war service as captain in the Royal Field Artillery and Royal Army Ordnance Division, followed by his distinguished research on combustion and, later, electron diffraction at Imperial College, London; and also on his extensive mountaineering and sailing activities.

Finch and his research team, since 1921, in the then 'Chemical Technology' Department of the Imperial College of Science and Technology, London, S.W.7, part of the University of London, had pursued many researches on the nature and control of combustion in gaseous mixtures, promoted by arcs, sparks and electrodeless discharges, or with catalytically active surfaces such as metal films sputtered in gas at low pressure.

Soon after G. P. Thomson came to the chair of Physics at Imperial College, Finch realised that his powerful and practical photographic technique for electron diffraction could throw much light on the nature of thin films and surfaces and their catalytic action, and they collaborated in results on sputtered Pt films (Finch, Murison, Stuart and Thomson 1933).

Finch saw that a greatly increased beam intensity and resolution could be obtained by applying the action of a coaxial current-carrying coil to focus the electron beam, as had recently been shown by Wierl, and also to scan specimens by tilting the coil. When I joined him in September 1932 to carry out research for the Ph.D. degree, he and A. G. Quarrell had already spent a year in putting together a first design of an electron diffraction apparatus or 'camera', and the ancillary high vacuum system and d.c. high-voltage supply giving 1–2 mA in the cold-cathode gas discharge tube producing a narrow intense beam of electrons at 40–70 kV.

In view of the extensive contributions of the Finch school it will be clearest to deal with this under the headings: apparatus; interpretation of diffraction patterns; and applications of electron diffraction.

1. CONTRIBUTIONS TO ELECTRON-DIFFRACTION APPARATUS AND TECHNIQUES

A description of the first Finch camera, made at Imperial College, was given by

Finch and Quarrell (1933) together with photographs from Mg, Zn and Al films condensed in vacuum, and sputtered Pt. The first commercially-produced version, made by the Cambridge Instrument Co., was in use in Finch's laboratory in 1933–4 (Finch and Quarrell 1934; Finch, Quarrell and Wilman 1935) and had a pair of interchangeable central sections for use in electron diffraction and for high-speed cathode-ray oscillography, respectively. Sauterne wine bottles were found to be suitable for use as discharge tubes, cut and sealed into a brass base, with polished Al (later stainless steel) cathodes, 0.5–0.7 in. in diameter, centred in the neck. This robust and practical apparatus was soon used in many research laboratories, both academic and industrial, including some abroad. Both in this form and the post-war form manufactured by Edwards and Co. Ltd., which was basically the same but had oil-diffusion pumps with separate pumping for the discharge tube, the Finch cameras gave extremely effective service and have continued to do so right up to the present time. A substantially similar camera with highly flat demountable brass portholes and specimen-carrier joints, and with oil pumps, was also designed by Finch in conjunction with Imperial Chemical Industries Ltd. and one was supplied to his laboratory (see Finch and Fordham 1936).

About 1935, S. Fordham and I introduced the use of much finer diaphragm apertures ($\sim$0.03 mm or less) in the anode block (Finch and Wilman 1936b), which much increased the resolution, for example allowing clear identification of the 100, $10\frac{2}{3}$, 101 and $10\frac{4}{3}$ graphite rings which show partly rhombohedral stacking of the layers (Finch and Wilman 1936a). I also introduced an extra set of movable diaphragm apertures about 15–20 cm below the anode diaphragm (in addition to the usual one close above the specimen, which reduced background scattering), thus limiting the beam to a still finer pencil when desired, to increase resolution still more, the equal of what, in recent years, was called elsewhere ultra-high resolution (e.g. see Finch and Wilman 1936b, 1937c).

Split-shutter and back-reflection arrangements were described by Jackson and Quarrell (1938). A modified central section for arc-heating specimens to $\sim$1000°C was developed by A. G. Quarrell and used by him in Finch's laboratory and later at Sheffield, to study oxidation of metals (Jackson and Quarrell 1939). Arrangements for liquid-air cooling of specimens were also used in Finch's laboratory.

Soon after the war Finch, J. F. Brown and C. E. Challice made an electron microscope (Finch and Wilman 1948, Challice 1950, Brown 1949) which allowed electron diffraction from selected areas of the specimen, and they emphasised the great importance of this combination. Shortly before Finch's retirement in 1952 to become Director of the National Chemical Laboratory at Poona, India, a high-voltage electron diffraction apparatus (to 200 kV) was designed and constructed in the laboratory (Finch, Lewis and Webb 1953). It was shown to be very worthwhile to use voltages of 100–140 kV for: (1) giving greater penetration of thicker films in transmission, and (2) virtually eliminating troublesome electrostatic charging of insulating reflection specimens, thus allowing the nature of the immediate surface region to be investigated at

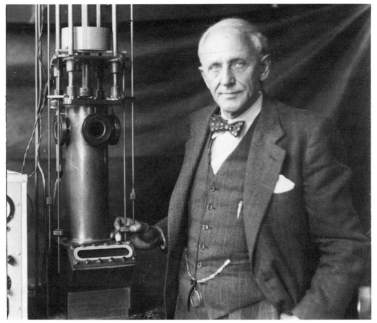

Fig. 1. Professor Finch standing beside his 200 keV electron microscope.

extremely low grazing incidence. This HV camera was also used later to study the α–γ transformation caused by abrasion on iron single-crystals (Agarwala and Wilman 1955) (see Figure 1).

Specimen-carrier assemblies were also developed and made in the workshops for special purposes, such as rotating a 'reflection' specimen about an axis in its surface (Finch and Wilman 1937c), and for rotating a transmission single-crystal film in its plane so that rotation of the specimen support about its axis could then give a rotation pattern about a selected lattice row, useful for example for organic crystals (Tyson 1938).

Finch and Quarrell (1934) suggested ZnO smoke as a convenient comparison standard for determining lattice dimensions, and later gold foil and colloidal graphite were also proposed (Finch and Fordham 1936). About 1946, Garrod (1952) in Finch's laboratory, and Lu and Malmberg (1943) elsewhere, found that the a spacing of beaten gold varied appreciably. Finch therefore pointed out that the best primary standard is graphite, for which both X-ray diffraction and electron diffraction gave consistent results, both finding $c/a = 2.726$ (e.g. Finch and Wilman 1936a). This standard is in effect based on the C=C spacing in the graphite (001) plane.

The Bunn–Bjurstrom-type charts ($\sim 46 \times 58$ cm) which I constructed in 1946 for indexing in tetragonal and hexagonal systems have been much used in our laboratories and in those of others. They were on a scale convenient for electron-diffraction work, and also for building up plots of comparison data, of d-spacing and relative intensities, such as those of the ASTM Index of X-ray Powder-Pattern Data, for identifying materials.

Wilman (1957b) described a stationary-crystal method of determining the inner potential of a crystal from photographs with the electron beam grazing two faces at once (cf. Finch and Wilman 1937c, Figure 45), and cited G. E. Thomas's (1959) estimates for FeS_2 pyrites crystals.

A. Charlesby (1940) made a considerable contribution to the theory of the blackening of photographic emulsions by electrons and X-rays, and its use for intensity measurement. The strong abnormalities of electron-diffraction ring intensities which can be caused by anisotropic absorption in, for example, needle-shaped crystals, as in ZnO smoke, were pointed out by S. Fordham (1940).

2. CONTRIBUTIONS OF THE FINCH SCHOOL TO THE INTERPRETATION OF ELECTRON-DIFFRACTION PATTERNS

Many types of pattern have been recognised and have required explanation in the wide-ranging work of the Finch school, and more effective ways of analysis of already known types had to be developed.

A. Kikuchi Line Patterns

Part of my contribution to one of our earlier papers (Finch, Quarrell and Wilman 1935) consisted of providing an analytical method for indexing these, and illustrating it with patterns from CaF_2 [100] and [111] faces. E. J. Whitmore later devised a formulation in terms of a vector treatment, relating the line equation to axes along and perpendicular to the trace of some identified main plane, i.e. line-pair median (Thirsk and Whitmore 1940a).

It seemed to me that one should be able to use a pattern by itself to give the lattice-point arrangement, symmetry and suitable axes. I therefore developed a method of constructing from the pattern a projection of the reciprocal-lattice points and their calculated heights, corresponding to the line pairs (Wilman 1948b). This was shown to work well for patterns from an arbitrary ground and etch-polished NaCl face, and from a concoidal fracture surface of an orthorhombic sulphur crystal, which has a fairly large unit cell and gives many narrow diffuse-edged bands. Other methods were also developed (Wilman 1948c), an especially neat one being in terms of the intercepts of the line-pair medians on the projections of the three crystal axial directions on the photographic plate.

B. The Diffuse-Area 'Thermal Molecular' Pattern from Molecular Crystals

This was first noticed by me in transmission patterns I recorded from thin single-crystal anthracene (Finch, Quarrell and Wilman 1935, Finch and Wilman 1937c). From further photographs, obtained with A. Charlesby, I was able to show (Charlesby, Finch and Wilman 1939) that the diffuse maxima corresponded well to the diffraction patterns which would be obtained from the

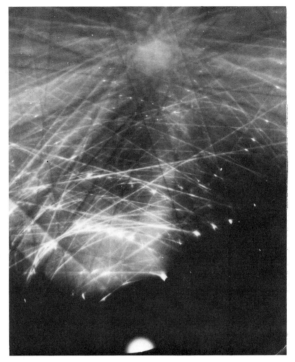

Fig. 2.   A reflection pattern from a conchoidal fracture of a sapphire crystal.

isolated molecules diffracting as separate units but oriented as in the crystal. We extended the Debye theory to the case of a molecular crystal (mainly by Charlesby, with a little tidying up by me). This theory broadly fitted in with the observations, though the effects of the binding forces being unsaturated at the surface were uncertain. Later I pointed out that elongated streaks and diffuse spots such as Preston had recently discovered in X-ray patterns from Al, very much less diffuse and extensive than our 'thermal molecular' maxima, are also present in our patterns from anthracene and other aromatic and aliphatic (long-chain) crystals; and we published a note stating this (Charlesby and Wilman 1942). The 'thermal molecular' diffuse patterns give valuable direct indications of the molecule orientations in the unit cell, when the atomic arrangement in the molecule is at least largely regular.

C. Patterns from Rotated Crystals, and Stationary Cylindrically Bent Crystals

The Laue-zone method of defining the expected diffractions on the ring positions was defined and illustrated by Finch and Wilman (1937c), and the straight-line, hyperbolic or elliptic row-line loci by Finch and Wilman for graphite, MoS_2, mica and CdI_2 (1936a,b, 1937b, resp.). The multi-rotation patterns of spots and lines associated with tears along, and cylindrical curvatures about, well-defined lattice rows in thin epitaxially oriented films isolated from their substrates were discussed and illustrated for films prepared on NaCl (001), from twinned Ag (Finch and Wilman 1937c, Goche & Wilman 1939), Ag halides (Wilman 1940) and later from many other compounds. Care is thus needed in use of *transmission* patterns to determine epitaxial relationships, and preferably *reflection* patterns from thin films in situ on the substrate should also be used.

D. Patterns from 'One-Degree-Oriented' Polycrystalline Specimens

Explicit expressions were derived, apparently previously lacking, for the indices [*uvw*] of the normal to a plane (*hkl*) for any crystal type, so that the Laue-zone method (Finch and Wilman 1937c) of constructing the diffraction positions expected from an orientation with an (*hkl*) plane parallel to a substrate, could be used for crystals of any lattice type, even if this involves the Laue index h_{uvw} being in general non-integral (Wilman 1952). Other methods were also described, and examples were given there and by Evans and Wilman (1952).

E. 'Directed Disorientation' of Epitaxial Deposits on Their Substrates

This phenomenon, apparently associated with misfit stresses and dislocation movements, was first observed by Evans and Wilman (1950) in ZnO formed on a zinc blende (110) cleavage face in air at 520°C, the ZnO showing a range of rotation about a main lattice row that is parallel to the substrate. Arcing in patterns from epitaxial deposits may often be due to such a 'directed disorientation' but is masked by the presence of symmetrically equivalent deposit crystals. One possible explanation might be in terms of rotational slip (e.g. Whapham and Wilman 1956). Further examples were found in Cu_2O on Cu (Goswami 1950; Wilman 1956); Cd electrodeposited on to Cu (Goswami 1950, see also 1969); in electrodeposited Cu, under tensional stress of epitaxy, in parallel growth on Ag (110), though there was no disorientation in Ag on Cu (110), where the Ag must be under compression due to the epitaxy (Wilman 1957a).

A 'one-sided' directed disorientation range or sometimes a discrete rotational displacement was observed by Reddy and Wilman (1959) in (211) Fe and Cr electrodeposited on to Cu (110) and (100) faces.

F. Patterns from Crystal Groupings of Rotational-Slip Type

Various types of pattern were recognised and accounted for and lamellar rotational slip was established as a newly-found slip process in crystals under torsion, at least in more or less layer-lattice crystals such as graphite, MoS_2, long-chain paraffins and their derivatives (Wilman 1950, 1951), and in gypsum and $K_4Fe(CN)_6 \cdot 3H_2O$, where high atomic cohesion exists within certain planes of atoms and much lower cohesion between the planes. The new types of electron diffraction pattern included spot patterns from discrete crystal pairs, triplets, etc. at regular azimuthal intervals corresponding to a low potential energy of fitting together of crystal lamellae, although in some cases unsymmetrical multiplets were observed. Patterns were obtained which showed double scattering, indicating extensive superposed crystals (Figure 66 of Finch and Wilman 1937c shows a striking example). Other variations included spots merging into continuous arcs, tailing off in intensity in one direction; or the same phenomenon but with one or more well-defined gaps. 'Laterally adjacent' crystal contacts could give the same set of possible azimuth intervals as contacts of superposed lamellae and might account for the observations in certain cases such as in some specimens crystallised from solution.

In 1951 (at the International Congress on Crystallography) at Stockholm I outlined these results and pointed out that a moiré pattern (analogous to that from two superposed sheets of 'perforated zinc') arising from two (001) lamellae in contact, superposed but azimuthally displaced, accounted best for the hexagonally arranged small dark areas on electron micrographs which Mitsuishi, Nagasaki and Uyeda (1951) had recently observed from a graphite flake. Finch had already written to Uyeda to tell him of my explanation, and I believe Uyeda later agreed that his photograph represented the first observation of a crystal moiré pattern.

My observation of many possible definite rotational-slip azimuths of crystal lamellae in superposed contact gave also interesting implications on the conditions under which epitaxy can occur.

3. APPLICATIONS OF ELECTRON DIFFRACTION BY THE FINCH SCHOOL

In the following outline only a limited number of publications are cited.

The Initial Period, from 1932 to About 1936

This was largely exploratory, but already important bases were established in our knowledge of crystal growth and orientation in the formation of various kinds of deposits.

Results on Al vacuum-condensed on oriented Pt (Finch and Quarrell 1933) and on the oxidation of Zn films (Finch and Quarrell 1934) suggested evidence of a 'basal-plane pseudomorphism' of the deposit crystals. Such a 'pseudomorphism' represents the tendency of crystals of the deposit material to be built up by the deposit atoms being at first constrained to fit into the potential troughs of the substrate crystal surface, so that the lattice spacings match at the interface, though as the crystals grow thicker the lattice dimensions must tend towards the normal ones. This concept of the nature of epitaxy was clearly reasonable, as indeed calculations by Frank and van der Merwe in 1949 and Smollett and Blackman in 1951 later confirmed; and the 'directed disorientations' we later observed (see above in Section 2) are in accord with this.

Some early results on vacuum-condensed and sputtered metals, electrodeposits, colloids, aerosols, abraded surfaces, oxide layers and organic materials were illustrated by Finch, Quarrell and Wilman (1935), and showed the widespread tendency for preferred orientations to be developed, especially when heating of the substrate occurs, as my earliest photographs had shown for sputtered Pt (Finch and Ikin, 1934). Finch and Wilman (1934) measured the lattice dimensions of ZnO and Finch and Fordham (1936) those of some other materials, the values tending to be slightly lower than those previously shown by X-ray diffraction, so they suggested this might be associated with the small crystal size. Finch, Quarrell and Roebuck (1934) and Finch and Quarrell (1936) investigated the nature of the 'Beilby layer' on polished metals.

From 1936 to About 1940

Many detailed studies were made in various fields. Warren's (1936) patterns from a wide range of organic materials were very useful in showing their nature and behaviour in forming crystalline films. Finch and Sun (1936) studied electrodeposits of many metals and showed that the orientation is at first affected by the nature of the substrate, epitaxy being observed e.g. for (110) Cu, Ni and Co on (110) Pt; and (100) Cu and (100) Fe on (100) Pd. Even on amorphous polished Cu, deposits soon showed preferred orientation with the most densely-populated atomic plane parallel to the substrate, e.g. (110) Fe, and (111) Au. Finch & Williams (1937) studied Ni electrodeposits and emphasised the need for care in use of optical microscopy as a guide to grain size. Williams also investigated photosensitive layers of Cu_2O on Cu, and Se on Ni, and he discovered a cubic $NiSe_2$ which had not been known previously (analogous to pyrites).

Quarrell (1937) studied vacuum-condensed metal films and concluded a slight rhombohedral deviation from fcc structure tended to occur in very thin films. Jackson and Quarrell (1939) later studied thermal oxidation of iron. Further results on the structure of polished or abraded surfaces of metals and non-metallic single crystals, and on the related problems of wear, lubrication and surface structure of sliding surfaces, were obtained (Bryant 1937; Finch 1938; Bailey 1938).

Combined electron diffraction and optical microscopy were used to study the epitaxy of $NaNO_3$ from solution on polished or otherwise treated calcite faces, and of NaCl on variously treated PbS (001) faces (Finch & Whitmore 1938; Beukers 1939). Thirsk and Whitmore (1940a,b) studied the epitaxy of NiO condensed on hot Al_2O_3 crystal faces, and of the spinel formed from Fe_2O_3 on hot MgO (001). The epitaxy of Pd vacuum-condensed on hot NaCl cleavage faces, and the effect of heating the isolated films in various gases, were studied by Fordham and Khalsa (1939).

My own work during this period included observations on SiC surface structure and oxidation, with use of long-chain hydrocarbon monolayers for oxide thickness estimation (Finch and Wilman 1937a); the behaviour and structure of colloidal graphite and the structure of larger graphite flakes, of MoS_2, and CdI_2 (Finch and Wilman 1936a,b; 1937b). I also spent time in assisting Professor Finch to install and use two Finch cameras at Brussels University during his year's tenure of the Francqui chair there in 1937–38. The study of electron diffraction by anthracene, especially the 'diffuse molecular' type of pattern has been referred to in Section 2. Studies of twinned epitaxial Ag condensed on hot NaCl cleavages (Goche and Wilman 1939) and the epitaxy of Ag halides formed by reaction on single-crystal films of Ag were also carried out (Wilman 1940).

1940 to 1945

In this war period electron diffraction work was carried out on electroplating β-brass on to iron and on the strong bonding of rubber on it, on wear and lubrication, oxidation of Cu alloys, Al, Mg and Al–Mg alloys, and films formed by various deposition processes (see Brummage 1942; De Brouckère 1945; Zentler-Gordon and Hillier 1944; Koenig 1946). My work for the Admiralty was on the structure of Ag deposits condensed in vacuo, the Ag_2O layers formed from them, and the caesiated oxide, in connection with production of infrared-sensitive photocathodes at E.M.I. Ltd. However, Finch soon was occupied as Scientific Adviser to the Ministry of Home Security, especially concerned with fire defence; and for the last two-thirds of this period I was entirely occupied at E.M.I. Ltd., Hayes, to assist in getting the Cs–O–Ag photocells ('picture transformers') into production, and then making detailed research on the optical properties of the photocathode layers and their relation to the photo-emission, leading to optimised preparation and much improved sensitivity.

1946 to 1952

After the war, electron diffraction studies were continued in Finch's laboratory. Extensive results were obtained on electrodeposit structure and growth, of many metals (Finch, Wilman and Yang 1947; Finch and Layton 1951; Layton 1952). Initial epitaxy was followed in general by a transition layer, then a characteristic 'lateral' type of growth with relatively smooth surface and preferred orientation with the most densely-populated net-plane parallel to the cathode, or 'outward' type of growth (rough or dendritic) with the densest row normal to the cathode, depending on the bath conditions (current density, temperature, concentration, stirring). A. Goswami (1950) investigated epitaxy of Fe, Cd and Zn plated on electropolished copper.

Various studies of epitaxy were made during 1948, in which large misfits occurred between the substrate and deposit crystals: e.g. for ZnO on Zn (0001) face, and for Sb_2O_3 on the Sb (111) face. The study of ZnO on ZnS surfaces (Evans and Williams 1950) provides a good example of this work. In the growth of $PbO \cdot PbSO_4$ and $PbCl_2$ (a chain structure) on PbS (Elleman and Wilman 1949) and as in that As_2O_3 (a molecular structure) condensed on FeS_2 or an NaCl and KCl faces the importance of the local atomic match at the interface rather than the match of the lattice spacings was stressed. It was concluded that epitaxy must conform to co-ordination bonding with the substrate, not just to lattice-translation similarity.

Fig. 3. A transmission pattern from oxidised tin-foil, $SnO + SnO_2$.

The structure of oxide films formed by heating smooth single-crystal faces of Cu, Fe, Al, Sn and other metals in air or oxygen was studied and the temperature at which amorphous oxide films gave place to crystalline was determined (see Wilman 1956 for summary).

K. G. Brummage (1942) had studied the structure of lubricated bearing surfaces in relation to their wear and lubrication, and later (1947a,b) he made extensive and fundamental experiments on the oriented nature and properties of films of homologous series of long-chain hydrocarbons and long-chain fatty acids. He made studies on their re-orientation by rubbing unidirectionally, analogous to my observations on stearic acid in 1937 (see Finch 1938). Fisher (1948) recorded patterns from strongly-oriented stretched natural rubber films, and used the diffuse-area ('thermal-molecular') pattern to compare with calculated intensity distributions, to show that an atomic arrangement in the molecule having normal bonds can give this pattern, without requiring the strained bonds which Bunn had concluded were present. She also observed some interesting cases of epitaxy of long-chain hydrocarbons on such rubber films when condensed on the films surrounded by a cooled screen. Her patterns from gutta percha films she concluded showed three crystalline forms, only two having been recognized previously (Fisher 1953).

Some basic results on the growth of vacuum-condensed films and development of preferred orientations were obtained by Evans and Wilman (1952), who showed the dependence on atomic mobility, substrate roughness, deposit thickness, and the quantitative relation of the orientation-axis tilt to the vapour angle of incidence. The work on 'directed disorientation' of epitaxial crystals (formed below the recrystallization temperature) covered in Section 2e above, was also carried out in this period.

Finch (1950), in his Guthrie lecture, discussed conditions at sliding surfaces, referring to a range of relevant electron diffraction results (a photograph of him by a Finch 'camera' is also shown). About this time the College acquired its first commercial electron microscope, a Siemens Elmiskop I, which was installed in Finch's charge as an interdepartmental facility. Later, Metropolitan-Vickers EM2, EM3 and eventually EM6 microscopes were in turn acquired and more recently a JEM 7 and a JEM 100B from Japanese Electron Optics Laboratory Co. Ltd. EM6s were also soon used in other departments of Imperial College. Finch retired in 1952 to become Director of the National Chemical Laboratory, Poona, India, for five years, and while there published results on Ni film structure (Finch, Sinha and Goswami 1955), on reactions in the solid state, and on the transformation of α-Fe$_2$O$_3$ to γ (Finch and Sinha 1957a,b).

In Finch's former laboratory in the Department of Chemical Engineering and Applied Chemistry (as it was then called) at Imperial College, much further electron diffraction work has been carried out since 1952 under my continuing direction, including both development of interpretation of diffraction patterns and also applications. Many systematic basic studies have been made on the growth of various types of crystalline films formed on single-crystal or polycrystalline materials and the associated friction and wear phenomena.

REFERENCES

Agarwala, R. P. and Wilman, H.: 1955, *J. Iron and Steel Inst.* **179**, 124–131.
Bailey, G. L. J.: 1938, Ph.D. Thesis, University of London.
Beukers, M. C. F.: 1939, *Rec. Trav. Chim. des Pays-Bas* (Holland) **58**, 435–447.
Blackman, M.: 1972, *Biogr. Mem. of Fellows of the Roy. Soc.* **18**, 223–239.
Brown, J. F.: 1949, Ph.D. Thesis, University of London.
Brummage, K. G.: 1942, Ph.D. Thesis, University of London.
Brummage, K. G.: 1947a, *Proc. Roy. Soc.* A**188**, 414–426.
Brummage, K. G.: 1947b, *Proc. Roy. Soc.* A**191**, 243–252.
Bryant, F. J.: 1937, Ph.D. Thesis, University of London.
Challice, C. E.: 1950, *Proc. Phys. Soc.* B**163**, 59–61.
Charlesby, A.: 1940, *Proc. Phys. Soc.* **52**, 657–700.
Charlesby, A., Finch, G. I., and Wilman, H.: 1939, *Proc. Phys. Soc.* **51**, 479–526.
Charlesby, A. and Wilman, H.: 1942, *Nature* (London) **149**, 411–412.
De Brouckère, L.: 1945, *J. Inst. Metals* **71**, 131.
Elleman, A. J. and Wilman, H.: 1949, *Proc. Phys. Soc.* A**62**, 344.
Evans, D. M. and Wilman, H.: 1950, *Proc. Phys. Soc.* A**63**, 298–299.
Evans, D. M. and Wilman, H.: 1952, *Acta Cryst.* **5**, 731–738.
Finch, G. I.: 1938, *J. Chem. Soc.*, 1137.
Finch, G. I.: 1950, *Proc. Phys. Soc.* B**63**, 465–483.
Finch, G. I. and Fordham, S.: 1936, *Proc. Phys. Soc.* **48**, 85–94.
Finch, G. I. and Ikin, A. W.: 1934, *Proc. Roy. Soc.* A**145**, 551–563.
Finch, G. I. and Layton, D. N.: 1951, *J. Electrodepositors' Tech. Soc.* **27**, 215–226.
Finch, G. I., Lewis, H. C., and Webb, D. P. D.: 1953, *Proc. Phys. Soc.* B**66**, 949–953.
Finch, G. I., Murison, C. A., Stuart, N., and Thomson, G. P.: 1933, *Proc. Roy. Soc.* A**141**, 414–434.

Finch, G. I. and Quarrell, A. G.: 1933, *Proc. Roy. Soc.* **A141**, 398–414.
Finch, G. I. and Quarrell, A. G.: 1934, *Proc. Phys. Soc.* **46**, 148–162.
Finch, G. I. and Quarrell, A. G.: 1936, *Nature* (London) **137**, 516–519.
Finch, G. I., Quarrell, A. G., and Roebuck, J. S.: 1934, *Proc. Roy. Soc.* **A145**, 676–681.
Finch, G. I., Quarrell, A. G., and Wilman, H.: 1935, *Trans Faraday Soc.* **31**, 1051–1080.
Finch, G. I. and Sinha, K. P.: 1957a, *Proc. Roy. Soc.* **A239**, 145–153.
Finch, G. I. and Sinha, K. P.: 1957b, *Proc. Roy. Soc.* **A241**, 1–8.
Finch, G. I., Sinha, K. P., and Goswami, A.: 1955, *J. Appl. Phys.* **26**, 250.
Finch, G. I. and Sun, C. J.: 1936, *Trans. Faraday Soc.* **32**, 852–863.
Finch, G. I. and Whitmore, E. J.: 1938, *Trans. Faraday Soc.* **34**, 640–645.
Finch, G. I. and Williams, A. L.: 1937, *Trans. Faraday Soc.* **33**, 564–569.
Finch, G. I. and Wilman, H.: 1934, *J. Chem. Soc.* 751–754.
Finch, G. I. and Wilman, H.: 1936a, *Proc. Roy. Soc.* **A155**, 345–365.
Finch, G. I. and Wilman, H.: 1936b, *Trans. Faraday Soc.* **32**, 1539–1556.
Finch, G. I. and Wilman, H.: 1937a, *Trans. Faraday Soc.* **33**, 337–339.
Finch, G. I. and Wilman, H.: 1937b, *Trans. Faraday Soc.* **33**, 1435–1448.
Finch, G. I. and Wilman, H.: 1937c, *Ergeb. exakt. Naturw.* **16**, 353–436.
Finch, G. I. and Wilman, H.: 1948, *Sci. Progress* **141**, 1–12.
Finch, G. I., Wilman, H., and Yang, L.: 1947, *Discuss. Faraday Soc.* **1**, 144–158.
Fisher, D. G.: 1948, *Proc. Phys. Soc.* **60**, 99–114.
Fisher, D. G.: 1953, *Proc. Phys. Soc.* **B66**, 7–16.
Fordham, S.: 1940, *Nature* (London) **146**, 807.
Fordham, S. and Khalsa, R. G.: 1939, *J. Chem. Soc.* 406–412.
Garrod, R. I.: 1952, *Proc. Phys. Soc.* **A65**, 292–293.
Goche, O. and Wilman, H.: 1939, *Proc. Phys. Soc.* **51**, 625–651.
Goswami, A.: 1950, Ph.D. Thesis, University of London.
Goswami, A.: 1969, *Indian J. Pure and Appl. Phys.* **7**, 232–233.
Jackson, R. and Quarrell, A. G.: 1938, *Proc. Phys. Soc.* **50**, 776–782.
Jackson, R. and Quarrell, A. G.: 1939, *Proc. Phys. Soc.* **51**, 237–243.
Koenig, R. T.: 1946, Ph.D. Thesis, University of London.
Layton, D. N.: 1952, *J. Electrodepositors' Tech. Soc.* **28**, 239.
Lu, C. S. and Malmberg, E. W.: 1943, *Rev. Sci. Instr.* **14**, 271–273.
Mitsuishi, T., Nagasaki, H., and Uyeda, R.: 1951, *Japan Acad. Proc.* **27**, 86–87.
Quarrell, A. G.: 1937, *Proc. Phys. Soc.* **49**, 279–293.
Reddy, A. K. N. and Wilman, H.: 1959, *Trans. Inst. Metal Finishing* **36**, 97–106.
Thirsk, H. R. and Whitmore, E. J.: 1940a, *Trans. Faraday Soc.* **36**, 565–574.
Thirsk, H. R. and Whitmore, E. J.: 1940b, *Trans. Faraday Soc.* **36**, 862–863.
Thomas, G. B.: 1959, Ph.D. Thesis, University of London.
Tyson, J. T.: 1938, Ph.D. Thesis, University of London.
Warren, J. B.: 1936, Ph.D. Thesis, University of London.
Whapham, A. D. and Wilman, H.: 1956, *Proc. Roy. Soc.* **A237**, 513–529.
Wilman, H.: 1940, *Proc. Phys. Soc.* **52**, 323–347.
Wilman, H.: 1948b, *Proc. Phys. Soc.* **60**, 341–360.
Wilman, H.: 1948c, *Proc. Phys. Soc.* **61**, 416–430.
Wilman, H.: 1950, *Nature* (London) **165**, 321.
Wilman, H.: 1951, *Proc. Phys. Soc.* **A64**, 329–350.
Wilman, H.: 1952, *Acta Cryst.* **5**, 782–789.
Wilman, H.: 1956, *J. Chim. Phys.* **53**, 607–619.
Wilman, H.: 1957a, *Acta Cryst.* **10**, 842.
Wilman, H.: 1957b, *Acta Cryst.* **10**, 857.
Zentler-Gordon, H. E. and Hillier, K. W.: 1944, *Sci. J. Roy. Coll. of Science* (London) **14**, 140.

HANS BOERSCH

II.17. The Physical Institute–I of the Technical University of Berlin

1. INTRODUCTION

After several years of studying physics in the Technical High School and the Humboldt University in Berlin with G. Hertz and M. Vollmer, and with M. von Laue, W. Nernst, M. Planck, E. Schrödinger and others, I took up an appointment with H. Mark at the University of Vienna. Mark wished me to continue the research, initiated jointly by Mark and Wierl in 1930, on electron diffraction from gases. Two years later a project was completed with the presentation of my dissertation (Boersch, 1935), and about the same time a paper was written jointly with L. Meyer entitled 'An Investigation of the Influence of Oxygen on Graphite at Elevated Temperatures by Means of Electron Diffraction' (Boersch and Meyer, 1935). From this period with Mark I gained a lasting interest in the principle of 'interference' and (together with Meyer) an appreciation of the real meaning of the expression, 'try, try, try again'.

Returning to Berlin in 1935 I took a post-graduate research position at the A.E.G. Research Institute under the direction of C. Ramsauer. Geometric electron optics was being developed there by E. Brüche. Similar ideas were being followed at the nearby Berlin Technical High School by M. Knoll, B. v. Borries and E. Ruska, but there was only occasional contact between the two groups.

2. FURTHER DEVELOPMENT

In 1935 the particle concept was still the accepted one in electron optics, although Houtermans – as far as I remember, about 1932 – had proposed [the idea] that the de Broglie wavelength should set the limit for electron-optical resolution. The move to AEG made it possible for me to carry out experiments on image formation. Using a simple single-stage electron microscope with Gabor lenses and a specimen of polycrystalline gold film, I managed to establish the existence of a primary image, having the form of an electron diffraction pattern, in the back focal-plane of the objective lens (Boersch, 1936a, b). As a consequence, earlier ideas of image contrast – by the processes of diffusion and absorption – were no longer tenable. Instead, the ideas of 'elastic-coherent and inelastic-incoherent scattering' were substituted in the theory of thin-object image contrast. With chromatic lens aberration included in the theory, it was predicted that elastic scattering would predominate in the image by virtue of being the single significant and sharply defined velocity-group in the transmitted spectrum.

Absorption contrast was not important for thin specimens. An absorption-like process occurs, however, due to the removal of all electrons scattered outside the objective aperture (i.e. scattering-absorption). An increase of contrast, and hence resolution, is thus afforded by decreasing the lens aperture, so long as the limit set by diffraction-error is not reached. These ideas were demonstrated by taking electron micrographs of thin gold films using a range of aperture sizes, both with a centered bright-field beam, and with centered and inclined dark-field beams (Bragg reflexions). The appearance of new details in the bright-field image with decrease of objective aperture was explained as arising from the removal of successive Bragg reflexions. This interpretation was supported by evidence from selective dark-field imaging (i.e. extinction). In order to complement the information from electron micrographs, a procedure was proposed and demonstrated whereby small specimen areas could be isolated for diffraction (i.e. selected-area diffraction) (Boersch, 1936b).

The experience and insight gained from this work provided a sound basis for planning future experiments. Direct evidence for 'diffraction contrast' arising from the excitation of crystalline reflexions was provided by the imaging of a thin mica crystal (Boersch, 1942), using the previously-developed 'shadow microscope' (Boersch, 1939). In another experiment (Boersch, 1943b) the phenomenon of 'ghost images', arising from Bragg reflexions transmitted by the objective aperture but not focused on the Gaussian plane, due to defect of focus and spherical aberration, was demonstrated. A measure of the contrast available in dark-field microscopy was obtained by the 'imaging' of a vapour-jet of CCl_4 having a scattering cross-section approximately the same as a condensed monomolecular film (Boersch, 1937). Furthermore, using the 'schlieren' technique, the existence of the Young–Sommerfeld edge-wave in electron optics was demonstrated (Boersch, 1943a). The edge-waves of light optics existed – not only for amplitude – but also for phase-objects (Boersch, Geiger, Kluge, 1963). The intensity symmetry observed in the electron-optical case is caused by a phase discontinuity at the boundary of the 'absorbing' half-plane, exactly as in the light-optical analogue.

3. TWO WAVELENGTH MICROSCOPY

The conclusion from those years was that the secondary image of a crystal lattice could be achieved only by a very considerable and concerted effort, while its primary image was already provided by the simplest possible electron diffraction apparatus. From that time the proposition emerged that the image-forming process should be divided into two stages. In the first stage, the pattern character should be recorded in both amplitude and phase in a plane of reciprocal space. In a second stage, the simulated optical distribution should be placed in the primary image plane of a light microscope. By means of coherent interference the secondary image of the crystal lattice could then be formed. I had carried out a test of this procedure of 'two wavelength microscopy' using light (Boersch, 1938), but by the following year Bragg had succeeded in imaging the diopside lattice by essentially this method, which he called 'X-ray microscopy' (Bragg, 1939). More recently I attempted to show the close relationship which exists between the two wavelength microscopy and holography (Boersch, 1967). In this context I had received a note from Gabor (dated 18.7.1948) in which he ascribed a 'genealogical kinship' to the two procedures.

Because the diffraction experiments give only the intensities of the crystal lattice reflexions, the necessary phase relationships must be obtained in another way – for example, by direct considerations of the structure as is done in normal X-ray analysis. To examine the problem of direct phase determination, model experiments were carried out with light and linear test objects, whose diffraction diagrams were superimposed on a coherent reference wave (Boersch, Geiger and Raith, 1960). From the resulting two-beam interference, the diffraction phases for the test object could be directly inferred. For the linear test lattices the intensity of the reference wave (represented by the central refraction maximum of a small slit) was too weak: at that time no lasers were available. Therefore, instead, an extended linear comparison lattice, with known refraction phases and with lattice constants matching those of the test object, was used to provide a more intense reference wave (Boersch and Raith, 1962). Here two-beam interferences occurred in the overlapping reflexions of both lattices, thereby providing phase information.

4. FRESNEL DIFFRACTION

The wave–corpuscle dualism of light was extended to matter by L. de Broglie in 1924–25. Duane (1923) and Compton (1923) sought to avoid this dualism by describing the characteristic interference-pattern by means of a particle picture. For this purpose they applied the phase integral of classical quantum mechanics to the internal motions of an infinitely-extended lattice. The resulting impulse is considered as a reaction to the light corpuscle, and leads to the same deflection as that given by the diffraction of the corresponding light wave. Epstein and Ehrenfest (1924) extended this assertion concerning the Fourier theorem and the correspondence principle to a finite lattice. Hence wave and corpuscle pictures lead to the same scattering distribution. The same authors (1927) pointed out that this statement held only for Fraunhofer diffraction, and not for Fresnel diffraction.

According to these little-studied papers, Fresnel interference effects represent the only true proof for the essential wave nature of light.

By the time the electrostatic electron lens was developed into a high-potential lens operating at 50–60 kV, the Fresnel diffraction could be predicted as being the main resolution-limiting factor in the shadow microscope (Boersch, 1939). The first demonstration of Fresnel electron diffraction at an edge was achieved experimentally soon after this prediction (Boersch, 1940). According to the Epstein–Erhenfest philosophy this was the first proof of the wave nature of electrons.

5. CURRICULUM

In the meantime I had been at the University of Vienna as an assistant and, after my doctorate (1942), as a lecturer [until 1946]. By the end of 1946, after several side-trips to Tettnang in South-Western Germany, I was employed as a laboratory head in the French-occupied Kaiser-Wilhelm Institute for Metallurgy at Stuttgart.

At the expressed wish of M. von Laue I took up a position as director of the newly-founded Physikalisch-Technischen Bundesanstalt at Braunschweig, and was there given, at the same time, an honorary professorship at the Technical University. In 1954 I was appointed successor to Ramsauer as head of department at the Technical University, and as director of their Physical Institute-I, until my retirement in 1974.

6. THE SETTING-UP OF ANALYTICAL PROJECTS AT THE BERLIN INSTITUTE

As the main theme of the Physical Institute-I of the Technical University of Berlin, I selected the interaction of light, electrons and ions with matter and electromagnetic fields. For the work of the Institute the development of experimental equipment was of the greatest importance. An advantage which we had at the beginning of our work was a developed electron-optical bench (Boersch, 1951), which combined versatility with immediate availability. Furthermore, we had two techniques for measurement of film thickness, developed at the Institute. Sauerbrey's (1959) method was based on the frequency change of an oscillating quartz crystal caused by the evaporated film. Niedrig, Just and Yersin's (1968) method was based on the back-scattering of fast electrons which, for thin layers, depended on the thickness and was relatively insensitive to the degree of order–disorder within the layer.

Apart from these, I placed great emphasis on the development of energy-analytical equipment, as follows.

I. Geiger (1961) altered the Möllenstedt (1949) analyser to an energy-angle-analyser in which the differential losses of scattered electrons are shown simultaneously as a function of energy loss and of the scattering angle. The angular resolution was about 10^{-4} radian in the range of 10^{-2} radians.

II. Gegenfeld analysers of the high-potential type were developed with large apertures, for a higher sensitivity with fast electrons (≤ 40 keV) (Boersch and Miessner, 1962).

III. In order to improve the resolution, an energy spectrometer was constructed in which the primary beam was first monochromatized and then, after interaction with the object, analysed (Boersch, Geiger & Hellwig, 1962). Wien filters were developed as the dispersing element in the gegenfeld filter. Thus the short-term resolution could be improved at first to 0.05 eV and later to 0.0017 eV (Geiger, Nolting and Schroeder, 1970).

7. ENERGY LOSSES IN GASES

Since 1959 the electron spectroscopy of gases has become an auxiliary technique to light absorption spectroscopy. It has proved itself of value in investigations of atomic and molecular excitations from the ground state.

Basic to these experiments was the theory of Bethe (1930), who calculated the differential elastic and inelastic effective cross-sections of atomic gas for fast, non-relativistic electrons in the first Born approximation, and made quantitative specifications for atomic hydrogen. The experimental development of energy analysers at high resolving power allowed a successful investigation of the results of Bethe to be published. For this purpose atomic hydrogen was produced by the thermal dissociation of molecular hydrogen (a dissociation value of 70% is obtained with an oven temperature reaching 2800 K) and was irradiated with 15 keV electrons. Its energy spectrum was obtained using spectrometer I (Boersch, Geiger and Reich, 1961). The analysis of spectra obtained at different temperatures confirmed Bethe's results, according to which the energy-loss spectrum for fast electrons corresponds to that from light absorption. Continuing this work, Geiger demonstrated in his thesis (1962) that Bethe's statement also held for helium, and furthermore that the small-angle inelastic scattering from other rare gases could likewise be described by means of Bethe's dipole approach. Helium was later re-examined at high resolution with spectrometer III (Boersch, Geiger and Schroeder, 1967). This time the singlet resonances in the region $1\,^1S$–$2\,^1P \ldots 11\,^1P$ were reproducibly analysed, and from these relative oscillator strengths in agreement with theoretical predictions were obtained.

The lack of suitable light sources in the vacuum UV prevented the investigation of the corresponding light-absorption spectrum for helium. These energy losses, as well as a (weakly) 'forbidden' transition $(1\,^1S \rightarrow 2\,^1S)$, and losses due to a double scattering and to double excitation, have been observed, but the more strongly forbidden intercombination transitions were not observed.

The analogy between light absorption and energy-loss spectra is not confined to atoms but extends also to molecules. In the energy-loss spectrum of molecular hydrogen, investigated using the spectrometer III at a resolution of 0.02 eV, the oscillator terms for the simultaneous excitation of the B-, C- and D-series were separately resolved (Boersch, Geiger and Stickel, 1964).

With further improvement in resolving power to 0.007 eV, the rotational structure in the $V' = 5$ Lyman-band of *para*-hydrogen could be analysed and compared with theoretical predictions (Boersch, Geiger and Topschowsky,

1965). The optical transition probabilities were also valid for fast-electron-induced molecular transitions in the infrared. Consequently infrared-allowed transitions corresponding to energy losses ($\Delta E < 0.4$ eV) were detected in CO_2, N_2O and ethylene, but not in the nonpolar molecules H_2, N_2, O_2 (Geiger and Wittmaack, 1965). Closely neighbouring transitions from the ground state, and particularly from initially excited states (e.g. in gas-discharge plasmas), could be examined (Boersch, Geiger and Topschowsky, 1971). In this work with spectrometer III the loss spectrum for 25 keV electrons in a helium gas discharge was measured with a resolution of 0.08 eV. Apart from the well-known energy losses occurring above 20 eV, six different losses appeared in the region from 1.15 to 3.15 eV. They were identified with transitions either within the singlet- or triplet-systems, and from their intensities the electron occupation of certain levels could be determined. Corresponding to the energy loss at 1.15 eV ($2^3S \rightarrow 2^3P$) a matching energy gain ($2^3P \rightarrow 2^3S$) ('superelastic' collision) was observed.

8. ENERGY LOSSES IN SOLIDS

In the energy range 0–50 eV, investigated by us, many processes occur. These processes govern the frequency-dependent complex dielectric constant of the solid. The dielectric theory referred to here is that given by Ritchie (1957), by Ritchie and Eldridge (1962), and by Hattori and Yamada (1963). Measurements made on Al and Ag layers with analysers I and II were compatible with this theory (Geiger, 1961; Boersch, Miessner and Raith, 1962), but it was only with the introduction of spectrometer III and its additional resolution, that the theory could be said to have been confirmed in a meaningful way (Geiger and Wittmaack, 1966).

The dielectric theory, which was originally conceived for the excitation of electron-plasmon oscillations, was then taken over the the energy-loss bands in the infrared region. These were detected from the transverse optical vibrations of ionic crystals (LiF, NaF, MgO, Al_2O_3) with spectrometer III (Boersch, Geiger and Stickel, 1968). In this case also there was a satisfactory agreement between experiment and theory.

As well as energy-loss bands (phonon production), corresponding energy-gain bands (destruction of phonons) were observed with LiF, NaF and MgO. These disappeared when the preparations were cooled from 300 K to 80 K.

9. IN CONCLUSION

The work I have reported above covers only a small part of our field of interest; much of the work was carried out at the Berlin Institute, though the themes had been developed much earlier.

In addition to the above topics I should like to mention the following in summary.

(a) Investigations of dynamical electron diffraction theory, culminating in our developing of a 'phase grating approximation' (Boersch, Jeschke and Raith, 1964).

(b) Considerations of thermal scattering, both theoretically and experimentally over a long period.

(c) The experimental verification of the Aharonov–Bohm effect (Boersch, Hamisch, Grohman and Wohlleben, 1961) in the Fresnel case.

(d) Investigation of the trapped and quantized magnetic flux in the bore of superconducting hollow cylinders by means of the Aharonov–Bohm effect and Fresnel diffraction (Boersch and Lischke, 1970).

(e) Lorentz microscopy by the dark-field method (Boersch and Raith, 1959), and application to a number of magnetic domain problems.

(f) Energy losses of elastic scattered electrons (Boersch, Schoenebeck, and Wolter, 1967).

(g) Investigations on the contrast of Thomas–Fermi atoms in dark- and bright-field by scattering absorption, phase contrast and filtering (Boersch, 1947).

(h) Image contrast, doses and object-destruction (Boersch, 1951).

(i) Investigation on anomalous distribution in electron rays (Boersch, 1954).

REFERENCES

Bethe, H. A.: 1930, *Ann. Phys.* **5**, 325; *Handbuch d. Physik* Bd XXIV/1.

v. Blankenhagen, P., Boersch, H., Fritzsche, D., Seifert, H. G., and Sauerbrey, G.: 1964, *Physics Letters* **11**, 296–297.

Boersch, H.: 1935, *Bestimmung der Struktur einiger einfacher Moleküle mit Elektroneninterferenzen.* PhD Thesis Univ. Wien; *Monatshefte f. Chemie* IIb, **65**, 311–337.

Boersch, H.: 1936a, *Ann. Physik* **26**, 631–644.

Boersch, H.: 1936b, *Ann. Physik* **27**, 75–80.

Boersch, H.: 1937, *Z. Phys.* **107**, 493–496.

Boersch, H.: 1938, *Z. techn. Phys.* **19**, 337–338.

Boersch, H.: 1939, *Z. techn. Phys.* **20**, 346–350.

Boersch, H.: 1940, *Naturwiss.* **28**, 709–711.

Boersch, H.: 1942, *Z. Phys.* **118**, 706–713.

Boersch, H.: 1943a, *Phys. Zeitsch.* **44**, 32–38.

Boersch, H.: 1943b, *Z. Phys.* **121**, 745–753.

Boersch, H.: 1947, *Z. f. Naturforschung* **2a**, 615–633.

Boersch, H.: 1951, *NBS Semicentennial Symp. Electron Physics*, 127–144.

Boersch, H.: 1954, *Z. Phys.* **139**, 115–146.

Boersch, H.: 1967, *Phys. Blätter* **23**, 393–404.

Boersch, H., Geiger, J., and Hellwig, H.: 1962, *Phys. Letters* **3**, 64.

Boersch, H., Geiger, J., and Kluge, W.: 1963, *Optica Acta* **10**, 21–31.

Boersch, H., Geiger, J., and Raith, H.: 1960, *Z. Phys.* **160**, 66.

Boersch, H., Geiger, J., and Reich, H.-J.: 1961, *Z. Phys.* **161**, 296–309.

Boersch, H., Geiger, J., and Schroeder, B.: 1967, *Abhandlungen der Deutschen Akad. d. Wissenschaften zu Berlin* **1**, 15–23.

Boersch, H., Geiger, J., and Stickel, W.: 1964, *Phys. Letters* **10**, 285–286.

Boersch, H., Geiger, J., and Stickel, W.: 1968, *Z. Phys.* **212**, 130–145.

Boersch, H., Geiger, J., and Topschowsky, M.: 1965, *Phys. Lett.* **17**, 266.

Boersch, H., Geiger, J., and Topschowsky, M.: 1971, *Z. Naturforschung* **26A**, 198–203.

Boersch, H., Hamisch, H., Grohmann, K., and Wohlleben, D.: 1961, *Z. Phys.* **165**, 79–93; 1962, **169**, 263–72.

Boersch, H., Jeschke, G., and Raith, H.: 1964, *Z. Phys.* **181**, 436–452.

Boersch, H. and Lischke, B.: 1970, *Z. Phys.* **237**, 449–468; and Lischke, B.: 1970, *Z. Phys.* **237**, 469–474 and **239**, 360–378.

Boersch, H. and Miessner, H.: 1962, *Z. Phys.* **168**, 298–304.

Boersch, H., Miessner, H., and Raith, W.: 1962, *Z. Phys.* **168**, 404–410.

Boersch, H. and Meyer, L.: 1935, *Z. phys. Chem.* **B29**, 59–64.

Boersch, H. and Raith, H.: 1959, *Naturwiss.* **46**, 574.

Boersch, H. and Raith, H.: 1962, *Z. Phys.* **167**, 152–162.

Boersch, H., Schoenebeck, H., and Wolter, R.: 1967, *Z. Phys.* **199**, 124–134.

Bragg, W. L.: 1939, *Nature* **149**, 470.

Compton, A. H.: 1923, *Proc. Nat. Acad. Wash.* **9**, 350.

Duane, W.: 1923, *Proc. Nat. Acad. Wash.* **9**, 138.

Epstein, P. S. and Ehrenfest, P.: 1924, *Proc. Nat. Acad. Wash.* **10**, 133.

Epstein, P. S. and Ehrenfest, P.: 1927, *Proc. Nat. Acad. Wash.* **13**, 400.

Geiger, J.: 1961, *Z. Phys.* **161**, 243–251.

Geiger, J.: 1962, *Streuung von 25 keV-Elektronen an Helium, Neon, Argon, Krypton und Xenon* Dr. Ing. Thesis, TU Berlin.

Geiger, J., and Wittmaack, K.: 1965, *Z. Phys.* **187**, 4333–443.

Geiger, J., and Wittmaack, K.: 1966, *Z. Phys.* **195**, 44–62.

Geiger, J., Nolting, M., and Schroeder, B.: 1970, *VII Congrès Int. de Microscopie Electronic* Grenoble, Vol. II, 111.

Hattori, M. and Yamada, K.: 1963, *J. Phys. Soc. Japan* **18**, 200.

Möllenstedt, G.: 1949, *Optik* **5**, 499.

Niedrig, H., Just, T., and Yersin, H.: 1968, *Z. f. angew. Phys.* **25**, 89–90.

Ritchie, R. H.: 1957, *Phys. Rev.* **106**, 874.

Ritchie, R. H. and Eldridge, H. B.: 1962, *Phys. Rev.* **126**, 1935.

Sauerbrey, G.: 1959, *Z. Phys.* **155**, 206–222.

II.18. Electron Diffraction in Danzig and Tübingen: Activities of G. Möllenstedt

In 1936 G. Möllenstedt started his post-doctoral studies with Kossel at the Physikalische Institut of the Technical High School in Danzig. The subject proposed for the thesis was the experimental investigation of the scattering of electrons from single crystals, with the motivating idea that the K-patterns of Kikuchi might be obtained in a more controlled way by using an incident cone of electrons instead of a single beam direction. Kikuchi's patterns were produced by electrons which were scattered through large angles within the crystal by inelastic scattering. Kossel had just published his first report of the equivalent phenomenon using X-rays (Kossel, Loeck and Voges, 1935) and he was quick to appreciate the advantage of using electrons, for which there existed a focussing lens and a relatively strong and monochromatic source.

Möllenstedt's abilities as an experimentalist were amply demonstrated by the speed with which results were obtained. The first note on 'Electron Interference in a Convergent Beam' was published in September 1938 (Kossel and Möllenstedt, 1938). Here, the details of interactions, resulting from the coexistence of more than two beams in the crystal, were displayed with surprising clarity. A more complete report appeared in 1939. An electromagnetic lens with a 2-cm focal length was used to produce a convergent beam, with 50 kV electrons, on a single crystal of mica. The single-lens camera produced a relatively small beam convergence. By 1939 Möllenstedt had already constructed a two-lens diffraction camera, in order to increase the convergence angle and to give some control over the incident beam direction by manipulation of an aperture. In the pictures resulting from this work many of the phenomena which occupied theoreticians for the following decades, relating to dynamic interference, and inelastic scattering, were observed and recorded.

The laboratory in Danzig was to continue operating through most of the war, until finally evacuated in 1945. Something of the two personalities involved is revealed by the following reported discussion. At the beginning of the war the young Möllenstedt, restless at the thought of any inactivity, was anxious to join the army. Kossel did his best to dissuade him, no doubt not wishing to lose his best experimentalist, by explaining that there would be plenty of time and that the war would wait. In the event the two worked on corrosion and other such defence-related problems and hence could keep the department functioning. Papers continued to appear on convergent beam diffraction from mica: for

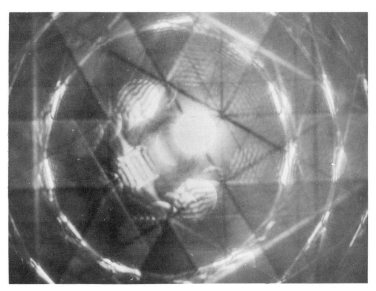

Fig. 1.   A pioneering (1938) convergent beam pattern from muscovite, taken at 50 keV.

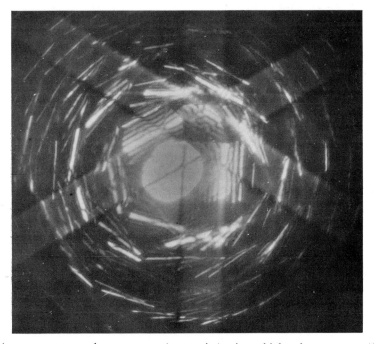

Fig. 2.   A 1942 convergent beam pattern (muscovite) using a high voltage camera (600 keV).

example, Kossel and Möllenstedt (1942) in which they compared the dynamic dip in the diffraction profile of the first-order reflexion to the phenomenon of anomalous dispersion in light.

After the war Möllenstedt moved to the University of Tübingen in Western Germany, where he became director of the Physics Department.

Any visitor to that Department when it was still housed in the old building near the centre of the town (in my case in 1963) would remember the large range of electron-optical experiments which had been established under difficult conditions by a man of outstanding abilities, imagination, and unbounding enthusiasm. Among these were the well-known Möllenstedt (1949) analyser, itself based upon the operation of an off-axis cylindrical electrostatic lens, working in a diffraction camera which had an incident-beam monochromator; electron interferometry based on the earlier invention of an electrostatic biprism (Möllenstedt and Düker, 1956); and also the beginnings of an electron micro-writing technique.

As a later visitor to the new building of the Physics Department at Morgenstelle, many years later, one was struck by the continuation of the same

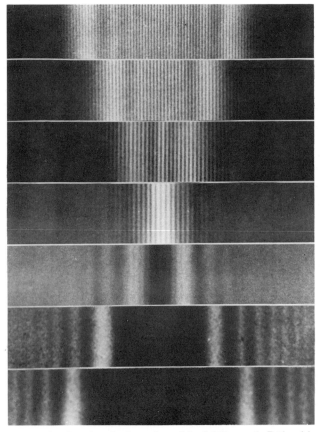

Fig. 3.   A series showing the focussing of the Möllenstedt-Düker biprism.

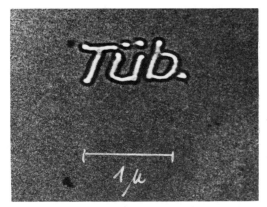

Fig. 4.   A monogram demonstration of the micro-writing technique developed at Tübingen University.

active personality shortly before Möllenstedt's retirement as Head of the Department in 1979. His main interests in this later period had changed slightly; the emphasis was now more on the industrial application of his electron-writing techniques to produce microcircuitry. His early interest in electron interferometry survived as a major activity, but his interest in the detailed interpretation of electron diffraction patterns had declined. Under construction at that time, however, was a large high-voltage (500 keV?) convergent beam camera.

His main laboratory interests were then in electron optical equipment and its application in industry. He had, after all, allowed others to pursue the structure-analytical implications of his discovery (e.g. MacGillavry, 1939; etc.).

The astonishing thing remains, in reviewing his early achievements, that the convergent beam technique had to remain dormant, and become overshadowed in its popularity by many other less powerful techniques for so long, before eventually becoming, with the evolution of the present-day microscope, the most accepted single crystal diffraction technique; and for the establishment of a Physics Department (in Bristol) which had this theme as its main subject.

From any point of view Möllenstedt's contribution to experimental electron diffraction was an outstanding one. He came into the field very near the beginning, when the discovery was only 9 years old. Together with Kossel he put single-crystal diffraction within practical reach, long before there was any hope of interpreting the results. He then turned his attention to other aspects of electron scattering. His later inventions in the fields of electron spectroscopy and interferometry are equally well known. His personality bridges the gap between the initial probing experiments and contemporary practice in a way which can be said of no other individual.

REFERENCES

Kossel, W., Loeck, V., and Voges, H.: 1935, *Zeit. f. Phys.* **94**, 139.
Kossel, W. and Möllenstedt, G.: 1938, *Naturw.* **26**, 660.

Kossel, W. and Möllenstedt, G.: 1939, *Ann. d. Phys.* **36**, 113.
Kossel, W. and Möllenstedt, G.: 1942, *Ann. d. Phys.* **42**, 287.
MacGillavry, C. H.: 1940, *Physica.* **7**, 329.
Möllenstedt, G.: 1949, *Optik* **5**, 499.
Möllenstedt, G. and Düker, H.: 1956, *Zeits. f. Phys.* **145**, 377.

CAROLINE H. MacGILLAVRY

II.19. Personal Reminiscences

In the early twenties, when I was a chemistry student at Amsterdam University, I soon found out that two of my teachers were really outstanding. One was the astronomer A. Pannekoek, to whom the extra task was given to instruct all first-year chemistry students in 'higher' mathematics. He did this in such a way that I later followed for my pure personal enjoyment all his courses in astrophysics, really intended for B.Sc.s in physics and astronomy. In this way I learned about such things as spectral inversion and plasmas, long before lasers were invented, or nuclear fusion was considered as an experimentally possible device for solving an energy crisis.

After obtaining my B.Sc. degree I majored in physical chemistry, following the advice of J. M. Bijvoet, then senior assistant in that department. After some years of postgraduate research I asked Bijvoet to be my supervisor for a thesis in crystallography. By that time Bijvoet was Reader in Crystallography and Chemical Thermodynamics. He was the inspiring and critical leader of a very small department with a submarginal budget but with an enthusiastic and eager group of coworkers. I took my doctor's degree in 1937, and shortly afterwards I became Bijvoet's assistant. We worked together with N. H. Kolkmeijer (Utrecht) on a book on 'X-Ray Analysis and its Results'. We were no narrow specialists – the idea of narrowing down to a single topic in the field would have seemed absurd, even if it had occurred to us! – and we tried to keep abreast of the whole relevant literature. We tried to read P. P. Ewald's original *Ann. d. Physik* papers on 'Dynamic Theory of X-Ray Diffraction', but we found that our chemist's stock of mathematics was not really adequate for Ewald's atomistic approach (Ewald, 1917). Later we fared better when studying H. Bethe's fundamental paper on electron diffraction in crystals (Bethe, 1928), and Laue's subsequent treatment of dynamic X-ray diffraction (v. Laue, 1931). Their approach, treating the electric field and the dielectric constant, respectively, as continuous functions, periodically fluctuating inside the crystal, was more congenial to a chemical crystallographer, familiar with the concept of Fourier analysis of such periodic functions. We studied these and other papers in a small seminar group in our free time, viz. on Saturday afternoons, bringing our sandwich lunch, and with the heat being turned off in winter. A young theoretical physicist, M. Biederman, helped us greatly in solving mathematical puzzles. (He fell a victim to racist fury shortly afterwards, in World War II.)

In 1939 Bijvoet was called to the Chair of General and Inorganic Chemistry at the State University of Utrecht. I tried to carry on Bijvoet's great tradition as well as I could, working my way, mostly all by myself, through the lengthy and laborious articles of that time. Then I came across Kossel and Möllenstedt's paper on transmission patterns of a converging electron beam through a thin mica foil (Kossel and Möllenstedt, 1939). These beautiful patterns intrigued me, especially the fringes and the kinks they suffered when crossed by a Kikuchi line. Kossel and Möllenstedt ascribed the fringes to diffraction effects of the atom row perpendicular to the foil surface. They noted that there were deviations from first approximation Fraunhofer scattering, and suggested that these might be due to 'Richtungsdispersion des Brechungsindex'. This expression somehow evoked Bethe's and Laue's treatment of dynamic scattering. So I tried to apply Bethe's results to the case in hand. Bethe had only fully developed the 'Bragg' case, that is, the diffracted beam emerging from the crystal surface hit by the primary beam. However, the 'Laue', or transmission, case had been fully treated for X-ray diffraction by Ewald and others. So it was easy to adapt Ewald's results to the simpler case of electron diffraction. I was quite excited to find that the interference minima could be quantitatively identified with those of Ewald's 'Pendellösung' for the transmission case. And even more so when I realized that this allowed one to determine the absolute values of the corresponding Fourier coefficients (structure factors) of the diffracted beam, purely by geometric measurement of these fine structure minima. The hardest part was perhaps the conversion of the electron density Fourier coefficients into those of the electric potential, as they come into the perturbation term of the Schrödinger equation.

Shortly after my results were published (MacGillavry, 1940), The Netherlands were occupied by the German invasion troops. We were, for five years, cut off from all international scientific contact. I did have a hunch about the shifts in the fringe pattern, occurring when a Kikuchi line indicated double reflection. I thought that the phase relation between the two simultaneous reflections might come into this shift, but I was unable to solve the three-beam problem. Also, during World War II we had other problems to worry about. But I look back with satisfaction to working on this paper which was for me the natural result of these Saturday afternoon seminar sessions under Bijvoet's inspiring supervision. And I am well aware that I could not have done it without Pannekoek's introduction into the world of thought of mathematics and physics.

REFERENCES

Bethe, H.: 1928, *Ann. der Phys.* **87**, 55.
Ewald, P. P.: 1917, *Ann. der Phys.* **54**, 519.
Kossel, W. and Möllenstedt, G.: 1939, *Ann. der Phys.* **36**, 113.
v. Laue, M.: 1931, *Der exakt. Naturwiss.* **10**, 133.
MacGillavry, C. H.: 1940, *Nature* **145**, 189.
MacGillavry, C. H.: 1940, *Physica* **7**, 329.

Subsequent Development at Various
Centers of Research
Period 3
1945–Present

The re-birth of LEED, due largely to Germer's return to the subject after about 30 years, is here symbolised by his colleagues at the Bell Labs.: A. U. Mac Rae (left) and J. J. Lander observe the pattern produced in a post-acceleration diffraction tube (see Articles II-20 and II-21).

Some of the personalities involved in the post-war development of electron diffraction. Clockwise from upper left: L. H. Germer, J. M. Cowley, L. Sturkey, and P. B. Hirsch with H. Watanabe on the right.

Professors S. Miyake and S. Ogawa (upper), and Professor H. Raether and A. F. Moodie, photographed during the Melbourne conference in 1965.

Upper: Professor Lawrence Brockway, seen here with Isabella and Jerome Karle, in the fall of 1945, at Ann Arbor. *Lower*: With Professor A. J. Guinier, on the left, are: (from left to right) Jon Gjønnes (University of Oslo), Gunter Lehmpfuhl (Fritz-Haber-Institute, Berlin), Colin Humphreys (University of Oxford) and Andrew Johnson (CSIRO, Melbourne). Active collaboration between these four centers over an extended period has been very largely responsible for a substantial improvement in the accuracy of electron diffraction measurements (see article III-6, by Gjønnes, this volume). The group is shown here at the IUCr. Kyoto Congress, 1972.

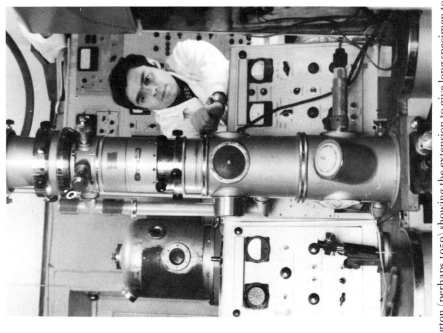

Left: A later version of the electron diffraction instrument built at N.R.L., Washington (perhaps 1950) showing the extension to give long specimen to plate distances and a gas injection nozzle on the right side, controlled by a solenoidal valve so that gas would only enter the camera when the beam was on. (see Article II-25). *Right*: A. V. Golubinski with the E. G. 100A diffraction camera at the Chemistry Department of Moscow State University. (see Article II-36).

E. G. McRAE

II.20. Low Energy Electron Diffraction at Bell Labs in the 1960s

Davisson and Germer (1927) foresaw that what we now call low energy electron diffraction (LEED) could be used to study the microscopic structure of crystal surfaces. This potentiality came to be widely exploited some thirty years later. By that time, improvements in the ability to control the chemical purity and crystalline order of solid surfaces, coupled with a growing awareness of the technological importance of surfaces, had set the stage for the emergence of surface science at the microscopic level. With the subsequent upsurge of activity in surface science in the 1960s came the 'renaissance' of LEED as the chief means of surface structure determination.

Before the beginning of the decade, many of the techniques and important properties of LEED had been established through the efforts of H. G. Farnsworth and his associates at Brown University (Schlier and Farnsworth, 1957; Farnsworth and Tuul, 1959; Farnsworth and Madden, 1961; Park and Farnsworth, 1960). For example, they worked out methods of preparing atomically clean and well-ordered surfaces by a combination of sputtering and annealing, and they showed that for many crystal surfaces – even including some atomically clean surfaces – the two-dimensional periodicity differs from that of the ideally terminated substrate. In all of this work the reflected electron current was measured with a moveable Faraday collector. This method of measurement was basically the same as used by Davisson and Germer but incorporated many technical improvements due to Farnsworth.

As far as Bell Labs is concerned, the renaissance of LEED started in 1960 with the construction of a display-type apparatus (Scheibner, Germer and Hartman, 1960). The idea of the apparatus was a modification by E. J. Scheibner, L. H. Germer and C. D. Hartman of a much earlier design by W. Ehrenberg (1934) of the Technischen Hochschule in Stuttgart. Instead of measuring the reflected electron current with a Faraday collector, as Davisson and Germer and subsequent workers had done, the electrons were accelerated after diffraction so as to make the diffraction pattern visible on a flat fluorescent screen perpendicular to the direction of incident electrons. Grids were used to repel inelastically scattered electrons, so that only the diffracted ones hit the screen. The great advantage of the display system was that the entire LEED pattern could be followed continuously under changing surface conditions such as its annealing or exposure to gases.

The display system did not immediately reveal any effect that could not have been anticipated from the work of Farnsworth and associates. However, the variety of diffraction patterns and the ease with which they could be observed with the display system caught the imagination of many workers. Within a year or so after the appearance of the display system, LEED efforts were established in many laboratories around the world.

The advent of the display system marked the return of L. H. Germer to the field that he had helped to establish decades earlier, and he subsequently played a leading role in the application of LEED to the study of chemisorption at metal crystal surfaces.

Germer's work on chemisorption at Bell Labs was done in collaboration first with Hartman and later with A. U. Mac Rae. Germer and coworkers studied the low-index surfaces of nickel crystal and its reactions with gases such as oxygen. (Germer and Hartman, 1960; Germer, MacRae and Hartman, 1961; Germer and Mac Rae, 1962; Germer and Mac Rae, 1961; Germer, Scheibner and Hartman, 1960.) These investigations, ultimately completed by Mac Rae (1964), centered on the dimensions of the unit mesh (two-dimensional unit cell) of surfaces formed by such reactions. In the conventional terminology of LEED, as definitively stated by Wood (1964), these unit mesh dimensions are expressed as multiples of the corresponding substrate unit mesh dimensions. For example, (5×2) denotes an enlargement of 5 times in one surface crystallographic direction and of 2 times in the other. The catalog of oxygen–nickel structures identified and described by Germer *et al.* was extensive; for example, it contained (2×1), (5×2), (3×1) and (5×1) for the Ni(001) surface alone. Together with the pioneering work of other groups, such as that of Farnsworth at Brown University, this Bell Labs effort helped to unveil the great variety and complexity of surface adsorption structures.

In addition to his work on surface reactions and surface periodicities, Mac Rae (working initially with Germer) made important observations of the intensities of LEED beams and their dependence on the temperature of the crystal. These observations were the first to show that the vibrational amplitudes of surface atoms exceed those of atoms in the bulk crystal and to make possible the determination of surface Debye temperatures (Mac Rae and Germer, 1962).

Meanwhile, J. J. Lander and co-workers F. C. Unterwald and J. Morrison had improved the display system by replacing the flat system of grids and fluorescent screen by a spherical one (Lander, Unterwald and Morrison, 1962). This arrangement (with the addition of an extra grid following Caldwell (1965) was later used in commercial equipment and is now a standard adjunct to surface experimentation.

Using the new display system, Lander very energetically set out to widen the realm of application of LEED. He and Morrison studied the surfaces of semiconductors including especially the cleavage surface (the (111) surface) of silicon (Lander, 1965). They confirmed in detail an earlier report by Farnsworth and co-workers of the (7×7) unit mesh of that surface. Then, together with G. W. Gobeli, Lander and Morrison went on to observe the metastable (2×1) surface formed by cleavage of silicon and of the transformation of this metastable surface to the stable (7×7) surface upon annealing (Lander, Gobeli and Morrison, 1967). These observations and Lander's (1965) speculations on them have been the focus for much of the subsequent effort to understand semiconductor surfaces.

Another of Lander's initiatives was the application of LEED to physisorption. By using a low-temperature crystal holder, Lander and Morrison (1967) were able to observe LEED patterns for such systems as bromine molecules absorbed on graphite. This pioneering study revealed a variety of systems for which both ordered and disordered surface phases existed, depending on the temperature and pressure. The possibilities thus suggested for the study of surface order–disorder transitions, for example, remain to be fully explored.

Lander's interests in LEED extended to the scattering processes governing reflection intensities. When I joined Bell Labs in 1963 he encouraged me to look into this question.

At that time most people thought that LEED intensities could be described reasonably well by the kinematic (single-scattering) type of theory. However, experiments that I did together with C. W. Caldwell on LiF surfaces showed that the specular reflection intensity depended strongly on the azimuthal angle of incidence – a result that could only be accounted for by a many-beam dynamical (multiple-scattering) theory (McRae and Caldwell, 1964). Guided by M. Lax and G. H. Wannier, I made some calculations demonstrating conclusively the dominant role of multiple scattering in LEED and the qualitative nature of the effects deriving from multiple scattering (McRae, 1966). Later developments from this work included the 'double-diffraction picture' (McRae, 1968a) – a first-order correction to the kinematical description – and an efficient scheme for computing LEED intensities usually called the 'layer multiple scattering' scheme (McRae, 1968b). P. J. Jennings used the layer multiple scattering scheme in a detailed calculation of electron reflection at the tungsten (001) surface (Jennings and McRae, 1970).

Another result of the experiments by Caldwell and me on LEED intensities was the discovery of electronic surface resonances or quasi-stationary surface states at energies above the vacuum level (McRae and Caldwell, 1967). We did not realize it at the time but now recognize that these resonances are quite analogous to the 'enhancement' effect known in high-energy electron diffraction. The ability to study electronic surfaces resonances was limited by the rather poor energy resolution of the LEED apparatus (0.4 eV). The phenomenon remains to be explored in detail using high-resolution equipment.

With the end of the decade came the end of the period of vigorous exploration of LEED at Bell Labs. One outcome of this exploration as carried out at Bell Labs and elsewhere can be seen in almost every surface science laboratory. The methods and phenomena of LEED are now largely understood. Consequently, LEED can be used reliably, and even routinely, as an aid in surface characterization. But from another viewpoint the most interesting outcome of the Bell Labs work of the 1960s lies in the phenomena that were observed but not fully explored because of technical limitations that have since been overcome. These phenomena include order–disorder transitions in physisorption systems and electronic surface resonances − topics that are now being taken up again in a number of research centers.

I am pleased to thank A. U. Mac Rae for useful comments and suggestions.

Bell Laboratories
Murray Hill
New Jersey, U.S.A.

REFERENCES

Caldwell, C. W.: 1965, *Rev. Sci. Instrum.* **36**, 1500.
Davisson, C. J. and Germer, L. H.: 1927, *Phys. Rev.* **30**, 705.
Ehrenberg, W.: 1934, *Phil. Mag.* **18**, 878.
Farnsworth, H. E. and Madden, H. H., Jr.: 1961, *J. Appl. Phys.* **32**, 1933.
Farnsworth, H. E. and Tuul, J.: 1959, *J. Phys. Chem. Solids* **9**, 48.
Germer, L. H. and Hartman, C. D.: 1960, *J. Appl. Phys.* **31**, 2085.
Germer, L. H. and Mac Rae, A. U.: 1962, *J. Appl. Phys.* **33**, 2923.
Germer, L. H. and Mac Rae, A. U.: 1961, *Robert A. Welch Foundation Research Bulletin* **11**, 5–26.
Germer, L. H., Mac Rae, A. U., and Hartman, C. D.: 1961, *J. Appl. Phys.* **32**, 2432.
Germer, L. H., Scheibner, E. J., and Hartman, C. D.: 1960, *Phil. Mag.* **5**, 222.
Jennings, P. J. and McRae, E. G.: 1970, *Surf. Sci.* **23**, 363.
Lander, J. J.: 1965, *Progress in Solid State Chemistry* (H. Reiss ed.) Pergamon, Vol. 2, Ch. 3, p. 2.
Lander, J. J. and Morrison, J.: 1967, *Surf. Sci.* **6**, 1.
Lander, J. J., Gobeli, G. W., and Morrison, J.: 1963, *J. Appl. Phys.* **34**, 2498.
Lander, J. J., Unterwald, F. C., and Morrison, J.: 1962, *Rev. Sci. Instr.* **33**, 784.
Mac Rae, A. U.: 1964, *Surf. Sci.* **1**, 319.
Mac Rae, A. U. and Germer, L. H.: 1962, *Phys. Rev. Lett.* **8**, 489.
McRae, E. G.: 1966, *J. Chem. Phys.* **45**, 3258.
McRae, E. G.: 1968a, *Surf. Sci.* **11**, 492.
McRae, E. G.: 1968b, *Surf. Sci.* **11**, 479.
McRae, E. G. and Caldwell, C. W.: 1964, *Surf. Sci.* **2**, 409.
McRae, E. G. and Caldwell, C. W.: 1967, *Surf. Sci.* **7**, 41.
Park, R. L. and Farnsworth, H. E.: 1960, *Appl. Phys. Letters* **3**, 167.
Scheibner, E. J., Germer, W. H., and Hartman, C. D.: 1960, *Rev. Sci. Instr.* **31**, 112.
Schlier, R. E. and Farnsworth, H. E.: 1957, *Advances in Catalysis* **9**, 434.
Wood, E. A.: 1964, *J. Appl. Phys.* **35**, 1306.

G. A. SOMORJAI

II.21. LEED: A Fifty-Year Perspective

In 1927 Davisson and Germer reported on the diffraction of low energy ($\sim$ 100 eV) electrons that were backscattered from one of the single crystal faces of nickel. Fifty years later, in 1977, the surface structures of ordered monolayers of CO and of C_2H_4 deposited on metal single crystal surfaces were analyzed in our laboratory by using the low energy electron diffraction (LEED) beams and employing multiple scattering dynamical calculations. Both the experiment and theory of surface crystallography reflected the scientific sophistication of modern surface science of the late 1970s. It is our purpose to review this important development in the history of surface science and to pay tribute to the many hard working and imaginative scientists who participated in it and contributed so much to its success. Their work is reviewed and summarized in several publications.

Progress has been slow for the 25 years following the discovery of LEED. In the absence of modern vacuum technology it was exceedingly difficult to maintain a clean surface or one with a constant composition for times long enough (hours) to carry out the experiments. At 10^{-6} torr partial pressure, which was about the best obtainable in that period, the surface becomes covered by a monolayer of gas in seconds, assuming a sticking probability near unity. The 10^{-10} to 10^{-9} torr ambient pressures needed to maintain the clean surface condition during the experiments have become available only in the 1950s along with the gauges necessary to measure such low pressures routinely. The electron detector in the early days was usually a Faraday cup in most cases, and it required the art of glassblowers to fabricate the delicate instruments that permitted the measurement of the electron beam intensities at various angles of scattering. Pioneering work in the United States in this period has been carried out mostly by H. E. Farnsworth at Brown University and his students, although other laboratories have also contributed. There are several notable achievements attributed to Brown University in this period. The surface ion bombardment technique was developed using 100–1000 eV inert gas ions to sputter-clean the surface (see McRae's article, this volume). This technique is commonly used today to strip the surface of unwanted impurities that are deposited from the surroundings or diffuse to the surface from the bulk of the samples to be investigated. They have also reported for the first time the phenomenon of surface reconstruction – that the atoms on the surface of germanium crystals have a different periodic arrangement than expected from the projection of the bulk unit cell to the surface. The continued LEED research by this group during this period, under the most difficult circumstances, has kept the field alive, and it well deserves our admiration and respect.

The rise of the semiconductor device technology and the aerospace industry during the 1950s has provided the conditions for the major changes that occurred in the field and for the explosive growth that followed. Semiconductor devices of ever smaller size (or increasing surface-volume ratio) have been developed to provide increasing computer speeds and less expensive electronic devices of various types. The electronic properties became more and more influenced by the structure and composition of the surface monolayer; therefore, it became essential to develop surface-sensitive techniques to investigate and control the surface behavior. In the same time, ultrahigh vacuum technology became available at a reasonable cost, which made it possible to produce and maintain clean surfaces for minutes or even longer. Germer, who had not remained an active researcher in LEED for many years, decided to return to it. By utilizing the idea of post acceleration of the electrons which were diffracted onto a fluorescent screen (Ehrenberg, 1934), he constructed a new type of instrument that permitted the detection of the diffraction beams visually *during* the LEED experiment (Scheibner, Germer and Hartman, 1960), thus the surface structure and its variations during gas exposure as a function of time could be monitored throughout the experiment. As a result the research effort of Bell Telephone Laboratories in this field has dominated the development of LEED. The renewed studied of nickel surfaces and oxygen chemisorption and oxidation of this surface, which played such an important role in the discovery of LEED, uncovered the rich surface structural chemistry of this system. Many ordered oxygen surface structures have been found by Mac Rae and Germer (1962) as a function of oxygen coverage and oxygen content in the bulk metal, and the oxygen inducted reconstruction of nickel was suggested. An active group of researchers, including McRae, Gobeli, and Estrup – under the direction of J. J. Lander – explored the structural chemistry of semiconductor surfaces of Si, Ge, GaAs, InSb, etc. (see Somorjai and Farrell, 1971). The phenomenon of surface reconstruction of semi-conductor surfaces was confirmed by the Bell group and found to be more of a rule than an exception. Wood (1964) developed a simplified notation of the two-dimensional unit cells of ordered adsorbates or reconstructed surfaces. The first experiments on a weakly adsorbed monolayer (bromine on graphite) and monolayers of condensable vapors (potassium and iodine on germanium) were performed. LEED studies of alkali halides were initiated by McRae and Caldwell. Investigations of the temperature dependence of the diffraction beam intensities yielded the surface Debye–Waller factor. Careful measurements showed that the mean square displacement of surface atoms perpendicular to the surface is appreciably larger than for atoms in the bulk.

In the early 1960s, the first commercial LEED apparatus became available from Varian. This set the stage for the wider dissemination of the research and its appearance in several universities and research laboratories. My research in LEED was started in January 1965 and employed platinum single crystal surfaces. We were immediately successful in discovering the surface reconstruction of the Pt(100) crystal face. The surface reconstruction of other metal crystal faces was detected shortly thereafter by Palmberg and Rhodin (1968) and Jona (1967).

In Berkeley we embarked on a study of the surface structures of hydrocarbon monolayers on transition metal surfaces. Blakely studied the surface structures of carbon which segregated from the bulk of metals onto the surface. The detection of many ordered surface structures on metal and semiconductor surfaces revealed the richness of surface structure chemistry that was unknown and unexpected. These observations focused increased attention on the need for a theory of low energy electron scattering.

A theory was needed to interpret the diffraction beam intensities to determine the location of surface atoms, their distances, and angles to their nearest neighbors. In short, surface structure analysis needed a detailed understanding of the LEED process. For this purpose the intensities of many diffraction beams, from alumina and other metal crystal surfaces, were measured as a function of electron energy (I-eV curves). The complexity of the I-V curves, the presence of "fractional order" peaks indicated that the kinematic, single scattering diffraction theory (so useful in X-ray crystallography) is inadequate to explain LEED.

As early as 1965, E. G. McRae pointed out the importance of multiple scattering in the surface diffraction process. Because of the high scattering cross section of low energy electrons (as compared to X-rays), there is a large probability that the electron that is scattered once elastically will scatter again before leaving the surface region. Shortly thereafter, Beeby developed a Green's function formalism that provided the framework for dynamical surface structure calculations that take into account the multiple scattering. Pendry (1971, 1974) and Jepsen et al (1971) developed the atomic scattering factor that was suitable for LEED calculations. Lundqvist (1969), and Duke and Laramore (1970) pointed out the rapid attenuation of the diffracted low energy electrons. Not only the elastic, but also the inelastic scattering has large cross sections. Thus the electrons are likely to lose energy upon multiple scattering, which limits their penetration and attenuates strongly the higher order scattering. Subsequently, efficient calculational schemes were developed by Pendry (1971, 1974), Tait, Tong and Rhodin (1972) and Van Hove and Tong (1979).

The first structure analysis using the multiple scattering theory was carried out on the Al(100) crystal face (Hoffstein and Boudreaux, 1970; Tong and Rhodin, 1971; Jepsen et al., 1971). With the limited data base available and the uncertainties of the physical parameters that had to be inserted in the theory (atomic potential, Debye-Waller factor, screening lengths), the agreement between the experiment and calculated intensity curve was satisfactory. From this time on the development of LEED has followed two parallel paths. Researchers made continued discoveries of ordered surface structures of clean surfaces and adsorbed monolayers as a function of temperature and coverage and reported their two-dimensional unit cells and the conditions of preparation. In these investigations, surface structure analysis had not been attempted, however. Other investigators embarked on a program of carefully measuring the diffraction beam intensities from many of these ordered surface structures as a function of electron beam energy, angles, and temperatures, and then carried out calculations to determine the location of surface atoms. We shall review the developments in both of these research directions during the past 15 years.

A major improvement in the experimental science of LEED has been the introduction of Auger electron spectroscopy (AES) into the experiments that provided surface composition analysis with about 1% of a monolayer sensitivity. Using AES, much of the ambiguity of interpretation as to the causes of the appearance of ordered surface layers with unit cells different from the bulk unit cell has been removed. Many of the controversies of interpretation of surface reconstruction have been resolved. The surface restructuring of many metal and insulator crystal faces have been discovered. Metastable structures of recon-structed surfaces were prepared by Bonzel and Helms. Work was extended to high Miller index clean surfaces. These have an ordered step structure where the steps are usually of monoatomic height. There are crystal surfaces with kinks in the steps, and their structures have also been explored using metals primarily. This research was carried out by Oudar and by Perdereau and Rhead (1971) in Paris, by my group (Joyner, Lang and Blakely) in Berkely (Joyner et al., 1972), at Sandia Laboratory in Albuquerque by Schwoebel and by Houston and Park (1971), at the Los Almos Laboratory by Ellis (Ellis and Schwoebel, 1968) and by Henzler (1970) in Aachen, West Germany. When we discovered that these surface irregularities, steps, and kinks have unique chemical bond-breaking abilities on various transition metal surfaces, their study took on special significance in investigations of surface reactivity. The work function changes and changes of sticking probabilities of gases associated with varying step densities were explored by Ibach (1975), Wagner (1979) and Besocke in Aachen and in Jülich.

Molecular crystal surfaces, ranging from butane to amino acids, have been investigated in my laboratory by Firment and Buchholz (see Firment and Somorjai, 1979). These, along with inert gas crystals, were grown epitaxially in the diffraction chamber. Inert gas crystals were studied for example by Shaw, Fain and Chinn (1978).

Between 1960 and 1979 over 700 ordered surface structures of adsorbed monolayers were reported. The substrates were, in most cases, low Miller index faces of monatomic solids, mostly metals. Ordered layers of atoms and molecules have been characterized as a function of coverage and temperature. Metal atoms adsorbed on crystal surfaces of other metals have been studied by LEED. Order–order and order–disorder transformations in the two-dimensional monolayers have been reported. These have been studied by Lagally, Ertl, and other groups. The ordering of molecules from diatomic (CO, O_2, N_2) to large polyatomic molecules (benzene, phthalocyanines) has been studied.

Along with these investigations, the interaction of the probing low energy electron beam with a clean surface and the adsorbed monolayers has been documented. Alkali halide crystals and some of the organic crystals are particularly sensitive to electron beam-caused damage. Some of the adsorbates, such as carbon monoxide, are found to be more sensitive to electron beam desorption and decomposition than others (π-bonded organic adsorbates). Certainly one of the future directions of research is to minimize the incident electron beam intensity in the LEED experiments to reduce the rate of electron beam-surface interactions. Studies of surface disorder, either in the clean surface or in the adsorbed monolayer, are in their infancy. Many researchers, (e.g. Lagally and Webb, 1969) have initiated work in this area. Since many phase transformations originate at the surface (melting, evaporation, crystal growth, nucleation), it is hoped that work will be expanded in this area in the near future.

New instrumentation has been developed in my laboratory that combines LEED studies of surfaces in ultrahigh vacuum with studies of the same surfaces in high ambient pressures. The samples can be enclosed in an isolation cell which can be pressurized. In this high pressure mode, catalytic studies or corrosion studies can be carried out, or the sample may be exposed to liquids for electrochemical or other solid–liquid interface studies. When the isolation cells is opened in the middle of the UHV chamber, the surface can be studied by LEED and by other surface diagnostic techniques before and after the high pressure experiments. This combination of UHV and high pressure studies permits us to couple the atomic scale scrutiny of the surface structure and composition with investigations of many macroscopic surface phenomena: heterogeneous catalysis, corrosion, radiation, adhesion, lubrication, and studies of the electrical properties of surfaces. The samples can also be cleaved or fractured in the UHV chamber before or after exposure to the high pressure environment. In this way, changes in the mechanical properties as a function of changes of surface structure amd surface composition can be investigated (Somorjai, 1979).

It would be of importance to combine electron microscopy studies of surfaces with atomic surface structure studies by LEED. It is hoped that these two techniques will be combined to their mutual benefit in the near future.

The more quantitative research direction in LEED has led to the development of surface crystallography. By joining forces, a handful of theorists (Pendry, Tong, Van Hove, Jepsen, and Marius) and experimentalists (Webb, Jona, Somorjai, Anderson, Mitchel, and Rhodin) developed this field rapidly during the past 10 years. By now over 100 surface structures have been determined by LEED structure analysis. There are many clean low Miller index surfaces of metals for which the atomic positions were verified. A contraction of the first atomic layer toward the second layer was uncovered as an important and dominant phenomenon as the coordination number of surface atoms decreases (relaxation). The structures of several reconstructed surfaces have been solved. Most surface structure analyses were carried out on ordered monolayers of adsorbed atoms. Although surface sites of the highest coordination number and symmetry are preferred in many cases, the adsorbed atoms may also occupy other, lower symmetry sites. There are reports of adsorbed atoms preferring sites *below* the topmost layer of substrate atoms in some instances. Strong chemisorption of oxygen can lead to a restructuring of metal atoms just as first predicted by Germer. Only recently there have been reports of surface structure determinations of adsorbed molecules of C_2H_2, C_2H_4, and CO on transition metal surfaces. This direction of research holds the promise of providing accurate bond angle and bond distance information for monolayers of molecules that will greatly help to verify their surface chemical bonds.

Adsorbed monolayers of molecules often undergo chemical rearrangement as the temperature is increased. Sequential bond breaking with increasing temperature is usually observed in strongly interacting systems. Recently, a combination of LEED and high resolution electron loss spectroscopy (HREELS) has been employed to verify the structure of adsorbed molecules, Ibach (1977). While LEED provides crystallographic data on ordered mono-layer systems, HREELS yields the vibrational spectrum of monolayer systems, whether ordered or disordered. The combination of these two surface structure determination techniques, when applied to the same surface adsorbate system, provides a more complete physical picture of the surface structure and its various transformations as a function of temperature and coverage.

As a result of LEED studies, the nature of the surface chemical bond is being increasingly better understood. LEED studies, when combined with other surface sensitive techniques such as HREELS, AES, X-ray and ultraviolet photoelectron spectroscopies (XPS and UPS), and ion scattering, reveal the structure and orientation of adsorbed atoms and molecules and their chemical transformations. This information is applied increasingly in heterogeneous catalysis, adhesion lubrication, corrosion, and semiconductor device technologies to mention a few important fields of surface science. Therefore, surface crystallography has a great impact on many modern technologies. Rapid-scanning instruments that use low incident electron currents, combined with more reliable and rapid computational methods for surface structure analysis, will likely enhance the utility and the application of LEED. The future is indeed very bright for this technique when viewed from a 50-year perspective.

Materials and Molecular Research Division
Lawrence Berkeley Laboratory and Department of Chemistry
University of California
Berkeley, U.S.A.

REFERENCES

Duke, C. B. and Laramore, G. E.: 1970, *Phys. Rev.* **32**, 4765.
Ehrenberg, W.: 1934, *Phil Mag.* **18**, 878.
Ellis, W. P. and Schwoebel, R. L.: 1968, *Surface Science* **11**, 82.
Firment, L. E. and Somorjai, G. A.: 1979, *Surface Science* **84**, 275.
Henzler, M.: 1970, *Surface Science* **19**, 159.
Hoffstein, V. and Boudreaux, D. S.: 1970, *Phys. Rev. Letters* **25**, 512.
Houston, J. E. and Park, R. L.: 1971, *Surface Science* **26**, 269.
Ibach, H.: 1975, *Surface Science* **53**, 444.
Ibach, H.: 1977, *Electron Spectroscopy for Surface Analysis* Springer, Berlin.
Jepsen, D. W., Marcus, P. M., and Jona, F.: 1971, *Phys. Rev. Letters* **26**, 1365.
Jona, F.: 1967, *Surface Science* **8**, 57.
Joyner, R. W., Lang, B., and Somorjai, G. A.: 1972, *Surface Science* **30**, 440.
Lagally, M. G. and Webb, M. B.: 1969, in *The Structure and Chemistry of Solid Surfaces* (ed. G. A. Somorjai) Wiley, New York.

Lundqvist, B. I.: 1969, *Phys. Stat. Sol.* **33**, 273.

Mac Rae, A. U. and Germer, L. H.: 1962, *Phys. Rev. Letters* **8**, 489.

Palmberg, P. W. and Rhodin, T. N.: 1968, *Phys. Rev. Letters* **20**, 925.

Pendry, J. B.: 1971, *J. Phys.* **C4**, 2501 and 2514.

Pendry, J. B.: 1974, *Low Energy Electron Diffraction* Academic Press, London.

Perdereau, J. and Rhead, G. E.: 1971, *Surface Science* **24**, 555.

Scheibner, E. J., Germer, L. H., and Hartman, C. D.: 1960, *Rev. Sci. Instr.* **31**, 11.

Shaw, C. G., Fain, S. C., and Chin, M. D.: 1978, *Phys. Rev. Letters* **41**, 955.

Somorjai, G. A.: 1979, *Surface Science* **89**, 496.

Somorjai, G. A. and Farrell, H. H.: 1971, *Advances in Chemical Physics* **20**, 215.

Tait, R. H., Tong, S. Y., and Rhodin, T. N.: 1972, *Phys. Rev. Letters* **28**, 553.

Tong, S. Y. and Rhodin, T. N.: 1971, *Phys. Rev. Letters* **26**, 711.

Van Hove, M. A. and Tong, S. Y.: 1979, *Surface Crystallography by LEED* Springer, Heidelberg.

Wagner, H.: 1979, *Physical and Chemical Properties of Stepped Surfaces*, Springer Tracts of Modern Physics, **85**.

Wood, E. A.: 1964, *J. Appl. Phys.* **35**, 1306.

VERNER SCHOMAKER AND KENNETH HEDBERG

II.22. Gas Electron Diffraction: Continuation at CalTech and Oregon State

I. THE CALTECH YEARS: VERNER SCHOMAKER

1. Introduction

Arriving at CalTech, in the Fall of 1935, I only knew a bit about Linus Pauling and Richard Tolman, and little more about Lawrence Brockway, who had preceded me from Nebraska by a few years and had been marvelously successful studying molecular structure by electron diffraction. Professor Pauling at first would not let me work on electron diffraction ("It's too complicated. Do something simpler first"), but after a year in which I found assigning Raman lines difficult and uncertain, I was allowed to join. Brockway's second apparatus (Brockway, 1936) was producing pictures that were leading to structure after structure, although getting a good enough vacuum to run was often difficult (finding a leak sometimes took days or weeks) and 'finding the beam' was a regular adventure. (There were occasional other adventures, usually explosions; e.g., John Beach told me of finding a trail of Brockway's blood leading from the apparatus in the second basement of the Astrophysics building to the street outside, the result of an explosion of fluorine nitrate. Brockway's note in the lab record runs simply "NO_3F. Exploded on turning stopcock. Three weeks spent in repairing the apparatus." No mention of the injury.) Pauling and Brockway's (1935) new electron diffraction radial distribution (RD) method had brought a fine feeling of excitement and achievement, but did not pinpoint the distances: to find the structure was still often a real puzzle. The intensity and RD summations were very laborious, even with the Cross–Brockway strips, and most of our time was spent pounding the calculator.

What follows is a much too brief, perhaps too personal view of gas diffraction as seen at the CalTech laboratory from nearly its beginning to the end. It touches on the experiment itself, the computations, a few of the structures, and the UF_6 problem, but it is mainly about the visual method, intertwined with the RD method.

2. The Experiment

Brockway's second apparatus, left to less skillful hands when he went to England in 1937, was used for nearly twenty years. Over the years, various small improvements were made in it: simple apertures replacing the original tubular collimator ended the problem of finding the beam; a thermocouple vacuum gauge vastly reduced leak-hunting time; fiduciary markers in the camera afforded corrections for exposure-to-measurement film expansion (as much as a percent, dependent on the humidity, and a partial cause of a troublesome apparent quasi-seasonal variation in wavelength); ZnO powder, replacing gold foil, improved the precision of the wavelength calibrations; crude magnetic shields reduced ring ellipticity (previously as much as a percent); and beam-stops improved picture quality. The most-used special 'nozzle' was Brockway and Palmer's (1937) monel-metal oven – with its troublesome lid gasket and needle-valve packing.

A vital feature of the apparatus was its large (for the time) aperture, about $s = 33$ rad $Å^{-1}$ with 40 keV electrons. This aperture and our continual effort to exploit it by manipulation of exposure time, beam intensity, sample pressure, and nozzle diameter were responsible for much of our success. At high s-values delicate distinctions often grow into obvious differences.

3. The Visual Method

When I had got my first pictures, Professor Pauling taught me how to measure them up on the comparator. He read a few ring diameters. Then I read them. Our values seemed to agree. He said something like "All right. You've got the hang of it." End of the course! Gradually, however, I came to know the visual method as subtle and taxing, rather than simple and easy.

Beginning early in 1939, David Stevenson and I made and measured up pictures of N_2, O_2, Cl_2, Br_2, I_2, CO_2, and CH_4, all of well known structure (Pauling and Brockway, 1934, had made similar tests). We treated our separate sets of measurements of each substance separately, averaging over both maxima and minima and always rejecting the first few rings and the outermost ring or two. The errors were 0.005 Å or less except for one set for O_2 (-0.015 Å), one long-camera set for Br_2 (-0.008), one set for I_2 ($+0.008$), and all the CH_4, which averaged C—H $= 1.112$ Å – this despite the mentioned difficulties with the wavelength calibrations. We concluded, in an unpublished manuscript, ". . . the interatomic distance in the molecule being investigated may be determined to within 0.01 Å . . . if the distance is the only parameter to be evaluated." The laboratory always adhered to Brockway's description of his error estimates as 'limits' of error. Unlike the estimated standard deviation both in quantity and in quality, logically absurd but somehow realistic, this description implied that if the actual error exceeded the estimated limit, one had *blundered*. Our meaning was sometimes misunderstood.

Pauling and Brockway (1934) had proposed that the visually perceived intensity is essentially $(I - I_{background})I_{background}$, with $I_{background}$ equal to the sum of the atomic and inelastic scattering intensities. We, however, assumed that the effective background at any point is fixed by the machinery of vision as an average of the neighboring local intensity so weighted as to follow the broad atomic part of the intensity rather completely, broad or doubled features less completely, and sharp single features least of all. This background would then be responsible not only for the main illusion of rings of maximum or minimum intensity, but also for the remaining clearly evident general outward decline, the greater apparent relative intensity of sharp than of broad rings, the electron-diffraction St John effect, and other effects not previously recognized. Varying the illumination, changing the viewing distance, using a large reducing lens, and the like, alter all these effects and were often helpful, but in any case an act of interpretation, not a literal matching of appearance to a theoretical intensity curve, was required.

We were much influenced by the RD method. Pauling had set me to changing the coefficients in the Pauling-Brockway RD sums to make the terms more like the sub-sums over individual rings that Charles Degard had been getting in his hand calculation of the complete integral of microphotometer intensity data for CCl_4. Factors of s^3 or s^4 along with an 'artificial temperature factor' (which Pauling also suggested to me) and the inclusion of properly balanced terms for both minima and maxima helped a lot, resolving closely spaced distances almost as well as the direct correlation of pattern with theoretical intensity curve. But raw intensity estimates of maxima and minima à la Pauling and Brockway did not always follow the $s^{-3.5}$ trend, and they remained ill-suited to representing important minor features, which tended to be lost in the general outward decline. This became obvious after the development in 1940–41 of our punched-card sine tables (Shaffer, Schomaker, and Pauling, 1946), which with a tabulating machine replaced the Cross–Brockway strips and the calculator and made it practical to do closely spaced summations to approximate the radial distribution integral (RDI). We carefully drew curves of 'visual intensity' ($V(s)$) as interpretations of the perceived patterns subject to the arbitrary convention that the general amplitude should be independent of s so as to follow our theoretical curves. Our punched cards, whether through wisdom or expedience, were tables of $\sin x$ rather than $(\sin x)/x$; it also implied a normalization to zero average mean-square amplitude of vibration. I was inspired by Bauer's early published visual curves which, however, were presented as absolute representations of visual appearance. (Our previous visual curves were only ring-number guides.) Drawing $V(s)$ curves, doing our 'integrals', and planning for an eventual scheme for least-squares refinement of both measured ring diameters and visual intensities (Hamilton, 1954) convinced us ever more strongly that the interpretation of each new pattern should be made by direct comparison with similar patterns or parts of patterns in our files of photographs and the corresponding 'final' (presumably correct) theoretical curves. Nor was $V(s)$ the best way to record our observations: the least-squares scheme therefore called for observation equations on differences of adjacent ring intensities, three-fold second differences, and the like.

All this contributed, I feel, to success in cases that tested our powers. Just as important was Pauling's example and assistance in checking the immediate evidence against whatever could be inferred from structural lore and theory. This checking was not always done well (stories could be told), but it usually served very well the primary interest in right answers rather than purity of method.

In any case, special interest in the visual method doubtless contributed to my reluctance to act on Pauling's often repeated suggestion that we build a sector apparatus. The visual method, I now know, also kept down the finished output: it is ultimately just too difficult; one sees a lot at a glance – and agonizes before committing to print. And hardly one co-worker in ten, however fine his other qualities, would ever carry through. The complaint was "But Verner, you can see the rings so much better than I can." Mostly not so! I just tried harder and got further behind, leaving all too much hard work unfinished and unpublished, except for Sutton's reviews (see his article in this volume).

The high resolving power of our RDI's greatly reduced but did not eliminate the incidence of puzzles, the details of the diffraction pattern often permitting distinctions that apparently couldn't be drawn from the RDI. (I held high hopes that Jürg Waser's (1944) 'modification functions', based on his extremely valuable general discussion in terms of *folding*, would equalize the situation, but we had no great special success with them.) Methyl acetate, methyl formate, and methyl chloroformate (O'Gorman, Shand, and Schomaker, 1950) were instructive: valid starting models and consequent good fits to our observations were obtained only by relying on the exact shapes of the incompletely resolved radial distribution peaks.

But even with a good starting model and the speed of our punched cards, it was still a trial to deal with more than three shape parameters. Worse, given an excess of zeal in the calculation and plotting of more and more curves for variations of a basic model, it becomes impossible to think. Hedberg formulated a rule about this: the more curves one calculates for a molecule, the less one learns about it. This rule and its implications, as well as early instruction by Henri Lévy and the enormously successful crystallographic example (Hughes, 1939) had already interested us in having a least-squares method of refining electron-diffraction structures, as mentioned earlier.

4. The Boron Hydrides

Our reinvestigations of the four simplest of these, all first investigated by Bauer at CalTech, are among the most interesting and difficult studies I have ever taken part in. All except diborane were still fascinating puzzles, and all had become controversial. Bauer had worked out an ethane-like structure for diborane, but other kinds of evidence had shown it to be almost certainly incorrect. I found myself answering inquiries about the diffraction problem, saying it would be difficult to distinguish between the bridged and ethane structures but, in view of the differing numbers of B—H and B . . . H distances in the two models, not impossible. Hedberg and I made pictures of ethane and diborane, saw that they are indeed different (to a degree that surprised me), and worked out the details of the ethane structure for ethane and the bridged structure for diborane. As I look back now I feel that I was much too conservative in estimating our chances of success and in stating the weight of our conclusion. "You might as well fall flat on your face as lean over too far backward" (Thurber, 1945).

After diborane, B_5H_9. For it, no automatic solution by a hydrocarbon structure had presented itself during the original study – as I recall, no promising structure at all, so weak still was the radial distribution method, until Pauling drew a characteristic far-reaching analogy to the boron octahedron in calcium boride. Given a few 90° angles there was considerable immediate success, but refinement led to something like methylene cyclobutane instead of the right structure. Later, Pauling again favored the square pyramid, but we were also impressed with Pitzer's protonated double bonds. His cyclopentadiene-like model didn't work at all, however, unless the ring were non-planar, and the pyramid didn't seem to fit for any ratio of the two B—B distances. Finally, however, knowing that Pitzer now also favored the pyramid but unexplainably oblivious of King and Lipscomb's (1949) apt suggestion of the hydrogen-atom arrangement and only after writing our abstract for the September 1950 ACS meeting in Chicago, we at last arrived at the correct structure by explicitly seeking an arrangement of the hydrogen atoms – we should have known enough already to respect the H's more – that would make the square pyramid fit the RDI. Hedberg described it at the meeting. This was our first example of singly hydrogen-bridged B—B bonds, encountered shortly before such bonds were established in the full report on the classical X-ray crystallographic study of $B_{10}H_{14}$ (Kasper, Lucht, and Harker, 1950) and by the independent microwave and X-ray determinations of B_5H_9 itself (see Lipscomb, 1954, for a comprehensive contemporary account of all the boron hydride structures).

Tetraborane (B_4H_{10}) and unstable pentaborane (B_5H_{11}) were further fine puzzles. Again, the old hydrocarbon-like structures were evidently wrong; in contrast to B_2H_6 and B_5H_9, however, no preconceived alternatives were at hand. We were proud to discover the B_4H_{10} structure in Spring 1951, well before the correct crystal structure had been worked out (Nordman and Lipscomb, 1953a, 1953b). In our B_5H_{11} study, never finished, the diffraction pattern, the RDI, and our experience with B_4H_{10} and B_5H_9 led uneventfully (Jones, 1953) to the correct multicoordinate boron framework, again quite unlike the original electron diffraction result, and to very nearly the crystal-structure hydrogen disposition (Lavine and Lipscomb, 1954) at about the same time that Lipscomb (1951) sent us his conjecture of the same framework and a hydrogen arrangement distinctly different in the placement of one atom.

5. Failure of the (First) Born Approximation and the UF_6 structure

A very early difficulty was the report of unsymmetrical structures for UF_6, WF_6, and MoF_6. Symmetrical octahedra might have been expected, as were in fact found for SF_6, SeF_6, and TeF_6. I remember that at a gathering of CalTech folk in Baltimore in connection with the 1939 ACS Spring Meeting nearly everyone argued that these unsymmetrical structures had to be incorrect because they violated Pauling's principles of quantum mechanical resonance among several valence bond structures. I argued correctly, but with little success, that symmetrical structures didn't necessarily follow: the patterns were surely unmistakable. However, although most of us didn't know about it until much later, the structure of uranium hexafluoride was soon to become an item of consuming importance. It was subjected to careful investigation and reinvestigation on all sides with the final conclusion that it is a symmetrical octahedron, the electron diffraction pattern, broadly confirmed in Bauer's reinvestigation, notwithstanding. After the war, when we became aware of this, we (especially Otto Bastiansen, who was spending a year with us, and William F. Sheehan, Jr.) photographed or rephotographed a series of relevant molecules. The unsymmetrical structures appeared to be all too true. We plagued Pauling for an explanation of the structures. None was forthcoming. Later, Sheehan worked to outline the possibilities for tunneling in such unsymmetrical molecules with otherwise equivalent bonds differing by only a few tenths of an Ångstrom. We also saw that if the Born approximation were inadequate, variations in more than the radial scale of the patterns might occur if the electron wavelength was changed, but such photographs were not made until later. Then, at the end of his Ph.D. oral, Sheehan asked Professor Robert Christie whether he thought the Born approximation should be trusted in our work. Christie counseled skepticism. Sheehan immediately left to take up his first job but very shortly thereafter, encouraged by Christie's remarks, I noticed that a differential phase shift approximately proportional to s and to difference in atomic number would explain all the patterns in terms of symmetrical structures. When I reported this to Christie, having failed to find direct support in 'Mott and Massey', he told me not to worry and put me in touch with Roy Glauber, who promptly worked out a comprehensive approximate theory of the phase shift (Glauber and Schomaker, 1953). Later, Jean Hoerni and Jim Ibers made more detailed calculations, and Ibers, Gary Felsenfeld, David Wong, and others additional structural studies, further supporting our interpretation.

6. The Last CalTech Apparatus

In 1951, Pauling finally talked me into building a new apparatus. It was agreed that we should copy Bartell's first Michigan apparatus (Brockway and Bartell, 1954), and this was done with a number of modifications. Leaks due to our unfortunate anachronistic use of brass plagued the new apparatus almost as badly as they had plagued the old. However, it eventually worked well. Of the studies done with it, I shall mention only the heavy-metal (W, Ir, Pt, U, Np, Pu), series of hexafluorides, done over a wide range of wavelengths and confirming the phase-shift interpretation as fully as could be hoped (Kimura, Smith, Schomaker, and Weinstock, 1968), and finally cuprous chloride trimer (Wong and Schomaker, 1957), the least volatile substance we ever studied. Pauling had long wanted it done. Our result was interesting (even shocking, in harking back to a rather notorious early derivation of the NO_2 bond angle from a failure to find *any* diffraction rings). Here we had almost no evidence of the copper–copper scattering, which ought to have been predominant. Our story is that the structure is a strange alternating 6-ring with the angle: Cl—Cu—Cl $\gtrsim$ 150° and the angle: Cu—Cl—Cu $\lesssim 90°$, the short Cu—Cu distances having a very large vibrational amplitude because the copper atoms can't decide whether to be metallic and multivalent or non-metallic and bivalent. I keep waiting for disproof or verification of our report.

Surely the main thing about the laboratory was that it was at CalTech, inspired by Pauling and abetted by many other contributing members of the Faculty, a continual stream of fine students, and dozens of distinguished visitors. Besides the comparatively few named in the text, fickle memory and the laboratory record books afford the following approximately chronological list of participants: L. E. Sutton, G. W. Wheland, C. C. Lauritsen, J. H. Sturdivant, R. M. Badger, F. T. Wall, A. J. Stosick, S. Weinbaum, D. M. Yost, H. J. Lucas, R. G. Dickinson, G. C. Hampson, A. Kossiakoff, K. J. Palmer, H. D. Springall, N. Elliott, H. O. Jenkins, C. D. Carpenter, J. A. Ketelaar, H. Sargent, H. J. Yearian, M. J. Schlatter, J. E. LuValle, S. M. Swingle, E. R. Buchman, H. G. Feldman, R. Spitzer, O. Hassel, Chr. Finbak, H. Russell, Jr., R. E. Rundle, D. A. Shaffer, Jr., R. A. Spurr, W. J. Howell, Jr., R. A. Cooley, W. West, H. Stücklen, W. Gordy, J. H. Saylor, Lu C-S, P. Giguère, J. Donohue, W. N. Lipscomb, D. P. Shoemaker, G. L. Humphrey, E. W. Malmberg, A. G. Whittaker, H. Pfeiffer, A. L. Wahrhaftig, J. M. O'Gorman, W. J. Shand, Jr., M. T. Rogers, H. Lemaire, S. Claesson, A. G. Biswas, J. D. Dunitz, J. R. Fischer, S. Naiditch, W. Nowacki, E. Heibronner, A. Perlis, B. Keilin, E. Goldish, M. E. Jones, W. C. Hamilton, J. N. Shoolery, R. G. Shulman, F. Dudley, J. D. Roberts, C-H. Wong, W. W. Wood, L. Hedberg, G. Nazarian, A. Berndt, R. E. Marsh, J. P. McHugh, D. W. Smith, M. Kimura, B. Weinstock.

Like others at CalTech, many (perhaps most) of those present more than briefly felt they had enjoyed the Golden Age. As an old hand there I came to know that it's *a* not *the* and that *my* Golden Age was neither first nor last.

The last entry in the laboratory record is for April 3, 1958. Ken and Lise Hedberg had moved to Corvallis about two years earlier and I left CalTech that summer.

II. DEVELOPMENT OF THE OREGON STATE LABORATORY: KENNETH HEDBERG

1. The Beginnings at Oregon State

The Oregon State electron-diffraction laboratory was started in 1956. I had arrived at OSU in January of that year with the notion of having all phases of the laboratory operational in three years. Alas, I reckoned poorly with the time required to handle simultaneously a full teaching load, the running of a small but increasing research program (based on data from elsewhere), and the design and construction of my own apparatus. It was not until 1963 that full-blown structure investigations based on Oregon State data could be routinely carried out.

Those early years were a trial, but there were bright spots too. One was that funding from the Research Corporation and the Alfred P. Sloan Foundation was available in sufficient amount to meet both the costs of constructing the apparatus and carrying on the research for several years. A second was an unexpected grant from the Air Force Office of Scientific Research for the purchase of an automatic recording microdensitometer. A third bright spot was that my first graduate student, Roger Crawford, was both skilled at electronics and eager to construct a high-voltage power supply from detailed drawings of the highly successful one at CalTech which had been given me by Holmes Sturdivant. Roger set to work immediately and completed the work before his graduation; however, because the electron gun and pumping system were not yet ready, it could not be thoroughly tested. When those tests were made later, some puzzling problems turned up which were eventually solved by no more than a few adjustments of bias voltages. The solutions, incidentally, were due to a crusty, retired Navy electronics technician who had a few salty remarks to make about the abilities of the more theoretical types who had tried earlier, failed and eventually given up. As he said afterward, "You told me it worked at CalTech. Then I figgered it oughtta work here."

2. The OSU Apparatus

The design of the apparatus itself was strongly influenced by my year in Norway in 1952–3 during which the new apparatus there, about which Otto Bastiansen has written in his article, was just becoming operational. The large size of the scattering chamber of this equipment dwarfed that of any other electron-diffraction apparatus of the time and permitted the recording of data at large scattering angles using a long distance between scattering point and photographic plate. Gas samples were bled into the apparatus at 'pushing pressures' of 10–15 torr, much lower than had been customary, through a nozzle tip of small diameter, with the result that exposure times much longer than usual were required. The results were spectacular: for cyclooctatetraene, data were obtained out to $s \sim 56$ rad $Å^{-1}$, nearly twice the range obtained for similar molecules in other apparatus. (When, in a high state of excitement, I wrote this to Lawrence Brockway, he gently suggested I examine the outer parts of the sector for unevenness. By the time of his reply, clearly different data at similar high angles from other molecules made that unnecessary.) It seemed that the combination of long camera distance, small scattering region (due to low run-in pressures which minimized gas spreading) and exceptionally clean electron beams must have been responsible for the ability to record data at these high angles. Because the first two of these items owed to the large size of the apparatus, I decided to incorporate that feature into the OSU design. Other features of the OSU apparatus, in particular the electron gun and the upper chamber in which it mounts, were adapted with modifications from Larry Bartell's Michigan design. His ideas had already found their way into the new CalTech sector apparatus where I had seen them work well. Figure 1 is a view of OSU apparatus.

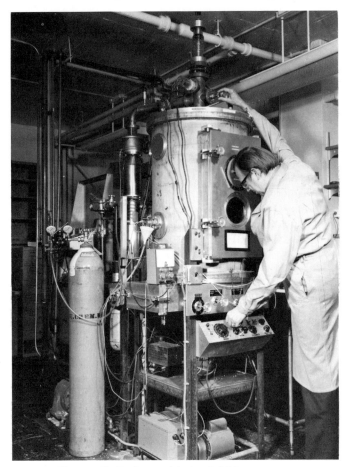

Fig. 1.   K. Hedberg with the O.S.U. apparatus.

3. Early Research Work

While the apparatus was under construction, research at OSU was continued with data from Norway. It is impossible to measure the debt I owe to Otto Bastiansen for the help he and other Norwegian friends provided during this period. Several graduate students at OSU wrote theses based wholly or in part on Norwegian data (David Barnhart on $AsBr_3$, Jerome Blank on NO_2, Robert Ryan on B_2Cl_4, Dorothea Gregory-Allen on cyclohexane-1,4-dione), and Machio Iwasaki as a post-doctoral fellow from Japan used Norwegian data for PCl_3 in part of his work. Most of these data were obtained during my stays in Norway in the summer of 1958 and the year 1962–3, and although I carried out most of the microdensitometry of the plates, others, particularly Arne Almenningen, did the diffraction experiments.

4. Development of Computer Programs

This period also saw the development of our computational techniques. Shortly after my arrival at OSU an ALWAC III–E digital computer was installed in the mathematics department and made available for use by others in the university – free in the beginning and later at a charge of $5 per hour. I immediately set myself to programming intensity and radial distribution calculations in machine language (no compiler existed), learned by reading instruction manuals. Roger Crawford also became interested and we very soon achieved our objective. For some reason I had never bothered to look at the machine; my first glimpse of it occurred on the occasion of using one of our completed programs. This machine was also the first programmed for our least squares calculations concerning which some historical remarks are in order.

The introduction of least squares into electron diffraction work began with Walt Hamilton's and Verner's application of it to visual data (Hamilton, 1954). When, in Norway in 1952, I saw the large amount of highly reproducible sector data on cyclooctatetraene mentioned above, it was immediately obvious that many more parameters could be accurately measured than was possible with the visual method and that our old visual ways of defining 'acceptable' regions of parameter space by comparing intensity curves couldn't possibly work. It occurred to me that least squares was a made-to-order tool for analysis of sector data. At that time, in Oslo, as in nearly all other laboratories, electron-diffraction calculations consisted largely of radial distribution curves, and were carried out with punch-card machines. The least squares calculation based on intensity curves was thus a formidable task. Fortunately, Otto had a young co-worker named Lise Smedvik who was persuaded to join the project, and after several months spent on densitometry with a non-automatic instrument, drawing backgrounds and combining the data (all by hand) we were ready to tackle the least squares problem. Using tables of trigonometric and exponential functions we first calculated an intensity curve for the trial structure, and then 15 others each reflecting a small change in one parameter. Derivatives were obtained from the 15 difference curves, the appropriate derivative products formed, and the normal equations solved on a small Monroe calculator. Each calculation was done at least twice to eliminate errors and the entire process required several weeks. At the end Lise and I were fed up with numbers, but the end result was exciting: we had determined values for several times as many parameters as had ever been done before, and it was clear we had seen the wave of the future. One important matter was lacking: the estimates of uncertainties. Despite several attempts, we had not managed to invert the 15×15 matrix on the Monroe. Later, back in Pasadena, I took my problem to an acquaintance at Consolidated Electrodata, a company which had developed a digital computer, and one evening we watched the lights on it blink for about 10 minutes as it did the job. I was filled with wonder at the speed of a machine which could do all those multiplications in a few minutes. The results, but not the sense of excitement generated by the process, are conveyed in our publication (Bastiansen, Hedberg and Hedberg, 1957).

Programming the ALWAC III–E for least squares calculations was a major undertaking not only because it had to be done in machine language, but because the machine had become unreliable through shortages of replacement parts, particularly vacuum tubes. The State of Oregon offered no direct support for the computer, which eventually was kept operational only by mathematics department graduate students who had learned how to permute dying vacuum tubes among racks of sockets in such a way as to maintain intermittent vestiges of life in the whole. At one stage I advanced a substantial sum of money to the destitute math department, against a credit for future computer use, for the purchase of vacuum tubes. At any rate, neither the programming difficulty nor the unreliability of the machine deterred Machio Iwasaki from attacking the problem in which he had become very interested, and his formidable talents led to rapid progress. At my suggestion the program was written to allow, at the end of each refinement cycle and at certain other points, a choice by the operator about whether to proceed, or in case of doubt about the reliability of the intermediate results, to recalculate them for comparison. Since a single cycle of refinement for a molecule of average complexity required about two hours, such choices were carefully made. One always approached ALWAC III–E with apprehension and frequently left it hours later discouraged and disappointed. Accounts of the method, program development, and an application are given respectively by Hedberg and Iwasaki (1964), Iwasaki, Fritsch, and Hedberg (1964), and Bastiansen, Fritsch, and Hedberg (1964).

These early beginnings were followed by a rapid expansion of the program library stimulated by the increased need for a larger variety of calculation and made practical by faster and more reliable computing hardware added by the university. The least-squares calculation has seen much effort by many people: Bob Ryan, Michael Gilbert, and most particularly, Lise Hedberg who, it may be fairly said, is responsible for the major part of the interactive version currently in use. Our turn to interactive computing was made possible by features of an operating system written by computer center personnel; it was attractive to us because our computing costs were increasing rapidly. Among these costs was waste incurred from least-squares refinements that failed to converge. Being able to change refinement conditions as necessary during the calculation allowed one to eliminate badly conditioned refinements, to do many successful refinements at a single sitting, and with the recent addition of a high-speed printing terminal to the laboratory, to have output immediately available. The handsome pay-off experienced for making this calculation led us to extend the interactive feature to nearly all the other calculations encountered in our electron-diffraction work. For example, curve-plotting and model-drawing for publications are done at a display terminal, arranged or ordered as desired, and only when the results are satisfactory are they routed to the plotter at the computer center. Probably our greatest single time-saving calculation is Lise's program for automatic background removal (Hedberg, 1974). Unfortunately it cannot be truly appreciated by one who has never faced the task of plotting large amounts of raw intensity data, positioning splines on the curves, and laboriously subtracting the results. Perhaps each new student should be required to analyze one structure by the old methods in order to appreciate better the power of the modern one.

5. Recent Laboratory Development

Shortly after my arrival at Oregon State I remember thinking that a functioning electron-diffraction apparatus, a densitometer, and a few computer programs were the only tools needed for a successful laboratory. (Good ideas were also important, but such was the power of the sector method that any study of a molecule – even of molecules previously investigated by visual techniques – was bound to produce interesting results.) I could scarcely have been more wrong: both apparatus development and program development and improvement have been continuing activities at Oregon State. The latter is certainly measurable in man-years, and the least-squares program alone has seen seven versions. Older and wiser, I now feel that change is the rule. So far as apparatus at OSU is concerned, it includes the recent addition of a quadrupole mass spectrometer to the diffraction unit and the design and construction of high-temperature nozzles by students Larry Eddy and Donald Danielson. A high-temperature oven for use with nonvolatile substances is contemplated, as well as the addition of a microprocessor for control of the mass spectrometer and the densitometer, and for mass-storage – the last to further reduce computing costs. Many other changes and additions are contemplated and in the aggregate will greatly expand the types of problems accessible to the gas electron-diffraction techniques.

6. Lines of Research

Out of the developments described in the preceding two sections has emerged a certain special capability that was foreseen some years ago, but only in part. That capability concerns the speed with which the structures of simple molecules may be worked out: experienced people have been able to do all the experiments and to solve such structures in a day or two. One of the consequences of this facility has been to turn us toward problems which can only be studied by doing a considerable amount of experimental and/or computational work. An example is the investigation of conformational equilibria which requires experiments at several sample temperatures in order to estimate the interesting thermodynamic quantities governing the equilibria. Our first study of this type (on oxalyl chloride) gave values for those quantities and at the same time revealed that the second conformer of this conjugated system was not the expected *cis* form, but a *gauche* form derived from the *trans* by a rotation about the central single bond of about 120°. This unexpected result has led to a series of investigations which is still being expanded and which raises new questions as fast as it answers old ones.

Dissociation equilibria, particularly the monomer–dimer type, comprise a second example of a large-scale experimental problem. Again, experiments must be done at several temperatures and, also, at different pressures to obtain thermodynamic information. Other examples include studies of a series of homologous compounds, and studies of large amplitude motion including internal rotation. Papers from this laboratory on all these subjects are to be found in the literature.

7. Concluding Remarks

Despite the larger than usual amount of experimental work required by the types of problems mentioned above, the diffraction experiments are short compared to the time used in analysis of the data. Thus, the apparatus can serve many more people than it is called upon to do. For that reason, we have encouraged visits from others who wish to do electron-diffraction experiments. Among such visitors who have spent times from one to several weeks at OSU have been Carl Aten, James Boggs, Thomas Borgers, István Hargittai, John Huston, Jean Jacob, Michael Kelley, Colin Marsden, Sushil Staija, Lothar Schäfer, my co-author Verner Schomaker, and Bernard Weinstock. We hope to be able to continue to offer the laboratory facilities to these and others who may wish to use them.

The foregoing account does not give proper credit to the many students and longer-term post-doctoral visitors who have brought so much to the program. In addition to those mentioned specifically or in references, these people are Vladimer Bondybey, Sven Cyvin, Gordon Goodman, Grete Gundersen, Bernhard Haas, Kolbjørn Hagen, Takao Iijima, Svein Samdal, Alan Robiette and Cuthbert Wilkins (longer-term senior colleagues); Virginia Cross, Robert DeMattei, Walter Emken, Roger English, Richard French, Dwayne Friesen, William Hartford, Anne Hedberg, Erik Hedberg, Robert Johnson, Bruce McClelland, John Neissus, James Patton, Vernon Plato, Celia Rockholt, and Quang Shen (students).

This brief history would be misleading without some extra statement of the enormous contributions made by my wife, Lise, to the Oregon State Laboratory. Her work on the computing side has been mentioned. Her efforts on behalf of students and visitors are continuous and often self-sacrificing; in a very real sense the efficient functioning of the laboratory work is due to her. As Otto Bastiansen may have meant to suggest in his article, the scientific contributions of the Oslo laboratory to Oregon State go much beyond those he chooses to document.

REFERENCES

Bastiansen, O., Fritsch, F. N., and Hedberg, K.: 1964, *Acta Cryst.* **17**, 538–543.
Bastiansen, O., Hedberg, L., and Hedberg, K.: 1957, *J. Chem. Phys.* **27**, 1311–1317.
Brockway, L. O. and Bartell, L. S.: 1954, *Rev. Sci. Inst.* **25**, 569–575.
Brockway, L. O.: 1936, *Rev. Modern Phys.* **8**, 231–266.
Brockway, L. O. and Palmer, K. J.: 1937, *J. Am. Chem. Soc.* **59**, 2181–2189.
Glauber, R. and Schomaker, V.: 1953, *Phys. Rev.* **89**, 667.
Hamilton, W. C.: 1954, Ph.D. Thesis, California Institute of Technology. For a published application of least squares to visual data see Jones, M. E., Hedberg, K., and Schomaker, V.: 1955, *J. Am. Chem. Soc.* **77**, 5278–5280.
Hedberg, K. and Iwasaki, M.: 1964, *Acta Cryst.* **17**, 529–533.
Hedberg, L.: 1974, Fifth Austin Symposium on Gas Phase Molecular Structure, Abstract 79.
Hughes, E. W.: 1941, *J. Am. Chem. Soc.* **63**, 1737–1752.
Iwasaki, M., Fritsch, F. N., and Hedberg, K.: 1964, *Acta Cryst.* **17**, 533–537.
Jones, M. E.: 1953, Ph.D. Thesis, California Institute of Technology.

Kasper, J. S., Lucht, C. M., and Harker, D.: 1950, *Acta Cryst.* **3**, 436–455.

Kimura, M., Schomaker, V., Smith, D. W., and Weinstock, B.: 1968, *J. Chem. Phys.* **48**, 4001–4012.

King, M. V. and Lipscomb, W. N.: 1949?, Unrecorded private conversation.

Lavine, R. and Lipscomb, W. N.: 1954, *J. Chem. Phys.* **22**, 614–620.

Lipscomb, W. N.: 1954, *J. Chem. Phys.* **22**, 985–988.

Lipscomb, W. N.: 1951, Personal communication.

Nordman, C. and Lipscomb, W. N.: 1953a, *J. Am. Chem. Soc.* **75**, 4116–4117.

Nordman, C. and Lipscomb, W. N.: 1953b, *J. Chem. Phys.* **21**, 1856–1864.

O'Gorman, J. M., Shand, W., Jr., and Schomaker, V.: 1950, *J. Am. Chem. Soc.* **72**, 4222–4228.

Pauling, L. and Brockway, L. O.: 1934, *J. Chem. Phys.* **2**, 867–881.

Pauling, L. and Brockway, L. O.: 1935, *J. Am. Chem. Soc.* **57**, 2684–2692.

Shaffer, P. A., Jr., Schomaker, V., and Pauling, L.: 1946, *J. Chem. Phys.* **14**, 659–664.

Thurber, James: 1945, Moral to 'The Bear Who Let It Alone' in 'The Thurber Carnival' (Harper and Brothers, New York and London) from his *Fables for Our Time and Famous Poems Illustrated*.

Waser, J.: 1944, Ph.D. Thesis, California Institute of Technology. See also Waser, J. and Schomaker, V.: 1953, *Rev. Modern Phys.* **25**, 671–690.

Wong, C.-H. and Schomaker, V.: 1957, *J. Am. Chem. Soc.* **61**, 358–360.

OTTO BASTIANSEN

II.23. Gas Electron Diffraction in Norway

In memory of Professor Christen Finbak
in gratitude for his contribution to our field

1. INTRODUCTION

A fairly large number of people have in various ways contributed to the work of the Norwegian electron-diffraction group during its nearly 50 years of history. It is impossible in a short presentation to do justice to all those who have taken part in establishing the group, in making it work as a unit of multifarious interest, and in solving a large number of molecular structure problems. So many of the results have been obtained through cooperation of several people that it has often been impossible to separate the contribution of the individual participants of the team. Several categories of people have contributed to the results: members of the research staff have guided the development; hard-working students have produced large numbers of molecular structure results; and a skilled technical staff has over the years helped to improve the procedure of the routine work and has been responsible for keeping the routine going. In a field where design of precision equipment is decisive for the quality of the work, the endeavour of dedicated and able workshop personnel has been essential.

Even though the total effort of the Norwegian electron-diffraction group is the result of many people, there are two persons whose individual contributions are easily separable. Without the contribution of these two persons there would hardly have existed a Norwegian electron-diffraction group worthy of a chapter in the present volume. They are the Nobel Laureate Professor Odd Hassel and the late Professor Christen Finbak.

2. THE EARLY HISTORY

Hassel was the initiator of electron-diffraction studies in Norway. It all started after a meeting of 'Deutsche Bunsengesellschaft' in Heidelberg, Germany, in the spring of 1930. After the meeting Hassel was invited by Professor H. Mark, with whom Hassel had been associated in the early 'twenties, to Ludwigshafen, where R. Wierl demonstrated his gas electron-diffraction unit. The potentiality of the electron-diffraction method challenged Hassel's creative imagination. He felt that the method probably would help in solving molecular structure problems

Fig. 1. Professor Odd Hassel.

that earlier had attracted his interest. Already before this time Hassel had dealt with a series of fundamental structure problems mainly using X-ray crystallography and dipole moment measurements. Hassel had also prior to that time taken an interest in the cyclohexane problem. He was well aware of the economic limitations of the University of Oslo, limitations that contemporary scientists can hardly imagine. Possessing an optimism that his colleagues at that time may have considered unrealistic, Hassel decided to try to start electron-diffraction activity in Oslo. Hassel's intention was realized a few years later. In 1934 he met Dr H. de Laszlo at a Faraday Society meeting in Oxford. De Laszlo had produced an electron-diffraction unit for sale, and Hassel was able to buy one of these units. Unfortunately it had been made of a rather porous alloy and was far from vacuum tight. However, after having been surface treated by a proper sealing wax, it performed satisfactorily for a few years. The first gas electron-diffraction study in Norway was on CI_4 and CBr_4 (Finbak and Hassel (1937)). Shortly afterwards the first electron-diffraction study of halogen derivatives of cyclohexanes provided firm support for Hassel's earlier finding, that the chair form of cyclohexane derivatives was prevalent (Gudmundsen and Hassel (1938)).

3. THE SECTOR METHOD

A decisive step towards the development of gas electron diffraction into a modern precision procedure was the introduction of the sector. The first person to suggest the use of a rotating sector for electron-diffraction studies was F. Trendelenburg (1933). His idea did not seem to have led to a sector apparatus, though he constructed a working sector device that was used for making contact copies of un-sectored powder electron-diffraction plates (Trendelenburg and Franz (1934)). The light used for making a negative copy of the diffraction plates was modified by a rotating sector, thus using a photographic trick to compensate for the steep descent of the background, and thus getting plates leading to photometer curves showing well-developed diffraction lines. Later, Chr. Finbak (1937) also suggested a rotating sector inside the vacuum system. The idea materialized a few years later. Finally P. P. Debye (1939) in two articles published the first experimental results of a working sector electron-diffraction machine. There seems to be a general agreement among those who were active in the field in the 'thirties, that the three inventors of the sector got the idea independently. It seems also obvious that the idea had been inspired by the use of rotating sectors for various purposes in spectroscopy.

Finbak's idea led to the first sector electron-diffraction equipment in Norway. It was in active operation for about 15 years. Finbak designed the apparatus, and it was built in the university workshop. It did not come into actual operation until 1940. During the first days of April that year the machine was ready to go. But so also was Hitler's war machinery. The day planned for the first pictures to be taken was April 9, the very day of the German assault on Norway. Not many bombs were dropped over Oslo that day, but strangely enough one bomb hit and exploded less than 100 meters away from the lab in which the electron-diffraction machine had been installed. It was demounted and brought to a safe place. After some time the university was back at work, and so was the electron-diffraction machine.

4. THE WAR YEARS

Already by the end of 1940 the first report based on experiments obtained by the new equipment was published, to no one's surprise on cyclohexane (Hassel and Taarland (1940)). That work put an end to a spectroscopically-based revival of the idea that the cyclohexane skeleton ought to be planar.

The hardship and suffering of wartime also hit the Norwegian electron-diffraction group. Concentration camp, prison, underground resistance, heavy sentences, death; all this was a daily part of life for the group. But as demonstrated so often during the war, the ability of people to perform efficiently under hard pressure is remarkable. As a matter of fact the nucleus of a permanent group was created during the war years. Hassel's famous and, of course, illegal lectures in the concentration camp, contributed to the recruitment. Sharing time between war goals and research activity, a series of structure problems of central importance was formulated, some of them were solved, and the sector method took shape, starting its development towards today's level. Among the contributors to the electron-diffraction work during the war years may be mentioned T. Taarland, B. Ottar, and A. Sanengen. H. Viervoll contributed significantly to the computing routine and also introduced the so-called 'normal-curve procedure' that helped in solving important structure problems at that time. The present author, who started out in Finbak's field of liquid studies, using monochromatic X-rays, shifted over to gas electron diffraction.

For obvious reasons many of the ideas and results from the wartime were never published. Scientifically, Norway was practically isolated, and Norwegian journals with very limited circulation were the only possibility for publication. However, a foundation for the post-war activity of the electron-diffraction group was laid. A summary of the results obtained during the war and a description of the procedure were presented by Viervoll (1947) and by Hassel and Viervoll (1947).

Hassel's important contribution that led to the Nobel Prize was also published during the war (1943) in Norwegian, later on (1971) translated to English by K. Hedberg. Other contributions concerning conformational analysis also date back to the wartime (Bastiansen and Hassel (1946)). Some significant steps were taken towards modern conformational analysis. Plans for further experimental improvements were made, involving both changes in the existing equipment and ideas for new design. In particular, the plans for new equipment took shape. It was decided that the new apparatus ought to be large in size, making it possible to take photographs at nozzle-to-plate distances varying from a few centimeters to half a meter, and also making space for possible auxiliary equipment inside the vacuum system. Finbak's exceptional talent for experimental design was above all decisive for the plans. The first ideas were perhaps more imaginative than realistic, but in retrospect one realizes that some of Finbak's ideas from that period are today part of the electron-diffraction method. Those first sketches led to the equipment that is still – more than 35 years later – producing high quality intensity data not only for the Norwegian electron-diffraction group, but also for foreign researchers. The design work done in Oslo during the war had also, apparently, an effect on the new generation of electron-diffraction machines to be built later outside Norway.

5. THE EXPERIMENTAL EQUIPMENT PRESENTLY IN USE

Well known post-war problems seriously interfered with the hopes for new equipment. All resources were channeled to more fundamental national needs than electron-diffraction machines. In fact it soon became extremely difficult to keep the existing sector equipment going due to lack of routine laboratory facilities. Experimental problems that under normal circumstances are irritating trifles, led to weeks of extra work. An irreparable leak developed in the main part of the apparatus. There was not sufficient workshop capacity to replace it. After two weeks of systematic bicycle hunting a small firm was found that took the job of replacing the defective part, and it was indeed delivered in a couple of months. At one time the experimental work came to a full stop as the milliammeter used for measuring the high voltage broke down. One of the American founders of gas electron diffraction, the late Professor L. O. Brockway, happened to visit Oslo shortly afterwards. A few months later a new meter arrived from his lab. Only two years ago the author accidentally learned that this precious gift had been purchased by our American friend out of his own purse. The road of science is thorny, but the nature of the thorns is different at different times.

Eventually, on the initiative of Professor Hassel and with economic support from two newly-created national research councils, concrete work on sector apparatus No. 2 started in the late 'forties. Finbak was the obvious leader of the design work, but in 1948 he took over a professorship in Trondheim, and about the same time his fatal disease started to develop. Engineer E. Risberg was hired to do the actual drawing of the pattern. Risberg has, through his skill, conscientiousness, and endless working hours, left his mark on the equipment. He was with the group until the new machine was in safe operation. In spite of Finbak's disease and his full-hearted dedication to his new position in Trondheim, he was a central person in the work until the machine functioned. Another person who played an important part was the head of the chemistry workshop Mr S. R. Sørensen, who not only personally machined most parts of the equipment, but also contributed valuable ideas to the design of the new research tool.

The new sector equipment was in operation the Summer of 1952. A few years before this a young physics student, A. Almenningen, joined the group. He became the first pilot of the machine, and it has since been under his supervision. He has in all these years been responsible for the quality of the pictures taken; he has also been responsible for maintenance of the machine and for the many new constructions and improvements that have been carried out.

Christen Finbak died on February 26 1954, after long suffering. He was happy to see that the product of his creative brain functioned satisfactorily, perhaps beyond expectations, but he was never able to participate in harvesting the scientific fruits of his efforts. Finbak died of chronic nephritis, the same disease from which Linus Pauling had been cured a few years earlier. Kenneth Hedberg was the one who drew our attention to the fact that Pauling had been given a special treatment that had saved his life. At our request Pauling hurried all details about his medical treatment to Norway, and Finbak's diet was regulated accordingly, but too late.

The main advantage of the new equipment was of course the quality of the intensity data obtainable. The first Norwegian sector camera gave data over the approximate range $s \approx 3\text{--}25$ Å^{-1}. In favourable cases, data going out to $s \approx 35$ Å^{-1} had been reported. The new equipment gave data from $s = 1$ Å^{-1} up to as much as 60 Å^{-1} in particularly favourable cases. Further, the reproducibility of the intensity data in the classical s range improved considerably compared to that obtained by the earlier sector apparatus. These experimental improvements increased the accuracy of the method, improved the power of resolution in cases of distance overlap, and opened the gas electron-diffraction method for larger and more complicated molecules. The technical features that shared the responsibility for the improvements were most probably better nozzle design, a larger and more precise sector construction, larger photographic plates, and the possibility of taking pictures at different camera distances. (For technical details see Bastiansen (1953) and Bastiansen, Hassel and Risberg (1955)). This second Norwegian sector machine is still in operation, and is probably going to be so for many years. Quite a few technical improvements have been made during the years.

In order to study intensity data at small s values, an apparatus of a few meters height has been built by A. Almenningen. Unfortunately the equipment has not yet been put to a crucial test.

In 1968 a Balzers Eldigraph KDG-2 was purchased. (Bastiansen, Graber and Wegmann (1969)). It has been functioning very reliably. About 240 compounds have been studied with that machine alone. Several new nozzles have been made in our lab for the Balzers machine. One nozzle system was made with two independent inlets to the same nozzle tip. This device is used for critical comparison of two compounds in successive experiments without making any change in the diffraction-point position. The device may also be used to study possible chemical reactions near or at the diffraction point. A possibility of heating the nozzle tip for conformational analysis was also included. Two new nozzles of Russian design have recently been introduced for use with the Balzers equipment. One of the two nozzles is a wide slit nozzle. The other one ends in a doughnut-shaped device with an inner slit all the way around. These nozzles have made possible an appreciable lowering of the temperature of the experiment $(30-40°)$. Molecules may thus be studied that, under normal conditions, would experience thermal decomposition. Engineer Ragnhild Seip has led the routine work and has been responsible for the new constructions of the Balzers equipment in our lab.

In order to obtain higher quality intensity data close attention must also be paid to the photometer equipment. In the first years of the sector method in Norway, a Zeiss 'Lichtelektrisches Registrier-Photometer' was used. The next photometer was a Leeds and Northrup design. In order to utilize as much as possible of the information on the electron-diffraction plates, the plates were oscillated during the photometer process about an axis perpendicular to the plate and through the diffraction center. Since the Oslo apparatus uses photoplates of 6×24 cm dimension, and since the sector support leads to a negative 'shadow' on the plates, spinning of the plates is not possible. An automatic digital photometer was therefore constructed, able to integrate photometer readings along an arc of a circle with the center in the diffraction point. The equipment was built in our lab by P. J. Molin under the supervision of Almenningen. This instrument, called 'Snoopy', has been in use since 1970. In 1971 a commercial photometer was added. It was a Joyce Loebl autodensidator type M. It is a double beam instrument that reads off data from individual small areas, systematically running through the whole plate. Through a computer program written by H. M. Seip, the readings are transferred to intensity data as function of s. The two last photometers are both still in use.

6. THE TWO CENTERS OF THE NORWEGIAN
ELECTRON-DIFFRACTION GROUP

When Finbak became professor in Trondheim, the field of molecular structure study was introduced to the Norwegian Technical University there. This started an intimate cooperation in this field between chemists in Oslo and Trondheim. During Finbak's first year in Trondheim an essential part of the second sector electron-diffraction machine, namely the plate box, was machined in the chemistry workshop under his supervision. In 1955, after Finbak's death, the author became his successor. During the following seven years an active electron-diffraction group was set up in Trondheim, basing its work on experimental material from Oslo. The idea was to concentrate an experimental base in Oslo in a national electron-diffraction lab that could also service institutions outside the University of Oslo. This was the beginning of the two Norwegian electron-diffraction groups that later on, with approval of the Norwegian Research Council for Science and the Humanities, came to be administered as a single group. In retrospect this appears to have been a wise economic and scientific decision. A distance of 500 km between two centers does not seem to hamper the functioning of this single group.

After 1962, when the author moved back to Oslo bringing with him four coworkers, electron-diffraction work in Trondheim slowed. It was quickly

Fig. 2.   Professor Christen Finbak.

renewed under the leadership of Marit Trætteberg who was appointed to the new section of the University in Trondheim. She became full professor in 1975.

Though the Norwegian electron-diffraction research operates in one group as far as equipment, computing facility, and support from the research council go, the group is divided into many sub-groups if the sub-groups are classified according to chemical problems involved.

7. COMPUTING TECHNIQUE

In the earliest days of the Norwegian electron-diffraction work, desk calculators were used even for the Fourier inversion. In 1948 the first IBM program was created by E. Amble, P. Andersen, and H. Viervoll, thanks to the generosity of the Norwegian IBM Corporation, who placed their facilities at the disposal of the electron-diffraction group, free of charge during night hours. The next computer was one initiated by Viervoll and built at the Central Institute for Industrial Research. A whole series of computer generations followed, and in many ways electron diffraction, together with X-ray crystallography and monochromatic X-ray liquid studies, helped to introduce modern computer technology to Norwegian science. These fields were the main users of computers for years.

An important improvement in electron diffraction work was the introduction of least-squares calculations on intensity data. In this field as in many others of importance for the development of Norwegian electron diffraction study, the contribution of K. and L. Hedberg was of great importance. The present least squares refinement program was originally written by H. M. Seip, who during the last 15 years has been the key person in establishing our present program library. The program was later modified in several ways by other members of the group. A series of other programs has been developed, improved, and adapted to new computers in good team work by B. Andersen, K. Kveseth, G. Gundersen, S. Samdal, H. M. Seip, R. Seip, T. Strand, R. Stølevik, and others.

Programs for calculating mean amplitudes of vibration and perpendicular amplitude correction coefficients were first developed by Cyvin and coworkers. A program with the same objective based on a work of W. Gwinn *et al.* was developed by H. M. Seip and R. Stølevik. Among other supplementary programs may be mentioned a program for conformational studies based on molecular mechanics and programs for calculating torsional barriers.

8. DIFFRACTION THEORY

The main emphasis of the Norwegian electron-diffraction group has been the study of molecules: their geometry, internal motion, conformational analysis, etc. However, members of the group have occasionally been engaged in activity of importance for the theory of electron diffraction.

A table of atomic scattering factors for the elements from H through Cu for

s-values up to 30 Å^{-1} was presented by Viervoll and Øgrim (1949). This table was widely used for many years.

During the year 1950 the author had the privilege to work at CalTech under the guidance and inspiration of Linus Pauling and Verner Schomaker. An electron-diffraction investigation of UF_6, WF_6, and MoF_6 formed the experimental basis of the recognition of the failure of the first Born approximation (Schomaker and Glauber (1952)). In order to study this effect quantitatively, H. M. Seip investigated a series of molecules containing atom pairs having particularly large differences in atomic numbers, including the three compounds mentioned above. It was demonstrated that the introduction of complex scattering amplitudes leads to a quantitative agreement between experimental and theoretical values. However, a small reproducible discrepancy remained, which was easily seen in the radial distribution curves. The disagreement concerns the double peak associated with an interatomic distance between a heavy and a light atom. Theory predicts a symmetric double peak, while the experimental peak complex is not symmetric, the inner component being slightly larger than the outer one. The effect was attributed to multiple scattering. This has later been confirmed. (For further details see Seip (1967)). With theoretical command of the effect of the failure of the first Born approximation, not only may the geometrical structure parameters be deduced with higher accuracy, but also the vibrational parameters. This question has been studied for the compounds involved in Seip's investigations by Cyvin, Cyvin, Brunvoll, Andersen and Stølevik (1967).

An obvious limitation of the accuracy of vibrational parameter calculations is due to uncertainties of the atomic scattering factors used in the electron-diffraction study. Before 1960 most investigations were based upon scattering factors obtained from Thomas–Fermi–Dirac fields in the first Born approximation. Introduction of Hartree–Fock atomic potentials (Strand and Bonham (1964) and Peacher and Wills (1967)), led to improvements of the quality of electron-diffraction studies in molecules containing atom pairs with large differences in atomic numbers.

9. INTERACTION AND COOPERATION WITH OTHER GROUPS

A characteristic feature of the gas electron-diffraction method is its potential for fruitful interaction with other methods aimed at solving molecular structure problems. For example, the breadth and accuracy of the results from electron diffraction increases considerably if results from one or several spectroscopic methods are included in solving a molecular structure problem. For this reason interaction with other methods has been part of the research policy since the electron-diffraction method was taken up in Oslo. This policy has been advanced in several ways, partly by interacting with existing groups inside and outside our country, and partly through taking initiative in establishing new fields in Norwegian labs.

In the earliest days of Norwegian electron diffraction the neighbouring fields that were of greatest importance were X-ray crystallography, dipole moment measurements, and monochromatic X-ray studies of liquids. Later, spectroscopic methods were established. The by now prosperous infrared and Raman spectroscopy group in Oslo, headed by P. Klæboe, was originally initiated by Hassel to make a partner for the electron-diffraction group. The microwave group was initiated for a similar purpose. Professor Cyvin's active group of molecular vibration studies in Trondheim branched off from the electron-diffraction group. So did Professor Skancke's quantum chemistry group in Tromsø. (Skancke's first publications were in the field of electron diffraction.) The interaction with quantum chemistry started, as a matter of fact, in the late forties with Professor Inga Fischer-Hjalmars in Stockholm, who helped to establish our first Norwegian group in the field. A number of other foreign groups have also helped to advance our policy of interaction with other methods. The spectroscopy group of Professor B. Bak in Copenhagen, and the combined microwave–quantum chemistry group of Professor J. Boggs in Austin, Texas, deserve special mention.

Interaction with other groups in our own field has been of vital importance for our own development. Our first foreign contact was the Caltech group already mentioned followed by Hedberg's group in Oregon, which in many respects inherited the CalTech tradition and responsibility. The Norwegian group has obtained invaluable inspiration and knowledge from Hedberg's group. In evaluating the benefit accrued to him through our cooperation a life partner and very gifted coworker obtained from us, should be included. Members of the Norwegian group have benefited from stays in other US electron-diffraction laboratories as well, for example that of Professor S. H. Bauer in Ithaca, Professor R. A. Bonham's in Indiana, Professor L. O. Brockway's and Professor L. Bartell's in Ann Arbor, Professor R. L. Hilderbrandt's in Fargo and the labs in Austin, originally started up by Professor H. Hanson and later taken over by Professors M. Fink and D. Kohl. Contact with British electron-diffraction activity was first established through Leslie Sutton, and for many years valuable cooperation has existed with Japanese groups, particularly those headed by Professor Y. Morino and Professor K. Kuchitsu. Contact and cooperation with other European groups like the Dutch, French, German, and Spanish have been in effect for many years. In recent years active cooperation of great reciprocal value has been established between the Norwegian group and the Moscow and Budapest groups. Concrete and continuous cooperation exists with such people as V. Spiridonov, L. Vilkov, A. Ischenko, and V. Mastryukov in Moscow, and with the two Hargittai's in Budapest.

The Norwegian electron diffractionists have had great help during the years from synthetic chemists who have supported us with material. Professor Lars Skattebøl, Dr. Else Kloster Jensen, and Professor Johannes Dale from the Chemistry Department of our university have helped us with syntheses of many useful compounds. Professor Otto Fischer's group in Munich helped in starting out our organometallic chemistry research and has later supported us with valuable chemicals. Professor J. Weidlein in Stuttgart and Professor Wolfgang

Lüttke in Göttingen and his group have supplied many of the organometallic and organic compounds we have been working with. Professor Heinrich Nöth in Munich and Professor Kurt Niedenzu in Lexington have provided the majority of the boron compounds.

We owe hundreds of colleagues and friends from many places in the world a debt of gratitude for their scientific contribution to the Norwegian group and for the lifelong friendship that has been established across national, linguistic, cultural, and political barriers.

10. MOLECULAR STRUCTURE PROBLEMS

A large number of people have at various times been associated with the Norwegian electron-diffraction group. How many molecular structure or molecular dynamic problems have altogether been dealt with is hard to guess. Certainly hundreds of molecules have been studied. As an indication of the research activity it may be mentioned that during the last 10 years about 250 articles have been published from our group, not including work done outside Norway with participation from members of the group. Space considerations allow only a naming of the broad areas of interest and those who have contributed to them: Conformational analysis (E. Astrup, G. Gundersen, K. Hagen, K. Kveseth, S. Samdal, R. Stølevik, M. Trætteberg), large amplitude motion (H. M. Seip, T. H. Strand), small ring systems (B. Andersen, E. Astrup), boron compounds (G. Gundersen, A. Haaland, H. M. Seip), and organometallic compounds (A. Haaland). Details may be sought in recent publications. (See for example Bastiansen, Boggs and Seip (1971), Bastiansen, Kveseth and Møllendal (1979), Haaland (1975), and Haaland (1979).)

11. FUTURE PERSPECTIVES

Gas electron diffraction has so far nearly exclusively been used in order to study stable and intact molecules. The structure determination by P. Andersen of free radicals, carbanions, and carbonium ions, drawing conclusions from combined electron-diffraction and X-ray crystallographic studies, ought to bring new and valuable knowledge. Combining electron diffraction and mass spectrometry for structure determination of unstable free radicals also presents promising possibilities. Combination of lasers and electron diffraction may also be useful. No doubt these fields are hampered by severe experimental difficulties, and it may be too optimistic to anticipate an immediate break-through.

An obvious advantage of the gas electron-diffraction method is its general applicability. Any molecule that can be brought into the gas phase with a certain gas pressure may, in principle, be studied. The necessary minimum gas pressure is a limitation, but as mentioned earlier our recent experience with the Russian nozzle design indicates an appreciable reduction of this limitation. But there is another limitation that has always been considered as serious, and that is the lack of resolution in determining internuclear distances of nearly the same value. This difficulty also seems to have been appreciably reduced during the last few years through combining electron diffraction and spectroscopy. Vibrational amplitudes deduced from vibrational spectra may be introduced and kept constant during the refinement of the electron-diffraction data. The conditions for success in this procedure are reliable vibrational amplitudes and particularly careful experimental technique when taking the intensity data. This approach has been successfully used for $H—C≡N—S—CO—O$, its methyl derivative, and for coronene.

Combination of X-ray crystallography and electron diffraction for studying molecular structure differences in solids and gases has a long tradition. This seems also to be important for the future. As a recent example we may mention work on the structure of cyanoformamide with emphasis on the study of hydrogen bonding carried out by S. Samdal *et al.*, combining electron diffraction, X-ray and *ab initio* calculations.

The future of gas electron diffraction also rests upon an obvious fact, namely the enormous effectiveness of synthetic chemistry. The rate of production of new structurally interesting molecules is by far higher than our ability to handle them by experimental structural methods.

Norwegian Technical University
Trondheim

REFERENCES

Bastiansen, O.: 1953, *Det 8. Nordiske kjemikermøte, Oslo 14.–17. juni 1953.* Beretning og foredrag, pp. 139–148. Oslo: Kirstes boktrykkeri.
Bastiansen, O., Graber, R., and Wegmann, L.: 1969, *Balzers High Vacuum Report* **25**, 1–8.
Bastiansen, O. and Hassel, O.: 1946, *Nature* **157**, 765.
Bastiansen, O. and Hassel, O.: 1946, *Tidsskr. Kjemi, Bergves. Metall.* **6**, 96–97.
Bastiansen, O., Hassel, O., and Risberg, E.: 1955, *Acta Chem. Scand.* **9**, 232–238.
Bastiansen, O., Boggs, J., and Seip, H. M.: 1971, *Perspect. Struct. Chem.* **4**, 60–165.
Bastiansen, O., Kveseth, K., and Møllendal, H.: 1979, *Top. Curr. Chem.* **81**, 99–172.
Cyvin, S. J., Cyvin, B. N., Brunvoll, J., Andersen, B., and Stølevik, R.: 1967, *Selected Topics in Structure Chemistry* (Ed. P. Andersen, O. Bastiansen and S. Furberg) pp. 69–89. Oslo: Universitetsforlaget.
Debye, P. P.: 1939, *Phys. Z.* 40, **66**, 404–406.
Finbak, C.: 1937, *Avh. Nor. Vidensk.-Akad. Oslo, Mat.-Naturvidensk. Kl.* No 13, 1937.
Finbak, C. and Hassel, O.: 1937, *Z. Phys. Chem., Abt. B.* **36**, 301–308.
Gudmundsen, J. G. and Hassel, O.: 1938, *Z. Phys. Chem., Abt. B.* **40**, 326–332.
Haaland, A.: 1975, *Top. Curr. Chem.* **53**, 1–23.
Haaland, A.: 1979, *Acc. Chem. Res.* **11**, 415–422.
Hassel, O.: 1943, *Tidsskr. Kjemi, Bergves. Metall.* **3**, 32–34.
Hassel, O.: 1971, *Top. Stereochem.* **6**, 11–17.
Hassel, O. and Taarland, T.: 1940, *Tidsskr. Kjemi, Bergves.* **20**, 167–169.
Hassel, O. and Viervoll, H.: 1947, *Acta Chem. Scand.* **1**, 149–168.
Peacher, J. L. and Wills, J. G.: 1967, *J. Chem. Phys.* **46**, 4809–4814.
Schomaker, V. and Glauber, R.: 1952, *Nature* **170**, 290–291.
Seip, H. M.: 1967, *Selected Topics in Structure Chemistry* (Ed. P. Andersen, O. Bastiansen and S. Furberg) pp. 25–68. Oslo: Universitetsforlaget.

Strand, T. G. and Bonham, R. A.: 1964, *J. Chem. Phys.* **40**, 1686–1691.
Trendelenburg, F.: 1933, *Naturwissenschaften* **21**, 173–176.
Trendelenburg, F. and Franz, E.: 1934, *Wiss. Veroeff. Siemens-Konzern* **13**, 48–55.
Viervoll, H.: 1947, *Acta Chem. Scand.* **1**, 120–132.
Viervoll, H. and Øgrim, O.: 1949, *Acta Cryst.* **2**, 277–279.

LAWRENCE S. BARTELL

II.24. Electron Diffraction at the University of Michigan and Iowa State University

1. LAWRENCE BROCKWAY

Although Lawrence Olin Brockway was not the first to do vapor-phase electron diffraction, he did more than anyone else to establish it as a flourishing field. Few are the diffraction laboratories around the world in which one encounters neither a professional descendant of his nor equipment bearing the mark of his ingenuity. His brief account of the electron diffraction method, dictated a few weeks before his death in 1979, is published elsewhere in this volume. Modesty inhibited him from reminiscing about his own history-making contributions. As a graduate student and Senior Research Fellow at the California Institute of Technology he not only put the electron diffraction method on a firm footing, but he also explored the structures of a large number of organic and inorganic molecules. His pioneering work to establish characteristic bond lengths and bond angles as well as shifts in these quantities induced by chemical substitution played a significant role in subsequent advances in chemistry and won him the coveted American Chemical Society Award in Pure Chemistry. After leaving the California Institute of Technology he spent a year at Oxford and The Royal Institution as a Guggenheim Memorial Fellow before accepting a permanent position at the University of Michigan in 1938.

In Ann Arbor he remained active in gas diffraction for a quarter of a century but turned his attention more and more to the study of adsorbed films and surface phases. Not many years after his arrival in Ann Arbor he applied his expertise in these areas to national defense projects as a consultant to the Navy, Air Force, the General Electric Company, and the Los Alamos Scientific Laboratory. His extraordinary vitality and relish for life made him a delightful teacher and an engaging colleague. As he grew older the emphasis in his career shifted from the technical to the more human aspects of science. As a counselor of students he became well known for his kindness, dedication, and unfailing common sense.

2. LIFE IN BROCKWAY'S GROUP

My first acquaintance with Lawrence Brockway was in 1943 when he was my instructor in the second term of elementary physical chemistry. Despite his puckish sense of humor he was a stern taskmaster; on his examinations it was not enough to get the correct answer – the style of the treatment was important, too. Being a rather lazy undergraduate student, I found this rather bitter medicine but, in retrospect, I am sure it was good for me.

He planted the diffraction seed in my consciousness when he invited me to carry out some research in his electron diffraction laboratory. Owing to the wartime depletion of universities, he needed assistance on his projects and was willing to try an undergraduate student, even a dilatory one. Despite the help of my father, F. E. Bartell, a distinguished surface chemist, and Brockway's assistants, Jerome Karle, Isabella Lagoski (later Mrs. Karle), and Elaine Glass, all of whom impressed me greatly, I was unable to get my electrons to distinguish between hydrophilic and hydrophobic samples of colloidal carbon. This inauspicious project was soon interrupted by a year of research on plutonium decontamination at the University of Chicago followed by a brief stint in the Navy.

Upon returning to Michigan, now as a graduate student, I was encouraged by Brockway to experiment with a new invention, the rotating sector device

Fig. 1.   L. S. Bartell with new Michigan diffraction unit, 1950.

introduced with notable success in the late thirties by C. Finbak and, independently, P. P. Debye. By putting odds and ends together, getting a small gear train constructed in the physics shop, and fashioning some sectors myself, I devised a sector machine that gave patterns so superior to those obtained before in Ann Arbor that Brockway asked Arthur Bond, who was nearly finished, to repeat his silane structure determinations. Unfortunately, this undertaking prevented Art from finishing his work before prior commitments required him to begin postdoctoral research in Minnesota. Then I designed and constructed a new diffraction camera, influenced strongly by Jerry Karle's successful new apparatus at the Naval Research laboratory. Brockway, who was very busy with other things, first saw my design the day before the deadline for getting it to the physics shop. Although he was quite skeptical, it was too late to make changes. Happily for me, the new apparatus performed considerably better than the older makeshift unit. Unhappily, it was completed at just about the time Art Bond returned to finish his thesis – just in time, as it worked out, for him to take his diffraction patterns all over again. I am not sure that he has ever forgiven me!

Although it was enjoyable to help the other students develop new procedures for molecular structure determinations, I wanted to make the most of my precious new hardware by doing something more original. Because kinematic electron scattering factors are simply related to X-ray scattering factors and, hence, to electron densities, I thought it should be possible to deduce the radial distribution function of atomic electrons from electron diffraction intensities. Electrons appeared to offer real advantages over X-rays in this application. Argon was selected for investigation. To avoid the Fourier termination errors so severe in X-ray studies it seemed a good idea to try to null out the leading terms by a differential method. Just as molecular studies concentrate upon the interatomic interference terms oscillating about the dominant atomic background, perhaps atomic structure studies could best analyze the intensity undulations about a smooth background calculated for the featureless Thomas–Fermi atom. These undulations should uniquely characterize the atomic shell structure. It turned out, however, to be simpler to use a more direct approach based on another good idea of Jerry Karle. Jerry had suggested representing X-ray elastic scattering amplitudes by certain analytical functions which, when transformed, generate probability distributions guaranteed to be non-negative. The 'witch functions' proposed by Jerry and Herb Hauptman needed modification to achieve adequate flexibility but the idea worked splendidly when applied to atom form-factors deduced from electron intensities. In fact, the electron diffraction results, published in 1953, yielded a far higher resolving power than prior X-ray work. They also yielded a Ph.D. degree.

Seeking to exploit the new unit, I stayed in Ann Arbor another year and a half to study molecular structure. It became apparent that our new apparatus yielded scattered intensities of greater precision than those provided by standard theoretical expressions. Indeed, it wasn't clear exactly what *was* being measured when structure parameters were refined by the procedures then in vogue. Therefore, quite innocent of theoretical physics but with a great faith in Peter Debye's 1941 scattering formalism, I began work on the crucial relationships.

One unattractive but inevitable outcome of this study was the proliferation of such symbols as r_e, r_g, $r_g(l)$, l_e, and l_g, characterizing molecular properties. Despite attempts by the referee to obstruct publication, the work caught on and served as a basis for all subsequent advances. A decade later an examination of Debye's formalism unearthed new and unexpected rewards.

While I was struggling to devise new equipment and new procedures, Verner Schomaker visited Ann Arbor several times. His visits were exciting times for me and they provided extra incentive for persevering. Near the end of this period Schomaker became convinced that the California Institute of Technology needed an apparatus with a rotating sector. Therefore he sent his young protegé, Ken Hedberg, to Ann Arbor to study the Michigan unit. It is absolutely impossible not to like Hedberg, and his droll sense of humor added zest to many occasions. Moreover, his sound judgement was impressive – usually – although I recall one time when, for a lark, he tried to see how far his Volvo would go on a tank of gasoline, only to have it run dry in the wilds of Oregon while transporting guests from Michigan! His efforts in Ann Arbor led to an apparatus similar to ours. It now operates in Lothar Schäfer's laboratory in Arkansas.

3. IOWA STATE UNIVERSITY

My first academic appointment began in 1953 at Iowa State University. Because little research support was available to me it was out of the question to start a diffraction laboratory. Therefore, my initial research concentrated on surface chemistry. This work was much more successful than my investigation of carbon black ten years earlier and led to several discoveries, inventions, and industrial consultantships. Unimpressed by all of this, in 1954 a bright, energetic, wildly imaginative student named Russell Bonham indicated a willingness to work with me if he could do something more interesting, such as electron diffraction. Needless to say, I quickly re-entered the field, functioning for several years by pilgrimages to Ann Arbor made possible by Brockway's kindness. Bonham and his undergraduate helper, Denis Kohl, bubbled over with ideas on computer analyses of data, and Kohl later devised for us the first digital microphotometer used in electron diffraction.

Bonham's structural results for n-alkanes and alkenes seemed strangely different from what we had expected from text-book conceptions of hybridization and steric effects. Fretting about this and other new results one night, I suddenly saw how everything fitted together simply if 'intramolecular van der Waals' interactions, especially geminal interactions (such as were ignored by almost everyone else except Robert Mulliken) were taken into account. This observation, together with my earlier experience with anharmonicity, suggested that more might be learned about molecular force fields by studying deuterium isotope effects on structures and vibrations. By this time my research was liberally supported by the Ames Laboratory of Iowa State University, and I had nearly finished construction of a new diffraction apparatus which eliminated the worst flaws of my Michigan design. By extraordinarily good luck, Kozo

Kuchitsu from Professor Morino's group was willing to begin his postdoctoral studies in this quiet rural center and to collaborate with an almost unknown diffractionist. Coming late in 1958, he was just in time to help complete the apparatus and to embark on our isotope explorations. He also shortly began his classic investigations of the effects of anharmonic vibrations on the interpretation of electron diffraction intensities and molecular spectra. These were much more thorough and rigorous than my first attempts. Also, by combining some of my inferences about nonbonded interactions with his remarkable analytical powers, he predicted (correctly, as recent events have shown), the equilibrium structure parameters of methane. Referees, again, were extremely negative but his work on this important prototype molecule gradually became accepted as standard.

My own ideas remained somewhat more qualitative and pictorial. Hydrogen atoms, being lighter, oscillated over large amplitudes of vibration and therefore acted as if they were larger than deuteriums. Rough orders of magnitude of the implied differences in nonbonded forces were not difficult to guess. The idea suggested that deuterium substitution, in a molecule such as ethane, should change not only the C—H (to C—D) bond length, but also the mean C—C bond length! I registered a prediction of the magnitude and still have the one word reply from the National Research Council in Ottawa – "Impossible!". But our experiments soon bore out our prediction and, moreover, suggested a variety of other properties including molar volume, surface tension, and rates of organic reactions that might be sensitive to such secondary isotope effects. A number of organic chemists embraced the idea (after, of course, protracted opposition by referees) and so began the active new field of 'steric isotope effects'. Years later, in 1974–75, Warren Hehre and I applied the powerful new *ab initio* molecular orbital methods he was developing and calculated the quadratic and the hitherto unknown key cubic force constants of ethane. This information allowed a rigorous computation of the secondary isotope effect on the C—C bond length. Results confirmed the original rough estimates as well as the interpretation.

Other implications of 'intramolecular van der Waals forces' in chemistry began to become apparent. Synthetic force fields embodying such interactions – modified Urey–Bradley fields – were found to work well in 'Molecular Mechanics' calculations in the hands of Jean Jacob, Brad Thompson, and later, Susan Fitzwater. Structures and thermodynamic properties of organic molecules could be accounted for by this approach with markedly fewer parameters than were invoked by other early workers in the field. Hans Burgi's 1971 diffraction study of the beautifully symmetric but exceptionally strained molecule tri-*tert*-butylmethane provided results that offered a powerful criterion for accepting or rejecting synthetic force fields. Modified Urey–Bradley fields fared well. Recently A. Robiette and G. Glidewell, and M. O'Keeffe extended our original table of 'intramolecular van der Waals radii' to show how nonbonded forces govern inorganic molecular and crystal structures.

In the early 'sixties I began to worry about seeming anomalies in scattering theory. Invariably, X-ray crystallographers averaged their scattered amplitudes over the quantum motions of their target particles (molecular electrons), then they squared the average amplitude to get the intensity. We electron diffractionists, on the other hand, followed Peter Debye's equations which square instantaneous amplitudes, then averaged the resultant intensities over the quantum motions of our target particles, the vibrating atoms. Whether first to average, then square, or to square first, then average: how to decide? After struggling with this question I began to realize that each sequence is correct, quantum mechanically. The first scheme yields the elastic scattering (that which crystallographers separate from the inelastic background) and the second gives the total scattering, elastic plus inelastic, which gas diffractionists measure. Striking inferences could be drawn. From the amplitude scattered elastically by an atom can be derived the (1-electron) distribution of electrons around nuclei. From the total scattering can be derived the (2-electron) distribution of electrons relative to other electrons, a concept new at that time. What was especially important was that such distributions reveal and characterize 'electron correlation'. Now, electron correlation poses a fundamental and difficult problem of great interest to quantum theoreticians. Therefore, the news that X-ray and electron diffraction intensities display characteristic evidence of electron correlation started a flurry of activity among the theorists. Although this idea was novel to quantum theorists, it is plain from Debye's early papers that he clearly understood the principles involved in 1915, a decade before wave mechanics! Appropriately, our development of the idea and Bob Gavin's nice illustrative calculations were announced at Debye's 80th birthday symposium in 1964.

During this time another project was unfolding in our laboratory. A number of years before I had accidentally discovered a fascinating paper while reading at random in the library. In 1932 Kapitza and Dirac proposed as a thought experiment a very different kind of electron diffraction called stimulated Compton scattering. It amounted to the diffraction of electron waves by 'crystals of photons' produced by reflecting coherent light waves from a mirror. In 1932 it was an impossible experiment to do by many, many orders of magnitude. But, in the early 'sixties, lasers were invented and one day it struck me that it should now be feasible to do the experiment. A series of coincidences suddenly brought me a windfall equipment budget, a new postdoctoral associate named Brad Thompson who was a real wizard with both hardware and quantum theory, and a talented graduate student with fierce determination, Roland Roskos. We developed the theory of stimulated Compton scattering and carried out the experiment. We had a headstart on other workers who entered the field and, in my opinion, we obtained the best results − though still rather crude − for this extraordinarily difficult experiment (1965–67). A few years later H. de Lang generalized our theory and corrected an error in the way we handled the degree of incoherence.

Also during this period, a chance encounter with Irving Sheft led to our diffraction study of XeF_6, the most remarkable and controversial substance we ever examined. The anomalous properties of this strange new molecule had perplexed chemists for several years. Although quantum chemists had predicted the molecule to be a regular octahedron, Bob Gavin's 1965 results (and, later, those of Ken Hedberg who acquired better and more extensive data) indicated, to the contrary, that the molecule is distorted from O_h symmetry. It isn't distorted greatly – just far enough for it to undergo a rather free, large amplitude pseudorotation that leads to its unusual properties. Although this interpretation was disputed for years it is now largely accepted, and our current *ab initio* pseudopotential SCF–MO computations strongly support it.

Our excursion into inorganic stereochemistry proved to be so enlightening that we carried out many investigations to test the popular 'Valence-Shell-Electron-Pair-Repulsion' (VSEPR) theory of R. J. Gillespie and to study pseudorotation and the concomitant anharmonic coupling of vibrational modes of different symmetry. We discovered that the VSEPR theory not only treats molecular structure easily and quite well, but it also has useful things to say about force fields and vibrations. With its aid we uncovered misassignments of vibrational spectra and a new 'secondary relaxation' principle in hexacoordinate stereochemistry. While this program was being carried out, my research group moved back to Ann Arbor. In 1965, the Ames Laboratory graciously made it possible to transfer the Iowa State Diffraction apparatus to Michigan.

4. RESEARCH AT THE UNIVERSITY OF MICHIGAN

An offshoot of our inorganic research was the observation that existing expressions could not account for details in the diffraction patterns of molecules with heavy atoms (Re, Xe, I). Similar discrepancies could be perceived in earlier patterns of uranium and tungsten compounds photographed in other laboratories, but Jean Jacob's careful 1968 study of ReF_6 documented these anomalies with precision. Such deficiencies in theory, of course, interfere with the derivation of structure parameters from diffraction patterns. We therefore initiated parallel experimental tests and theoretical studies to discover the source of the trouble. Theory won, but accidentally, at first. Gjønnes and Bonham had already shown a potential source of trouble – dynamic, or intramolecular multiple scattering – but their complex expressions were difficult to apply rigorously, leaving us uncertain about the interpretation. Tuck Wong and I, and, independently, Albert Yates set out on an alternative tack and applied Glauber scattering theory, which was much more tractable. We found that it accounted quite well for the dynamic effects in ReF_6. Later, to my embarrassment, I noticed that this agreement was fortuitous because we had all applied Glauber theory far beyond its limits of validity in momentum transfer. Then, in 1975, I discovered a remarkable scheme for correcting Glauber theory – a surprisingly simple way to take into account the intricate propagation of waves inside molecules between scatterings. Results showed that for 90° bond angles, by a quirk of geometry, the corrections to Glauber theory virtually vanish – and hence that ReF_6 fortuitously gave the right results – but that at other bond angles the corrections became large. Recent massive calculations by Denis Kohl indicate that our analytical scheme to correct Glauber theory is a fairly good approximation for 40 keV electrons and that the problem of multicenter potential scattering is now solved. This does not mean that scattering theory is now fully satisfactory for our purposes, however. Unpublished results obtained in several laboratories reveal unresolved discrepancies for certain molecules, no doubt associated with inelastic scattering.

An invitation to speak at the 1972 American Crystallographic Association meeting had serendipitous consequences. Asked to lecture on electron density, a topic to which I had not made any original contributions in several years, I cast around for fresh ideas. Always interested in optics, I had recently bought an inexpensive hologram to get a first-hand impression of how holography works. In a flash I saw how it was possible to devise a holographic electron 'microscope' capable of forming images of electron clouds in gas-phase atoms with a resolving power better than 0.1 Å. A very crude image was even ready in time for the ACA meeting. Now, it isn't widely recognized that Dennis Gabor, who invented holography in 1948, intended to apply it to electron beams, not laser beams (lasers had not yet been invented). He sought to enhance the resolving power of electron microscopes by circumventing their weakest links, their small-aperture objective lenses. His principle worked with dazzling success, when later put into practice with laser beams but technical problems prevented its successful application to electron beams. When Gabor heard that we had obtained atomic images he wrote, ". . . now what is your trick that has escaped others?" Before my answer reached him he figured it out by himself and used the 'trick' to design a holographic microscope which, if it worked, would be enormously more useful than my device. An enjoyable correspondence followed. The 'trick', of course, is to use scattering instead of lenses to get coherent reference beams. Unfortunately, Professor Gabor suffered a severe stroke before he was able to carry his ideas to completion. In my laboratory several undergraduate students developed our method to the point where we can photograph not only atomic images but also rotationally averaged molecular images of good enough quality to allow bond lengths to be measured with a ruler!

Some of the most beautiful structural investigations carried out in my laboratory were those by Colin Marsden in 1975–76. These uncovered, among other things, a principle of 'altruistic covalent interaction'. Too new to qualify among reminiscences are our current researches on extremely hot molecules, extremely cold molecules (supersonically chilled), and laser-irradiated molecules. At an Austen Symposium in 1974, Lawrence Brockway delivered a delightful reminiscence of early days in electron diffraction entitled, 'When Molecules were Fun'. Today, as we grope our way in new directions, subjecting molecules to indignities never before imposed in gas diffraction studies, we encounter surprises and disappointments, make mistakes leading to useful new tricks, and recapture a sense of wonder and fun. At middle age, gas-phase electron diffraction has matured to an enormously more powerful tool than it was in its youth. It offers so many unique advantages and untried possibilities that a stodgy senility is, as yet, nowhere in sight.

Department of Chemistry
University of Michigan
Ann Arbor, Michigan, U.S.A.

II.25. *Electron Diffraction at the Naval Research Laboratory*

1. OUR INTRODUCTION TO ELECTRON DIFFRACTION

We were both introduced to the field of electron diffraction by gaseous molecules as graduate students at the University of Michigan during the early 1940s. Except for being very fascinated with the field of physical chemistry, we did not enter graduate school with any strong convictions concerning a particular area of specialization for thesis research. Having been attracted to a bright young professor in the Chemistry Department, as a consequence of being students in his excellent physical chemistry course, we sought to carry out our research program under the direction of L. O. Brockway. The photograph on the picture page preceding this section shows us together in 1945, some time after our graduate studies were completed.

In those days, gas electron diffraction research was performed in circumstances that were quite different from today and undoubtedly unfamiliar to more recent students of the subject. High-frequency electronics had not been commercially developed and high potential was achieved by use of greatly overpowered equipment designed for use with X-ray tubes. Electronic control of the high potential, beam current and focusing system was quite excellent, but the circuits were home-designed and home-made. Pump design and vacuum measurement, while adequate and, in principle, similar to present systems, had not begun to attain current sophistication. Perhaps one of the greatest differences in the technical aspects of conducting gas diffraction research was in computing. Molecular intensity functions for comparison with visually estimated data were calculated by making use of tabulations of the function $\sin sr/sr$ for various values of r. The tabulations were in the form of paper strips with each strip associated with a particular value of r and a range of values for s. Calculations were performed in a fashion similar to those for electron densities in X-ray crystal structure analysis that employed Beevers–Lipson strips. With the selection of appropriate strips to correspond to the distances in a model structure and a simple estimate of their coefficients, the computation could proceed. The calculations were first performed with a hand-cranked calculator that was subsequently replaced with the type driven by an electric motor.

We both investigated the structures of numerous molecules by these means in the course of pursuing our graduate research. During this time, Jerome became quite interested in the theoretical paper of Debye (1941) that displayed a deep insight into the effects of molecular motion on the molecular intensity function and their expression in terms of the Fourier transform of this function, the radial distribution function. Although it did not appear that experimental data at the time would be readily applicable to this theory, it was thought that it might afford some insight into the internal rotation occurring in the molecule trifluoroacetic acid, that Jerome was investigating at that time. The paper of Debye made a profound impression and, as it turned out, had a strong influence on our research program after we joined the Naval Research Laboratory.

2. GETTING STARTED AT NRL

We completed our graduate studies in 1943 within some months of each other and went off to Chicago to work on a war-related research project. Some time later we returned to the University of Michigan where Jerome became associated with a research project sponsored by the Naval Research Laboratory. After the war had ended we needed to seek employment. Jerome's project ended and Isabella's teaching position at the University had no potential for future permanence. It is still difficult for a young married couple to find closely located scientific employment, particularly in academia. In 1946, it was all but impossible. The unique opportunity for us both to join a laboratory and pursue a research program in subjects of interest was offered to us by the Naval Research Laboratory and we joined this institution in 1946.

A valuable feature of this Laboratory was its extensive and very fine machine shop facilities. It was therefore not only readily possible to construct the necessary equipment for gas electron diffraction research but also to develop and introduce innovative features. As we entered into gas electron diffraction again after the hiatus that was filled by war-related research, we recalled the potential for advances in this field implied by an ability to implement the various theoretical aspects discussed in the previously mentioned paper by Debye (1941). The decision was soon made not only to construct a state-of-the-art diffraction instrument but also to develop the data reduction process and the theoretical tools in an attempt to fulfill this promise.

A main feature of the diffraction instrument that was built (Karle, 1973), was a double electromagnetic lens system permitting a beam crossover after the first lens, enhancing the quality of the beam passing through the second one before reaching the sample. The apparatus also had the first rotating sector used in the United States and a relatively large sample chamber. The construction proceeded quite readily. In fact, we built two instruments at the same time in order to obtain one suitably leak-proof instrument more quickly. This was achieved during 1947 and had proceeded smoothly except for one difficulty. For about a three month period, we experienced hair-raising electrical breakdowns, not to mention electronic damage, from time to time when the high potential was on for a while. Such nerve-racking breakdowns must be experienced to be appreciated. The strength of the induction effects from the breakdown of a standard X-ray high potential supply is illustrated by the fact that we had power

Fig. 1. Isabella and the electron diffraction instrument completed in 1948. The rotating sector was placed just above the horizontal plate holder whose flat vacuum compartment is seen extended on the left side.

outlets elsewhere in the room, not directly connected to the electronic circuits in the instrument, burn out during a breakdown. John Ainsworth who helped us with the instrument construction shared with us the intense discomforts. We recognized that the electron gun section was the source of the trouble, but were very confused because gauges indicated that the vacuum in the gun section was quite fine, close to 10^{-6} torr, and no changes in the gauges warned of an impending breakdown. The problem was finally solved when we recalled that the glass cylinder that insulated the gun section from the main body of the instrument was installed with a wax that was recommended to us. We replaced the wax with a rubber gasket and the problem disappeared. The wax had the property of sporadically emitting into the vacuum small amounts of gas that were enough to cause an electrical breakdown, but not enough to give a forewarning of such an occurrence in the gauges that we were using. No unusual incidents were encountered after this matter was resolved; the completed instrument is shown in Figure 1.

We made plans to use microphotometry for measuring the scattered electron intensities rather than depend upon visual estimates. The great capabilities of a well-trained eye in eliminating changing backgrounds, pinpointing positions of maxima and minima and approximating their relative levels made this a not-so-obvious decision at the time. We were aware of the use of rotating sectors and microphotometry already initiated in Norway. However, among our concerns was the relatively weak signals that occur with increasing scattering angle and the interference that arises in microphotometry from the granular character of the photographic emulsions. It occurred to us that spinning the photographic plate while tracing it should be quite helpful and we fitted out a Leeds and Northrup microphotometer with a rotating attachment that worked rather well. Before we submitted a little note on this subject for publication (Karle, Hoober and Karle, 1947), we learned that the idea had been pursued more than a decade earlier by Degard and colleagues (Degard, Piérard and van der Grinten, 1935) in Liège, in a laboratory that was under the direction of P. Debye.

Our diffraction instrument was completed before the rotating sector was installed and so the smoothing was initially applied to unsectored photographs. We had occasion to discuss this technique at an American Society for X-ray and Electron Diffraction meeting in early June of 1947 at St Marguerite, Canada, a beautiful mountain resort area some distance from Montreal. Getting there by bus was the most thrilling ride that we ever had. We were traveling on a Sunday and the two-lane highway was very crowded in each direction. Not to be deterred, our bus driver decided to make his own lane up the middle; and it was quite incredible how with horn blasting and foot on the accelerator he was able to sweep both oncoming traffic and also traffic flowing in the same direction away from his path.

P. P. Debye (son of P. Debye) who developed the concept of employing a sector (Debye, 1939) independently but perhaps a little later than Finbak (1938), was also at St Marguerite. Jerome recalls discussing the technique for smoothing microphotometer traces with P. P. Debye. His impression was that we intended to use the smoothed trace to replace the rotating sector in the sense that accurate enough data could be treated entirely mathematically without the aid of mechanical background correction. In a very friendly fashion he suggested that it would still be a good idea to employ a sector in addition, an opinion with which we indeed concurred. In fact, we were already experimenting with various sector designs. We settled on a design in which the rotor holding the sector rotated freely on stainless steel balls after being brought up to speed by a motor placed in the vacuum chamber and turned off during photographing to avoid spurious electrical effects. We later used a design in which the motor was outside the vacuum chamber.

The visit to St. Marguerite was somewhat early in the season and the water in the swimming pool was so cold that all we could manage was to dive in and quickly get out a few times. It was still enjoyable enough to do this every day, even on the morning that we left. We returned on a flight that made innumerable stops between Montreal and Washington. It made, in addition, an unscheduled stop on the Canadian–U.S. border in an attempt to check on suspected smuggling operations. We were all treated with great suspicion. The fact, for example, that Isabella was chatting with another colleague from NRL who was misidentified by the stewardess as her husband was treated as evidence that something sinister and deceptive was surely going on. We had to leave the plane and be present for a thorough search of our luggage, done with a general disregard for our belongings. The one satisfaction that we had from this unpleasantry came from witnessing the look on the inspector's face when he plunged his hand into Isabella's bathing cap where we had packed our wet bathing suits.

3. PROBLEMS TO BE OVERCOME

If we were to fulfill the hope of not only obtaining interatomic distances and bond angles of enhanced accuracy but also information concerning the average internal molecular motion, problems had to be overcome. If successful, it would be possible to obtain a radial distribution function that would represent accurately the probability distributions for the interatomic distances. The problems to be solved were:

1. The molecular intensity function, whose Fourier transform would produce the radial distribution function of interest, had to be obtained from the total intensity of scattering in such a way that the interference functions ($\sin sr_{ij}/sr_{ij}$, for example, where s is a function of the scattering angle and r_{ij} is the distance between the ith and jth atoms) would have coefficients that arise from essentially point atoms, i.e., the coefficients would be, ideally, independent of the variable s.

2. The experimental molecular intensity function is known only over a limited angle range, whereas the calculation of the Fourier transform of this function requires it to be known over an infinite range or, at least, to a large enough value of s so that the omitted region is negligible. This is ordinarily not attainable so that the calculation of the Fourier transform had to be modified to avoid 'termination errors'.

3. In the course of extracting the molecular intensity from the total intensity function, it is necessary to generate a suitable background scattering function. There were difficulties with the theory for the background scattering as well as added difficulties arising from the recording of the total intensities on photographic plates. As a consequence, adjustments were required for the theoretical background function in order to avoid serious errors in the derived molecular intensity function.

4. The microphotometry of the photographic images of the diffraction patterns had to be optimized. As mentioned previously, the rotating sector and plate spinning helped considerably. In addition, the densitometry and transformation of photographic density to intensity had to be made as accurate as possible.

These various problems were overcome. For example, by dividing the total intensity of scattering by an appropriate background function, a molecular intensity function could be obtained whose coefficients varied little over the experimental range of s (Karle and Karle, 1947 and 1949). A suitable damping function introduced by Degard (1937) and Schomaker (1939) took care of termination errors and a further analysis based on the known character of the probability distributions for interatomic distances took account of the spreading effect of the damping function on the probability distribution (Karle and Karle, 1950a).

The solution to the problem of developing a suitable background scattering function was facilitated by the use of the rotating sector. Adjustments to optimize the shape of this function were based on a principle that has found extended application to other areas of structure research. The principle concerns the introduction of constraints into the analysis of diffraction data to satisfy the non-negativity of certain structural characteristics. The non-negativity of the Fourier transform of the molecular intensity function representing modified probability distributions was used to produce an accurate background function.

It was therefore possible by various means to solve the four problems listed above and develop a molecular intensity curve whose Fourier transform could be interpreted in terms of probability distributions for the distances between pairs of atoms in a molecule. Along with this came enhanced accuracy that permitted not only the determination of interatomic distances, but also the possibility of investigating the internal motion of the molecules. The procedures for performing such analyses were described in 1949 and 1950 (Karle and Karle, 1949, 1950a, 1950b).

4. EARLY STUDIES

In the course of these developments, we began to make use of IBM punch card machines for our computing. The operation of addition proceeded with tolerable speed but multiplication in the early machines proceeded at the rate of about two multiplications per minute, too slow to be useful. Then about 1950, IBM came out with a multiplier that could perform 100 multiplications per minute. For those of us who had initially performed calculations with a crank-handle computer, computing paradise had indeed arrived.

Confidence in the increased accuracy with which interatomic distances could be determined, as well as the possibility of obtaining reliable values for root mean squared amplitudes of vibration, was enhanced by satisfactory comparisons with results from infrared and microwave spectroscopy. An incident that occurred soon after our first work using the new analytical techniques was published provided a poignant example of the value of such comparisons. One of the first molecules investigated by the new analytical techniques was CO_2 (Karle and Karle, 1949). We reported a value of 0.040 ± 0.007 Å for the rms amplitude for the $O \cdots O$ distance. However we had calculated a value of 0.029

Å from spectroscopic data which indicated a discrepancy somewhat larger than our reported limit of error for the electron diffraction experiment. It was soon shown by Yonezo Morino (1950) that the formula that we were using was in error and that after correction the computed value for spectroscopic data was 0.041 Å, in very good agreement with the result from the electron diffraction experiment.

The study of CO_2 (Karle and Karle, 1949) also provided a first experimental insight into the concept of 'shrinkage'. The value of the $O \cdots O$ distance was observed to be smaller than twice the value of the C—O distance. Owing to the early stage of the development of the new techniques, no particular notice was taken of the result in the publication. However, it did catch the attention of Lawrence Bartell and Jerome recalls discussing the matter with him during a visit to Ann Arbor. The observed shrinkage can be explained in terms of a bending mode such that in the course of the vibration, the $O \cdots O$ distance is equal to the sum of the C—O distances only when the molecule passes through the equilibrium position. It is otherwise always smaller. In a theoretical treatment of internal rotation about a single bond (Karle and Hauptman, 1950a) that soon followed the investigation of CO_2, it was noted that motions of this type could cause large shrinkage effects for *trans* distances. An experimental application of the theory was soon made to 1,2-dichloroethane (Ainsworth and Karle, 1952).

Internal rotation has remained a problem of interest theoretically and experimentally over the years and has been extended to investigations of double rotors (Karle, 1980). Theoretical studies coupled with electron diffraction experiments afford rough insights into potential barriers that hinder internal rotational motion.

5. IMPLICATIONS

Our work in electron diffraction of gases provided the foundation concepts that led us into other fields of structural research of which the investigation of crystalline materials has occupied the largest part of our time. The broad concept concerns the observed importance in structure determination of imposed mathematical and physical constraints on the analytic procedure. The possibility of making advances in gas diffraction was realized because such constraints were appropriately implemented. Perhaps the most fruitful one in its implications to other fields of structure analysis was the non-negativity criterion. It arose in the attempt to obtain an appropriate experimental function for the background intensity on the basis that the resulting radial distribution function had to be non-negative since it was interpretable in terms of probability functions.

Non-Negativity

The non-negativity criterion was impressively useful in analyzing gas diffraction data to such an extent that it motivated the search for other areas of application.

One such area that attracted our interest was atomic structure in which there was the possibility of obtaining more accurate electron distributions around atoms on the basis that the electron density is a non-negative function. Since the electron density is expressible as a Fourier transform of the diffraction data, the possibility exists that the non-negativity criterion could be used to improve the experimental data. This was done by developing an extrapolation procedure for the data, which are terminated at some finite scattering angle, in such a way that the Fourier transform of the measured and extrapolated data combined is non-negative. A discussion of this matter and illustrative calculations were published (Hauptman and Karle, 1950). This was followed by a very fine electron diffraction investigation of the electron distribution in argon by Lawrence Bartell (1953) that made use of this non-negativity criterion.

The area that has found the most extensive application of the non-negativity criterion is crystal structure analysis. With the phase problem in mind, it seemed quite appropriate to consider the possibility that there would be constraints imposed on the magnitudes and phases of the structure factors by the fact that the electron density in a crystal had to be a non-negative function. Such constraints do indeed occur and are given in terms of an infinite sequence of determinantal inequalities that contain the structure factors as elements (Karle and Hauptman, 1950b). The implications of these inequalities for crystallography were presented in collaboration with Herbert Hauptman who was a colleague in our laboratory at that time. The third order inequality and its probabilistic characteristics have provided the main mathematical tools for phase determination procedures (Karle, 1978b). The higher order inequalities associated with their general joint probability distributions (Karle, 1978a) are of current interest as a potential means for facilitating further advances in structure determination and refinement. It is most fascinating that the procedures used for a very large number of routinely performed crystal structure determinations had their origins in the fact that there was a problem concerning an appropriate background intensity function in the analysis of gas electron diffraction patterns which was solved by means of the non-negativity criterion.

Additional Areas of Interest

Boundary lubrication on surfaces that rub against each other can be effected by the deposition of monomolecular films of long-chain aliphatic compounds having a polar end-group to form the attachment. Curiously, although these films project the nonpolar ends of the molecules into the bulk of a lubricating oil, the films are oleophobic. In an investigation that was a continuation of wartime research in Lawrence Brockway's laboratory at the University of Michigan, the energy of binding of the molecules in films was evaluated in a dynamic experiment involving electron diffraction. The films were heated and the rate of evaporation was determined by measuring the rate of disappearance of the diffraction pattern. From the measurements, energy estimates were made that gave additional insights into the character of the films and the mechanism of evaporation (Karle, 1949).

Some investigations were made at the end of the 1950s and into the '60s on the inelastic scattering of electrons from molecules, in collaboration with our colleague, David Swick. Some experimental and theoretical insights were developed (Swick and Karle, 1961; Karle, 1961), but the opportunity to pursue this field of study did not materialize. Russell Bonham came to our laboratory for a year of postdoctoral work at that time and soon interested himself in theoretical studies of inelastic electron scattering from atoms (Bonham, 1962) that developed into a long series of experimental and theoretical investigations that continue to the present day. Much of this is documented in his book written with Manfred Fink concerning the scattering of high energy electrons (Bonham and Fink, 1974).

Another research area that has been under study for some years in our laboratory has involved the use of X-ray and neutron diffraction, as well as electron diffraction. This area concerns the structures of noncrystalline solids, often called amorphous materials. The term noncrystalline is in our view much to be preferred over amorphous since a key insight that our studies have provided for us is the considerable extent to which numerous so-called amorphous materials are, in fact, fairly well ordered (Karle and Konnert, 1974; D'Antonio, Moore, Konnert and Karle, 1977). We have termed this ordering, which, of course, stops short of being periodic, 'structural ordering'.

The development of our analytical procedures in the studies of noncrystalline solids owed much to our experience in gas electron diffraction. There was a considerable carry-over of basic techniques and advantage was once again taken of the possibility of introducing mathematical and physical constraints into the analysis (Karle, 1977).

Structure determination by diffraction has, in general, benefited greatly in the past from opportunities to introduce into the analysis additional information obtained from physical considerations and other sources such as related structural information or other techniques such as spectroscopy. This can be expected to continue and be considerably enhanced in the future. An increasing role will be played by calculations in theoretical chemistry. Perhaps we are on the verge of a great blossoming in this area.

6. POSTSCRIPT

Our early work in electron diffraction of gases and crystal structure analysis was done in the best of atmospheres. As young people we were well-supported and had the freedom to set our course and find our own way through our research problems. Very importantly, the support and the course of the work were not subject to the operational constraints currently imposed on scientific program managers or the opinions of fellow-scientists that govern the present funding systems. Our work in gas electron diffraction met with considerable opposition when we attempted to publish it and it required support from Lawrence Brockway to overcome the influence of the unfavorable comments. This was mild, however, compared to the unfavorable reactions generated by our work on crystal structure analysis. The negative responses were not directed toward the more superficial aspects of the research results but rather the fundamental concepts that governed our entire approach to structure analysis. This has made us wonder many times how our research program would have fared in today's atmosphere. Such a question actually has no specific answer because it is possible to conjure up an infinity of scenarios. However, the present weakened support for science and young people in science is certainly not conducive to the establishment of creative research programs. In particular, with so much of science support dependent upon the reviews and recommendations of those of presumed expertise, one cannot help but wonder what proportion of innovative research programs are suppressed because scientific peers do not have the vision or insight to comprehend their possibilities.

Laboratory for the Structure of Matter
Naval Research Laboratory
Washington, D.C., U.S.A.

REFERENCES

Ainsworth, J. and Karle, J.: 1952, *J. Chem. Phys.* **20**, 425.

Bartell, L.S.: 1953, *Phys. Rev.* **90**, 833.

Bonham, R. A.: 1962, *J. Chem. Phys.* **36**, 3260.

Bonham, R. A. and Fink, M.: 1974, *High Energy Electron Scattering*, ACS Monograph 169. Van Nostrand Reinhold, New York.

D'Antonio, P., Moore, P., Konnert, J. H., and Karle, J.: 1977, *Amorphous Materials*, in Proceedings of Symposium Commemorating Fifty Years of Electron Diffraction (ed. L. O. Brockway) *Transactions of the American Crystallographic Association* **13**, 43 (Polycrystal Book Service, Pittsburgh, Pa.).

Debye, P.: 1941, *J. Chem. Phys.* **9**, 55.

Debye, P. P.: 1939, *Physik. Zeits.* **40**, 66 and 404.

Degard, C.: 1937, *Bull. Soc. Roy. Sci. Liège* **12**, 383.

Degard, C., Piérard, J., and van der Grinten, W.: 1935, *Nature* **136**, 143.

Finbak, C.: 1938, *Avhandl Norske Videnskaps* – Acad. Oslo, *I. Mat. – Naturv. Kl.* No. **13**.

Hauptman, H., and Karle, J.: 1950, *Phys. Rev.* **77**, 491.

Karle, J.: 1949, *J. Chem. Phys.* **17**, 500.

Karle, J.: 1961, *J. Chem. Phys.* **35**, 963.

Karle, J.: 1973, *Electron Diffraction*, in *Determination of Organic Structures by Physical Methods* (eds. F. C. Nachod and J. J. Zuckerman) Academic Press, New York.

Karle, J.: 1977, *Proc. Natl. Acad. Sci. USA* **74**, 4707.

Karle, J.: 1978a, *Proc. Natl. Acad. Sci. USA* **75**, 2545.

Karle, J.: 1978b, *Proc. Natl. Acad. Sci. USA* **75**, 3540.

Karle, J.: 1980, *Internal Rotation and Electron Diffraction*, in *Diffraction Studies on Non-Crystalline Substances* (eds. I. Hargittai and W. J. Orville-Thomas) Akadémiai Kiadó, *In press*.

Karle, J. and Hauptman, H.: 1950a, *Acta Crystallogr.* **3**, 181.

Karle, J. and Hauptman, H.: 1950b, *J. Chem. Phys.* **18**, 875.

Karle, I. L., Hoober, D., and Karle, J.: 1947, *J. Chem. Phys.* **15**, 765.

Karle, J. and Karle, I. L.: 1947, *J. Chem. Phys.* **15**, 764.

Karle, I. L. and Karle, J.: 1949, *J. Chem. Phys.* **17**, 1052.

Karle, J. and Karle, I. L.: 1950a, *J. Chem. Phys.* **18**, 957.

Karle, I. L. and Karle, J.: 1950b, *J. Chem. Phys.* **18**, 963.

Karle, J. and Konnert, J. H.: 1974, *Analysis of Diffraction from Amorphous Materials and Applications*, in

Proceedings of Symposium on Liquids and Amorphous Materials (ed. A. Bienenstock) Transactions of the American Crystallographic Association **10**, 29 (Polycrystal Book Service, Pittsburgh, Pa).

Morino, Y.: 1950, *J. Chem. Phys.* **18**, 395.

Schomaker, V.: 1939, presented to the April meeting of the American Chemical Society, Baltimore.

Swick, D. A. and Karle, J.: 1961, *J. Chem. Phys.* **35**, 2257.

GORO HONJO

II.26. My Thirty-Five Years in High Energy
Electron Diffraction

I began electron diffraction in 1943 in the naval technology research institute after graduating from the University of Tokyo. I learned the technique from Professor R. Uyeda, who sometimes visited the institute as a consultant. My job was to apply the technique to problems of corrosion in naval engines. Although wet corrosion was important, I started from dry corrosion or oxidation of iron and copper in air following the report of Miyake (1936). I got some new results but was soon involved in the confusion and general upheaval that followed our involvement in World War II.

In 1946 I got a post as research fellow in Dr S. Miyake's laboratory in the Kobayashi Institute of Physical Research. Besides continuing the study of oxidation (Honjo, 1949; 1953) I began, with the use of a simple camera of the third design of Uyeda (Uyeda, this volume), the study of fine-structure due to refraction, in Debye–Scherrer rings. By preparing a fine electron source formed by a very small amount of mixed strontium and barium carbonates on the tip of a platinum hair pin filament, I could observe not only the splitting of rings but also multiplets of reflections from individual particles. Their configurations corresponded nicely to the cubic habit of magnesium smoke particles, and also to the octahedral habit of antimony smoke particles (Honjo, 1947). The observation of the latter particles was made at the suggestion of Miyake on the basis of his early work on the same oxide formed on stibnite (Miyake, 1938). This was a prominent example of submicroscopic crystal morphology found by an analysis of the fine-structure of electron diffraction spots. Replica and reflection electron microscopy by Watanabe (1957) showed the habit very clearly, together with details of size and location of the oxide particles on the substrate.

In the Kobayasi Institute I made another observation, finding an anomalous surface structure, the $\sqrt{3} \times \sqrt{3}R$ structure in the recent terminology of surfaces, of silicon carbide (Honjo 1949). I learned the ABC of crystallography in silicon carbide and found by X-rays a very long stacking period of 594 SiC layers (Honjo, Miyake and Tomita, 1950).

In 1949 I moved to the Tokyo Institute of Technology, following Professor S. Miyake. With K. Mihama, one of our first students, I continued the work on fine structure for which Sturkey had suggested the double refraction effect due to the dynamical diffraction in 1948. We observed multiple doublets due to the refraction from individual particles as well as from the particles of a fibrous orientation (Honjo and Mihama 1951). We noticed that the outer spot of each doublet was always distinctly weaker than the inner spot (Honjo and Mihama, 1954). This was, I believe, the first observation of the effect of periodic absorption in crystals, the counterpart of the Borrmann effect in X-rays.

In the 1950's, I tried various applications of electron diffraction with my colleagues, N. Kitamura, K. Mihama, K. Shimaoka and J. Harada. We applied Boersch–le Poole's method of selected area diffraction to a study of clay minerals and observed 'rotation patterns' from individual stationary crystallites of kaolin and halloysite as evidence for their tubular morphology (Honjo and Mihama, 1954). We constructed liquid nitrogen-cooled specimen holders for an electron diffraction unit and an electron microscope, analysed the hydrogen positions in cubic ice (Honjo and Shimaoka, 1957), and found a low temperature tetragonal phase of hydrogen sulfide (Kitamura, Kashiwase, Harada and Honjo, 1961). We also observed whisker growth of *in situ*-deposited mercury crystals and polygonization of solidification of mercury droplets (1956). With Watanabe I constructed an improved liquid nitrogen stage for electron microscopy and, since the biological sample remained undamaged at low temperature (Honjo and Watanabe, 1958), found an unknown structure of Valonia microfibril. In the meantime, the Russian group reported many results of crystal structure analyses by electron diffraction taking advantage of the fact that electrons are relatively more strongly scattered by light atoms than are X-rays. In Japan where the dynamical pest was prevailing (Uyeda, this volume), it was impossible to undertake such an analysis without finding some way to manage or eliminate the dynamical effect or the primary extinction in the observed intensities. Two ways occurred to me. One was to evaluate the extinction by taking local averages of intensities following the method of Wilson (1949), devised to find absolute intensities of X-ray diffraction. This was applied by S. Nagakura to his analysis of the structure of nickel carbides (Nagakura, 1957). The second was to extrapolate observed intensities to zero wave length. Kitamura measured the change of intensities of Debye rings from evaporated aluminium particles up to 300 keV with use of an electron microscope with a van de Graaff accelerator in the Hitachi Central Laboratory, the first very high voltage unit in Japan, and found that the intensities tended to the kinematical values at zero wave length in accordance with the two-wave theory (Honjo and Kitamura, 1957). However, it was soon pointed out by Miyake (1959) that the many-wave effect becomes important at high voltage.

In the course of these studies, which I made with a purpose to extend the field of application of electron diffraction, I repeatedly felt sad: eventually being confronted with the limited capability of this method in comparison with X-ray and neutron diffraction. The one possible breakthrough I noticed was to take advantage of combination with electron microscopy. Recently it has become well-recognized that electron microscopes for crystallographers are nothing but tools to see electron diffraction effects in real space. It took a long time until this was recognised. As a matter of fact, early electron microscopes had no specimen goniometer. I stressed its importance but could not attract electron micro-scopists and engineers of electron microscopes. Then I began, in 1956 with

Kitamura, the construction of a special electron microscope having a 'eucentric' goniometer which could be inserted horizontally just above the top of objective pole pieces of an unusual shape. It took a few years to get started. In the meantime, M. Watanabe and coworkers in JEOL devised a lens with a wide upper bore (20 mm) of a pretty high quality and constructed, on my advice, a convenient goniometer. I got a new 150 keV electron microscope incorporating these systems by a project called 'An Electron Microscope Dedicated to Crystals' in 1961, the year when the International Conference on Magnetism and Crystals was held in Kyoto. Afterwards we constructed a high temperature goniometer (up to 900 °C) and a liquid helium-cooled low temperature goniometer for this microscope (Honjo, Yatsuhashi, Kobayashi and Yagi, 1970). Electron microscopies of metals and alloys, pioneered by Hirsch's group in 1954, had been progressing in various places in 1961. Therefore, I directed my attention to nonmetallic substances.

The first substance we studied was barium titanate, a typical ferroelectric. Känzig (1955) had reported that surface layers of about 100 Å thickness of this substance had a tetragonal deformation about 50% larger than that of the bulk crystal (1954–5). I suspected the general validity of the conclusion they claimed from line profile analyses of Debye rings of powdered samples. M. Tanaka, who was a graduate student, succeeded in preparing electron transparent thin single crystal specimens. Their transmission electron diffraction patterns showed spots of systematic splittings in accordance with the bulk tetragonality and their transmission electron micrographs showed domains with the 90° boundaries (Tanaka, Kitamura and Honjo, 1962). Domains with the antiparallel polarizations were also seen with a strong contrast in dark field images (Tanaka and Honjo, 1964). I believe that this was the first observation on the failure of Friedel's law of transmitted electrons. Fujimoto (1959) had shown by a many-wave theory that the failure can take place in transmission. Even by the two-wave theory the failure was expected to occur by the periodic absorption effect. The observed contrast, however, was much stronger than those calculated by these theories. The same kind of contrast was observed in the lower temperature orthorhombic phase, but it was not observed in the lowest temperature rhombohedral phase. Tanaka (1975), afterwards at Tohoku University, found a peculiar temperature dependence of the contrast in the ferroelectric phase of lead titanate. Then, he suggested that the contrast may be due to polarity not in the static structure but in the dynamic structure of the crystals (Tanaka, 1975). But no clear explanation has been given as yet of this interesting phenomenon.

Recently, N. Yamamoto, K. Yagi and myself observed domains of antiparallel polarizations and of antiphases in gadolinium molybdate ($Ga_2(MoO_4)_3$) crystals and analyzed the structure of the antiphase boundaries which play important roles in the ferroelectric and ferroelastic phase transition of this substance (Yamamoto, Yagi and Honjo, 1974, 1976, 1977). They also found a new superlattice structure of strontium tantalate ($Sr_2Ta_2O_7$) and an incommensurate superlattice structure of strontium niobate ($Sr_2Nb_2O_7$), both of which had been overlooked in X-ray analyses on account of the very weak intensities of the relevant reflections (Yamamoto, Yagi, Honjo, Kimura and Kawamura, 1980).

Thus, I found a fairly systematic field of application for our high energy electrons. Here, however, we met again a serious limitation due to the fact that we can apply the electrons only to substances which can stand up to the irradiation in vacuum. The above substances belong to one group of dielectrics which undergo the displacement type of phase transition. There is another important group of dielectric substances which make the order–disorder type of phase transition, almost all of which cannot stand the irradiation of high energy electrons.

The next substances we studied were germanium and silicon. They excited me by showing beautiful electron diffraction patterns of streaks which transform continuously from honeycomb-like hexagonal networks, for incidences around the [111] axis, to square networks, for incidences around the [001] axis (Honjo, 1962; Honjo, Kodera and Kitamura, 1964). Although less sharp and distinct, similar diffuse streaks were also observed from barium titanate, fcc metals and alkali halides. Diffuse streaks of the same character were also observed in monochromatic X-ray diffraction patterns of silicon (Kodera, Kitamura and Honjo, 1963). Then, I knew that the diffuse streaks were of the same kind as those observed by Lonsdale for ice and benzil in her early works on X-ray diffuse scattering (1941–3) and also observed by Bouman in the X-ray pattern of β-tin in 1946. I noticed that the diffuse streaks corresponded to intensity distributions on parallel planes in the reciprocal lattices which are perpendicular to the directions of chains of nearest neighbour atoms, and considered that they were caused by parts of the chains which displaced incoherently with respect to the matrix (Honjo, Kodera and Kitamura, 1964). Professor Guinier, who observed the X-ray streaks of silicon, gave a similar explanation. However, Komatsu (1964) pointed out that the streaks were due to soft mode vibrations in the crystals, the theory of which had been developed in 1960 by Cochrane and Anderson in connection with the displacive type phase transitions. Then, I observed with Tanaka and Harada that the diffuse streaks in three perpendicular directions appearing from cubic barium titanate vanished one after another, corresponding to the freezing-in of the soft modes successively by a series of transitions to the lower temperature phases (Harada, Tanaka and Honjo, 1966). From the intensities of streaks of different orders, we found that titanium atoms vibrate in antiphase with respect to the framework of barium and oxygen atoms (Harada and Honjo, 1965).

The third substance we studied was sodium chloride. Electron microscopy of this typical ionic crystal had already been undertaken in various places. In these works specimens were prepared outside the electron microscope. Surfaces of such specimens attacked by moisture were quickly damaged by electron irradiation. Yagi, who began the study as a graduate student, found that the surface layers were cleaved away by electron beam flashing and succeeded in observing the inner layers of the crystal. Professor Miyake had carried out, with a collaborator, an interesting work by X-ray diffraction to analyse the precipitation process of calcium chloride dissolved in sodium chloride (Miyake,

Kohra and Takagi, 1954). This made a nice simulation of precipitation processes in alloys which were one of the leading topics of X-ray diffraction at that time. Yagi, who pursued the process by electron microscopy, showed finer and more exact details of the process (Yagi and Honjo, 1967). He also observed *in situ* the transformation of straight dislocations into helical ones induced by electron irradiation (Yagi and Honjo, 1964). This was a simulation, made in advance in the ionic crystal, of the electron irradiation effects in metals and alloys which later became a topic of very-high-energy electron microscopy.

Yagi's method of electron beam flashing gives thin crystals with clean surfaces which are very appropriate as substrate crystals for *in situ* electron microscopy of the epitaxy of thin films. The *in situ* observations pioneered by Bassett and Pashley *et al.* in the early 1960s marked an epoch in the study of epitaxial thin films which had long been a favorite subject of electron diffractionists. I noticed, however, that their thin molybdenite and graphite substrates were prepared by cleavage with adhesive tape, and rinsed by solvents. I suspected the cleanliness of these substrates, because it was a common practice in Japan since Professor Kikuchi's time to thin down crystals of layer structures with use of a clean and sharp needle to avoid contamination. I began in 1966 (Honjo, Shinozaki and Sato, 1966) *in situ* observations of thin magnesium oxide crystals prepared by Yagi's method during my stay in Dr H. Sato's Laboratory at the Ford Motor Company. While there, I also collaborated with Sato in his study of long period stacking orders in alloys (Sato, Toth and Honjo, 1967), begun after his success in long period antiphase structures (1961–5). My *in situ* study has since made good progress due to the active collaborations of Yagi and K. Takayanagi, and the excellent technical assistance of K. Kobayashi. We developed some other methods for preparing thin and clean substrate crystals and also a versatile ultra-high-vacuum (UHV) 100 keV electron microscope (Takayanagi, Kobayashi, Yagi and Honjo, 1974; 1978).

We were not the first in UHV electron microscopes. In the late 1960s a number were constructed in various places, mostly for *in situ* deposition studies. Only a little was reported, however, on their applications. This was due to the fact that too many of the microscope amenities, e.g., specimen tilting and specimen exchange facilities, were sacrificed for the sake of UHV conditions. Also, they lacked techniques to prepare thin and clean substrates. Our electron microscope can work with an ambient vacuum of $10^{-8} \sim 10^{-10}$ torr around specimens, retaining the full electron optical performance and operational amenities of a commercial unit: in routine operation, the substrate crystal, two kinds of materials to be evaporated and 50 photographic films can be reloaded within 10 min.

Thus, we made *in situ* observations on growth processes, in clean conditions, of a large number of systems of deposit and substrate materials. The results showed many aspects of the growth processes which had not been known, but which had been inferred misleadingly from *in situ* observations in dirty conditions, and from post-deposition observations for a very limited class of systems (Honjo and Yagi, 1980).

Furthermore, the UHV microscope has given us new approaches to surfaces and thin films at monatomic layer levels both in the transmission (TEM) and the reflection (REM) modes. The TEM works started from our own research interest in preparing thin crystals *in situ* in the UHV microscope. Takayanagi first noticed extra spots in diffraction patterns and the corresponding 63 Å spaced stripes in TEM images of (111) gold platelets. During Takayanagi's visit to the Fritz-Haber-Institute in Berlin, Yagi continued the observations in Tokyo with graduate students, Y. Tanishiro, N. Osakabe and H. Kanamori, and proved that the stripes were a reconstructed surface structure (Yagi, Takayanagi, Kobayashi, Osakabe, Tanishiro and Honjo, 1979). They also observed monolayer nuclei of palladium deposits on (111) silver and of silver deposits on (111) gold. Takayanagi and co-workers in the Fritz-Haber-Institute made a comprehensive study of growth of a few monolayers of lead on (111) silver with use of low energy electron diffraction (LEED), Auger electron spectroscopy (AES) and some other surface techniques (Takayanagi, Kolb, Kambe and Lehmpfhul, 1980). Takayanagi succeeded recently in observing finer details of the process by TEM.

The REM works were motivated by S. Ino's successful work with UHV reflection high energy electron diffraction (RHEED), on the cleaning process of (111) silicon surface, the (7×7) structure on the clean surface, and its transformations to various structures by heating and by deposition of a few monolayers of gold (Ogawa, this volume). The first REM observation was made by Yagi and Kobayashi on clean (001) surface of magnesium oxide prepared *in situ* by electron beam flashing, but in this case cleaving perpendicular to a thin crystal. Yagi trained Osakabe and Tanishiro in surface studies with a home-made RHEED unit, and they soon succeeded in REM observations on clean and gold and silver deposited (111) and (001) silicon surfaces (Osakabe, Tanishiro, Yagi and Honjo, 1980a; 1980b).

TEM and REM show fine microtopographs of surfaces which cannot be shown by the other techniques (Yagi, Takayanagi and Honjo, 1980). We may say that they opened our eyes to 'see' surfaces.

Achieving these ends just around the year (1979) of my retirement from the Tokyo Institute of Technology, I am now very happy with all that was achieved, having regard to the following circumstances:

Professor Germer, who deeply impressed me by his pioneering work in LEED, which was reported in the 1961 Conference in Kyoto, visited us and asked me, after seeing my new 150 keV electron microscope, "Why don't you follow me instead of spending so much money on such a complicated and dirty machine!". At that time we had no UHV technique in Japan but I had some prospects of progress in our microscope work. Soon after that, however, I saw the great progress made by LEED and other techniques in studies of surfaces and thin films at monatomic layer levels, at which our dirty, high-energy electrons could do nothing at all. Now I can say confidently that electrons in UHV are able to add something at clean levels of vacuum. I spent a lot of money on the UHV microscope. To get this money, I repeatedly stressed the importance of epitaxy in electronic device technology. After my retirement I actually came into working contact with a device manufacturer, and realized then that our *previous*

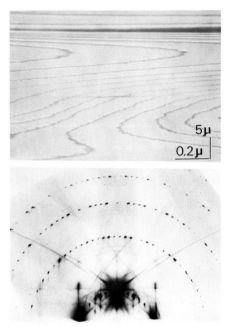

Fig. 1. A REM image of a clean (111) silicon surface. The corresponding RHEED pattern shows the 7 × 7 structure reflections. The image is foreshortened as indicated by the magnification scales. The wavy lines are atomic steps. The horizontal dark line at the center is the image of a screw dislocation, which indicates that the steps are monatomic steps.

studies on epitaxy (Honjo and Yagi, 1980), had nothing to do with the device technology. Present work on TEM and REM, on surfaces, however, will make certain contributions not only to the technology but also to studies of catalysis. Chemical reactions on solid surfaces, one field in which I began my electron diffraction, will be studied starting from clean surfaces and will reveal details of monatomic layer levels and yield electron micrographs.

Surfaces and thin films are the fields in which X-rays and neutrons cannot compete with electrons.

REFERENCES

Fujimoto, F.: 1959, *J. Phys. Soc. Japan* **14**, 1558.
Harada, J. and Honjo, G.: 1965, *Int. Conf. El. Diff. and Cryst. Def.* Melbourne, Pergamon IM-1.
Harada, J., Tanaka, M., and Honjo, G.: 1966, *J. Phys. Soc. Japan* **21**, 968.
Honjo, G.: 1947, *J. Phys. Soc. Japan* **2**, 133.
Honjo, G.: 1949, *J. Phys. Soc. Japan* **4**, 320, 352.
Honjo, G.: 1953, *J. Phys. Soc. Japan* **8**, 113.
Honjo, G.: 1962, *J. Phys. Soc. Japan* **17**, Suppl. II, 277.
Honjo, G. and Kitamura, N.: 1957, *Acta Cryst.* **10**, 601.
Honjo, G., Kodera, S., and Kitamura, N.: 1964, *J. Phys. Soc. Japan* **19**, 351.
Honjo, G. and Mihama, K.: 1951, *Acta Cryst.* **4**, 282.

Honjo, G. and Mihama, K.: 1954, *J. Phys. Soc. Japan* **9**, 184.

Honjo, G. and Mihama, K.: 1954, *Acta Cryst.* **7**, 511.

Honjo, G., Miyake, S., and Tomita, T.: 1950, *Acta Cryst.* **3**, 396.

Honjo, G. and Shimaoka, K.: 1957, *Acta Cryst.* **10**, 710.

Honjo, G., Shinozaki, S., and Sato, H.: 1966, *Appl. Phys. Letters* **9**, 23.

Honjo, G. and Watanabe, M.: 1958, *Nature* **181**, 326.

Honjo, G. and Yagi, K.: 1980, *Current Topics in Materials Science* (ed. E. Kaldis) **6**, 195.

Honjo, G., Yatsuhashi, K., Kobayashi, K., and Yagi, K.: 1970, *Proc. 7th Int. Cong. EM*, Grenoble, 109.

Känzig, W.: 1955, *Phys. Rev.* **98**, 549.

Kitamura, N., Kashiwase, Y., Harada, J., and Honjo, G.: 1961, *Acta Cryst.* **14**, 687.

Kodera, S., Kitamura, N., and Honjo, G.: 1963, *J. Phys. Soc. Japan* **18**, 317.

Komatsu, K.: 1964, *J. Phys. Soc. Japan* **19**, 1243.

Miyake, S.: 1936, *Nature* **137**, 457.

Miyake, S.: 1938, *Sci. Pap. Inst. Phys. Chem. Res. Tokyo* **34**, 565.

Miyake, S.: 1959, *J. Phys. Soc. Japan*, **10**, 1347.

Miyake, S., Kohra, K., and Takagi, M.: 1954, *Acta Cryst.* **7**, 393.

Nagakura, S.: 1957, *Acta Cryst.* **10**, 601.

Osakabe, N., Tanishiro, Y., Yagi, K., and Honjo, G.: 1980a, *Surface Sci.* **97**, 393.

Osakabe, N., Tanishiro, Y., Yagi, K., and Honjo, G.: 1980b, *Japan J. Appl. Phys.* **19**, 309.

Sato, H., Toth, R. S., and Honjo, G.: 1967, *J. Phys. Chem. Solids* **28**, 137.

Takayanagi, K., Kobayashi, K., Yagi, K., and Honjo, G.: 1974, *Japan J. Appl. Phys.* Suppl. 2, PT 1, 184.; (1978) *J. Phys.* **E11**, 441.

Takayanagi, K., Kolb, D., Kambe, K., and Lehmpfuhl, G.: 1980, *Surface Sci.* **100**, 407.

Tanaka, M.: 1975, *Acta Cryst.* **A31**, 59.

Tanaka, M. and Honjo, G.: 1964. *J. Phys. Soc. Japan* **19**, 954.

Tanaka, M., Kitamura, N., and Honjo, G.: 1962, *J. Phys. Soc. Japan* **17**, 1197.

Watanabe, M.: 1957, *J. Phys. Soc. Japan* **12**, 874.

Wilson, A. J. C.: 1949, *Acta Cryst.* **2**, 318.

Yagi, K. and Honjo, G.: 1964, *J. Phys. Soc. Japan* **19**, 1892.

Yagi, K. and Honjo, G.: 1967, *J. Phys. Soc. Japan* **22**, 610.

Yagi, K., Takayanagi, K., and Honjo, G.: 1980, in preparation.

Yagi, K., Takayanagi, K., Kobayashi, K., Osakabe, N., Tanishiro, Y., and Honjo, G.: 1979, *Surface Sci.* **86**, 174.

Yamamoto, M., Yagi, K., and Honjo, G.: 1974, *Phil. Mag.* **30**, 1161.

Yamamoto, M., Yagi, K., and Honjo, G.: 1976, *J. Phys. Soc. Japan* **40**, 601.

Yamamoto, M., Yagi, K., and Honjo, G.: 1977, *Phys. Stat. Sol.* **41**, 523; **A42**, 257; **A44**, 147.

Yamamoto, M., Yagi, K., Honjo, G., Kimura, M., and Kawamura, R.: 1980, *J. Phys. Soc. Japan* **48**, 185.

H. HASHIMOTO AND K. TANAKA

II. 27. *Electron Diffraction in Kyoto and Hiroshima*

Electron diffraction in the Kyoto and Hiroshima schools may best be described jointly, in four periods, which are (1) before 1945, (2) after the end of the Second World War in 1945, (3) after construction of a universal electron diffraction microscope in 1953, (4) and from 1959, when the first research-worker visits to overseas laboratories were arranged, to 1977.

As is well known, Hiroshima University was heavily destroyed by an atomic bomb and became the worst place in the world, whereas Kyoto University had no such destruction. However these two laboratories developed many joint projects and enjoyed a good collaboration.

1. THE PERIOD BEFORE 1945

The first electron diffraction work in Kyoto was carried out in 1938 by Tanaka and Kano, who constructed a simple and useful electron diffraction apparatus, using a Shearer gas discharge X-ray tube as an electron source, a magnetic deflecting coil which allowed the selection of electrons of constant energy, a specimen goniometer stage and a viewing screen and camera box. Since the Shearer tube could produce rectified electrons at optimum air pressure (10^{-2} torr) even though an unrectified voltage was applied, the high voltage power supply consisted of only a transformer with no kenotron and condenser system. Using this electron diffraction camera, the relative orientation of oxide crystals formed on the surface of aluminium foil was studied. The transition temperatures of the arrangement and other properties of long chain molecules such as rubber film, paraffin, grease and fatty acids on a copper surface were also studied by Tanaka (1938, 1939, 1941, 1942) using a specimen heating stage.

The first electron diffraction work in Hiroshima University was an investigation, started in 1944 with a diffraction camera designed by Kikuchi and made by Rikagaku Kenkyusho (Institute of Physical and Chemical Research), on the surface materials of copper-based alloys produced by the reaction of hot air, acids, tap water and salt water. This experiment aimed to study the materials which were formed inside the water pipes used for cooling the engines of warships. The results were published after the end of the war (Tazaki and Hashimoto, 1949; Hashimoto, 1948). An electron lens was fabricated by them for this diffraction camera to focus the electron beam and produce diffraction patterns of high quality. This camera survived even under the atomic bomb explosion, though almost all rooms and apparatus in the Hiroshima University were destroyed and burnt down by fire. This room was protected by the small water pools devised by Hashimoto near the entrance and the windows, and also by his fire protecting activities. This camera was successfully used thereafter by S. Kuwabara, a student one year junior to Hashimoto. After the end of the war, Hashimoto left Hiroshima and joined Tanaka's group in Kyoto University.

2. THE PERIOD 1945–1952

For several years after the war, not only materials for constructing electron diffraction apparatus but also electricity and coal gas were not always available even in Kyoto: i.e., gas was supplied only in the daytime and electricity was supplied only from evening to morning. In Kyoto University, three diffraction cameras with a Shearer-tube gun were operated by Tanaka's students, and produced results for their theses even under the bad conditions stated above. In 1947, in Kyoto University, Hashimoto designed a new type of small diffraction

Fig. 1   Electron diffraction camera made of shell cases, designed and machined by Hashimoto in 1947.

camera, whose main column and window were made from anti-aircraft shell cases and every part of the instrument was machined by himself (Hashimoto, 1950) (Figure 1). Electrons from a heated tungsten filament were accelerated by rectified high voltage and focussed by an electron lens. Specimens could be evaporated and heated inside the vacuum. Using this instrument, Tanaka and Hashimoto (1950; 1951), and Hashimoto (1950) studied the structures of zinc films evaporated on copper, and on molten zinc surfaces, and noted that the evaporated zinc diffused into the substrate copper at elevated temperatures, and that zinc film formed on the melt showed two types of superstructure, of zinc and zinc oxide crystals.

Hashimoto (1951) investigated the corrosion processes of copper and its alloys by sulphur vapour, and noted that a few layers of different kinds of sulphides were formed discretely on the specimen surfaces, corresponding to different pressures of sulphur vapour and reaction temperatures. This phenomenon was named 'selective sulphuration'.

Fujiki (1949) noted that the transition temperature of paraffin film was altered by the presence of a substrate support. Nagata (1951) studied the relation between the orientation and length of aromatic compound molecules coated on the metal surfaces. Fujiki and Suganuma (1953; 1954) investigated the film structure of bismuth, deposited from the vapour appreciably below its melting point, and found an anomalous axial ratio of the unit cell, even with the substrate at room temperature.

In Hiroshima University, Tazaki and Kuwabara (1950), (1951), (1952) studied the structures of sulphide and oxide films formed on the surfaces of copper and its alloys.

3. THE PERIOD 1953–1959

In 1953, Tanaka and Hashimoto (1953) constructed a universal electron diffraction microscope, as they called it, which consisted of a three stage electron lens and two specimen chambers as shown in Figure 2 and functioned both as a three stage electron lens microscope and as a high-resolution electron diffraction apparatus. This instrument could obtain the following images: (1) electron microscope images of magnification up to 20 000 times; (2) electron diffraction patterns from selected areas as small as 3.3 μm in diameter; (3) dark-field electron microscope images; (4) high-resolution electron diffraction patterns; (5) electron diffraction patterns with a variable camera length from 0 to 250 cm; (6) shadow images; and (7) shadow micro-diffraction patterns. Similar instruments were constructed in Hiroshima University by Tazaki and Kuwabara at the suggestion of Hashimoto, and also the design was taken up by the Shimadzu and Hitachi companies for commercial production.

Using this instrument the following observations were carried out. Hashimoto (1954) studied the relation between the contrast of electron microscope images and diffraction patterns. He observed the subsidiary maxima of electron diffraction in the electron micrograph, high resolution electron diffraction

Fig. 2   Universal electron diffraction microscope in Kyoto University with seven operations possible, constructed in 1953.

patterns and shadow micro-diffraction patterns of bent MoO_3 thin film; this was discussed by the dynamical theory of electron diffraction. Subsidiary maxima of electron diffraction corresponding to the harpoon-like shape structure in MoO_3 crystals were also observed in the diffraction pattern. In the micrographs, Hashimoto observed many 'boundary lines' and noted that the Bragg reflecting condition varied on either side of the lines. These lines were actually the images of dislocations, as was pointed out clearly two years later by P. B. Hirsch, at Cambridge (see article this volume). Hashimoto showed that, from the analysis of the subsidiary maxima the thickness and inner potential, V_{000}, of the specimen film were obtained. Hashimoto and Tanaka (1956) observed the dynamic refraction effect of electrons in the high resolution electron diffraction pattern of a cupric sulphide crystal, and images of single dislocations and dislocation arrays in the moiré pattern. Hashimoto and Uyeda (1956) suggested that dislocations in moiré patterns were formed by double diffraction and corresponded to actual dislocations in one of the crystals of the rotationally superposed pair. Hashimoto (1958) observed stepped structures in the moiré patterns of cupric and zinc sulphide crystals and deduced that the steps in flat crystals were due to actual stacking faults while those in bend contours arose purely from dynamic interaction. A more complete dynamical-theory inter-pretation was published in 1961 (Hashimoto, Mannami and Naiki 1961).

Hashimoto, Yoda and Maeda (1956) made an energy analyser of the net-filter type developed by Boersch, inserted it between the second specimen holder and second projector lens in their Universal Diffraction Microscope, and found that Kikuchi lines from the cleavage surface of zinc-blende, and in transmission patterns from molybdenite, contained a certain amount of elastically-scattered electrons besides the inelastically-scattered part.

The tungsten oxide needle crystals formed on the surface of a heated tungsten filament were also investigated using the microscope (Hashimoto, Tanaka, Yoda and Araki 1958). By analysing both diffraction patterns and micrographs, the characteristic bundle-structure of needle crystals was discovered. In the lattice images of copper phthalocyanine, first observed by J. Menter, a spacing anomaly was found by Hashimoto and Yotsumoto, and explained by the dynamical theory (Hashimoto, Naiki and Mannami, 1958), but a more complete discussion was published in 1961. Edge dislocation contrast in thin crystals, in three types of imaging mode, was subsequently treated by the dynamical theory (Hashimoto and Mannami, 1959) and a full treatment later published by Mannami (1962).

In Hiroshima University, Kuwabara (1953) studied the double-refraction phenomena appearing in high resolution electron diffraction patterns of MgO, CdO and ZnO and also observed (Kuwabara 1954) subsidiary maxima in patterns from PbI_2.

Kuwabara (1957) measured the intensity of electron diffraction rings from evaporated films of Al, Ag, Au, NaCl, KCl and MgO with various thicknesses and various crystal sizes and noted that dynamical scattering and temperature factor cannot be disregarded. Using the difference Fourier syntheses from the electron diffraction ring intensity data Kuwabara (1959) determined the hydrogen position in NH_4Cl.

4. THE PERIOD 1959–1972

In 1959, the Kyoto and Hiroshima schools each sent one of their research workers overseas. For one year from 1959 to 1960, Hashimoto (Kyoto) and Kuwabara (Hiroshima) went to Cambridge and Melbourne, joining respect-ively Hirsch's, and Cowley's groups.

Hashimoto, with the collaboration of A. Howie and M. J. Whelan, studied the contrast of bend contour and stacking fault images using dynamical theory including absorption and showed the relation between the absorption coeffi-cients of component Bloch waves and their current flow vectors (Hashimoto, Howie and Whelan, 1960, 1962). During his stay in Cambridge, the work completed in Japan with the collaboration of his colleagues was published

(Hashimoto and Mannami, 1960; Hashimoto, Mannami and Naiki, 1961). These papers give the dynamical theory interpretation for edge dislocation images, and for crystal lattice fringes of perfect crystals and their moiré patterns. They discussed and gave appropriate explanations for the spacing anomalies and stepped structure in lattice fringes and moiré patterns.

Kuwabara, in collaboration with J. M. Cowley in Melbourne (Cowley and Kuwabara, 1962), made intensity measurements of electron diffraction patterns of thin crystals of bismuth oxychloride, and showed that the relative intensity could be explained by the phase-grating approximation theory developed by Cowley and Moodie.

The 1961 Conference and After

In 1961, the International Conference on Magnetism and Crystallography was held in Kyoto, and many important papers on electron diffraction were reported. Hashimoto, Tanaka, Kobayashi, Suito, Shimadzu and Iwanaga (1962) showed experimental evidence of the relativistic effect in electron diffraction and absorption phenomena, using the 300 keV electron microscope in Kyoto University. They showed that extinction distance is proportional to electron velocity, and transmissive power (inverse of absorption coefficient) is proportional to the square of electron velocity. These parameters are very important in determining electron diffraction intensities, and naturally are of consequence in image formation. It was now possible to test the theoretical prediction of improved image contrast, and improved visibility of defects, using this microscope (Hashimoto, 1962). A detailed description was published later (Hashimoto, 1964). Tanaka, Hashimoto and Mannami (1962) discussed the contrast of electron microscope images of dislocations and G.P zones by extending the dynamical theory to distorted regions, assuming these regions to consist of small blocks with a perfect-lattice structure (block approximation). Full accounts of both theory and development were published later (Mannami, 1962; Okabe, Hasebe and Mannami, 1969; Hasebe, Okabe and Mannami, 1971).

At this conference Kuwabara (1962) discussed the variation of the electron diffraction intensities from BiOCl with tilting angle and λH, where λ and H are the wavelength and thickness of crystal respectively. With this established, he turned to the intensity measurement of electron diffraction patterns using a 'Gegenfeld' energy filter after moving to Kure Technical college in 1964 (Kuwabara, 1966, 1967) and to Saga University in 1967 (Kuwabara and Takamatsu, 1969, 1970; Kuwabara and Cowley, 1973). After 1975, Kuwabara studied the phenomena appearing in the electron microscopic images of a crystal containing a dislocation loop (Kuwabara and Uefuji, 1975a), a bent crystal considering plasmon excitation (Kuwabara and Uefuji, 1975b) and considering both core and plasmon excitation (Kuwabara and Cowley, 1977).

Most of the work done by the Kyoto school after the Kyoto conference of 1961 concerned specific phenomena appearing in the electron microscopic images: for example, lattice fringes appearing outside the image of a crystal in defocus (Hashimoto and Mannami, 1962); the image contrast from crystals containing two stacking faults in different slip planes (Hashimoto and Marukawa, 1965); a lens-shaped cavity and a plate-shaped layer (Hashimoto, 1965; Hashimoto, Kamei and Endoh, 1971); and a dislocation in a grain boundary (Fisher, Hashimoto and Nogele, 1966). Moiré patterns appearing in the electron microscopic images of a grain boundary (Hashimoto and Kodera, 1969) were also discussed using the dynamical theory of electron diffraction. Intensities from selected area patterns of crystals containing a planar fault were investigated theoretically and experimentally by Mannami with the collaboration of Fitzgerald (Fitzgerald and Mannami, 1966). The effect of diffraction condition on the mean free path of a 50 eV electron for plasmon excitation was also investigated (Ishida, Mannami and Tanaka, 1967). Yoshida, a student of Tanaka, studied the structure of Mn–Al films in Kobe University with the collaboration of Nagata, using electron diffraction (Yoshida, 1972; Yoshida, Yamamoto and Nagata, 1974).

Hashimoto and his colleagues (1971) observed the image of single thorium atoms in thorium pyromeritate molecules and thorium oxide crystals (Hashimoto, 1971) using a transmission electron microscope operated in the dark field imaging mode, and discussed the contrast of the image by the kinematical theory

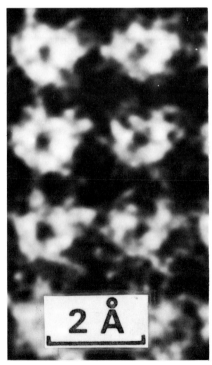

Fig. 3   Fine structure of the images of gold atoms in a crystal. Doughnut shape contrast is due to the flux flow of electron waves in crystal (1977).

of electron diffraction and image forming theory (Hashimoto, Kumai, Hino, Endoh, Yotsumoto and Ono, 1973).

1977

In 1977, the inner fine-structure of the images of atoms in a crystal, corresponding to the flux distribution of electron waves in gold, was photographed by Hashimoto, Endoh, Tanji, Ono and Watanabe (1977) using a conventional 100 keV electron microscope operated in the Aberration-Free-Focus condition, which is illustrated in Figure 3.

Also in 1977, the 5th International Conference on High Voltage Electron Microscopy was held in the Kyoto Kaikan Hall, which is the same place where the International Conference on Magnetism and Crystallography was held sixteen years before, and coincided with the fiftieth anniversary of electron diffraction. In this conference many important results were presented which where predicted sixteen years ago. The proceedings of this conference (Imura and Hashimoto, eds., 1977), and those of the Fiftieth Anniversary Meeting of Electron Diffraction in London in 1977, provide a good source of additional references.

Faculty of Engineering
Osaka University
Japan

REFERENCES

Cowley, J. M. and Kuwabara, S.: 1962, *Acta Cryst.* **15**, 260.
Fisher, R. M., Hashimoto, H., and Nogele, W.: 1966, *Proc. Sixth International Congress Electron Microscopy Kyoto*, **1**, 79.
Fitzgerald, A. G. and Mannami, M.: 1966, *Proc. Roy. Soc.* **A293**, 169.
Fujiki, Y.: 1949, *Memoirs Colleg. Sci. Kyoto Imp. Univ.* **A25**, 119.
Fujiki, Y. and Suganuma, R.: 1953, *J. Phys. Soc. Japan* **8**, 427.
Fujiki, Y. and Suganuma, R.: 1954, *J. Phys. Soc. Japan* **9**, 144.
Hasebe, H., Okabe, T., and Mannami, M.: 1971, *J. Phys. Soc. Japan* **30**, 417.
Hashimoto, H.: 1948, *Collection of Treatises of Kyoto Tech. College* 46; 53.
Hashimoto, H.: 1950, *Memoirs Kyoto Tech. College* **7**, 4.
Hashimoto, H.: 1950, *Memoirs Kyoto Tech. College* **7**, 9; 13.
Hashimoto, H.: 1951, *X-ray* **6**, 77.
Hashimoto, H.: 1954, *J. Phys. Soc. Japan* **9**, 150.
Hashimoto, H.: 1958, *J. Phys. Soc. Japan* **13**, 534.
Hashimoto, H.: 1962, *Proc. Fifth International Congress for Electron Microscopy*, **B–11**.
Hashimoto, H.: 1964, *J. Appl. Phys.* **35**, 277.
Hashimoto, H.: 1965, *Proc. International Conference on Electron Diffraction* Melbourne, 1 0-1.
Hashimoto, H.: 1971, *Jernkont. Ann.* **155**, 479.
Hashimoto, H., Endoh, H., Tanji, T., Ono, A., and Watanabe, E.: 1977, *J. Phys. Soc. Japan* **42**, 1073.
Hashimoto, H., Howie, A., and Whelan, M.J.: 1960, *Phil. Mag.* **5**, 967.
Hashimoto, H., Howie, A., and Whelan, M. J.: 1962, *Proc. Roy. Soc.* **A269**, 80.
Hashimoto, H., Kamei, S., and Endoh, H.: 1971, *Memoirs of Kyoto Technical University*, **20**, 1.
Hashimoto, H. and Kodera, M.: 1969, *Japan J. Appl. Phys.* **8**, 1390.

Hashimoto, H., Kumao, A., Hino, K., Endoh, H., Yotsumoto, H., and Ono, A.: 1973, *J. Electron Microscopy* **22,** 123.

Hashimoto, H. and Mannami, M.: 1959, *J. Electron Microscopy* **8,** 19.

Hashimoto, H. and Mannami, M.: 1960, *Acta Cryst.* **13**, 363.

Hashimoto, H., Mannami, M., and Naiki, T.: 1961, *Phil. Trans. Roy. Soc. London*, **253**, 459.

Hashimoto, H. and Mannami, M.: 1962, *J. Phys. Soc. Japan*, **17**, 520.

Hashimoto, H. and Marukawa, K.: 1965, *J. Phys. Soc. Japan* **20**, 1035.

Hashimoto, H., Naiki, T., and Mannami, M.: 1958, *Reprints Fourth International Conference on Electron Microscopy*, **1**, 331.

Hashimoto, H. and Tanaka, K.: 1956, *Proc. First Regional Conference in Asia and Oceania*, 292.

Hashimoto, H., Tanaka, K., Yoda, E., and Araki, H.: 1958, *Acta Metall.* **6,** 557.

Hashimoto, H., Tanaka, K., Kobayashi, K., Suito, E., Shimadzu, S., and Iwanaga, M.: 1962, *J. Phys. Soc. Japan*, Supplement, **BII**, 170.

Hashimoto, H. and Uyeda, R.: 1956, *Acta Cryst.* **10**, 143.

Hashimoto, H. Yoda, E., and Maeda, H.: 1956, *J. Phys. Soc. Japan* **11,** 464.

Imura, T. and Hashimoto, H. (eds.): 1977, *Proc. Fifth International Conference High Voltage Electron Microscopy, Kyoto*.

Ishida, K. Mannami, M., and Tanaka, K.: 1967, *J. Phys. Soc. Japan* **23,** 1362.

Kuwabara, S.: 1953, *J. Sci. Hiroshima Univ.* A**17**, 111; 221.

Kuwabara, S.: 1954, *J. Sci. Hiroshima Univ.* A**17**, 377.

Kuwabara, S.: 1957, *J. Phys. Soc. Japan* **12**, 637.

Kuwabara, S.: 1959, *J. Phys. Soc. Japan* **14**, 1205.

Kuwabara, S.: 1962, *J. Phys. Soc. Japan* **17**, 1414.; *ibid.* Supplement B-II, 115.

Kuwabara, S.: 1966, *J. Phys. Soc. Japan* **21**, 127.

Kuwabara, S.: 1967, *J. Phys. Soc. Japan* **22**, 1245.

Kuwabara, S. and Cowley, J. M.: 1973, *J. Phys. Soc. Japan* **34,** 1575.

Kuwabara, S. and Cowley, J. M.: 1977, *J. Phys. Soc. Japan* **42,** 1973.

Kuwabara, S. and Takamatsu, Y.: 1970, *J. Phys. Soc. Japan* **28,** 1031.

Kuwabara, S. and Uefuji, T.: 1975a, *J. Electron Microscopy* **24**, 137.

Kuwabara, S. and Uefuji, T.: 1975b, *J. Phys. Soc. Japan*, **38**, 1090.

Mannami, M.: 1962, *J. Phys. Soc. Japan* **17**, 1160–1171, 1423.

Nagata, S.: 1951, *Tech. Rep. Osaka Univ.* **1**, 167.

Okabe, T., Hasebe, H., and Mannami, M.: 1969, *J. Phys. Soc. Japan* **27**, 1245.

Tanaka, K.: 1938, *Memoirs Colleg. Sci. Kyoto Imp. Univ.* A**21**, 169.; 85.

Tanaka, K.: 1939, *Memoirs College. Sci. Kyoto Imp. Univ.* A**22**, 377.

Tanaka, K.: 1941, *X-ray* **2**, 7.

Tanaka, K.: 1941, *Memoirs Colleg. Sci. Kyoto Imp. Univ.* A**23**, 195.

Tanaka, K.: 1942, *X-ray* **3**, 72.; 133.

Tanaka, K. and Hashimoto, H.: 1950, *X-ray* **6**, 1.

Tanaka, K. and Hashimoto, H.: 1951, *J. Phys. Soc. Japan* **6**, 406.

Tanaka, K. and Hashimoto, H.: 1953, *Rev. Sci. Inst.* **24,** 669.

Tanaka, K., Hashimoto, H., and Mannami, M.: 1962, *J. Phys. Soc. Japan*, Supplement B-II, 166.

Tazaki, H. and Hashimoto, H.: 1949, *J. Sci. Hiroshima Univ.* A**14**, 1.

Tazaki, H. and Kuwabara, S.: 1950, *J. Sci. Hiroshima Univ.* A**14**, 251.

Tazaki, H. and Kuwabara, S.: 1951, *J. Sci. Hiroshima Univ.* A**15**, 67.; 133.

Tazaki, H. and Kuwabara, S.: 1952, *J. Sci. Hiroshima Univ.* A**15**, 263.

Yoshida, K.: 1972, *J. Phys. Soc. Japan* **32,** 431.

Yoshida, K., Yamamoto, T., and Nagata, S.: 1974, *Japan. J. Appl. Phys.* **13,** 400.

J. M. COWLEY

II.28. High Energy Electron Diffraction in Australia

1. THE EARLY STAGES

The first electron diffraction instrument in Australia was a Finch-type camera built by Edwards Co., England, with its cold cathode electron gun made from a wine bottle and vacuum seals made by abutting flat metal plates with a copious layer of grease. It was installed in the Physics Department of the University of Adelaide to support the program on surface physics of Dr Roy S. Burdon, the only one of the three faculty members in the Department who had had a research program. The first consistent user of the instrument was John M. Cowley who did his master's thesis during the years 1943–45 on RHEED studies of fatty acid layers on metal surfaces and on rectifying surfaces of crystals.

In 1945, at the age of 28, A. L. G. (Lloyd) Rees was put in charge of the newly created Chemical Physics Section of the Division of Industrial Chemistry of the Council for Scientific and Industrial Research (later to become the Division of Chemical Physics of the Commonwealth Scientific and Industrial Research Organization, the CSIRO). Lloyd Rees set about introducing to Australia a number of modern techniques in fields such as X-ray diffraction and spectroscopy, but the first major instrument to arrive for the Section was an RCA EMU electron microscope. To work with this John L. Farrant was employed for electron microscopy and John Cowley for electron diffraction.

The student who took over the Finch camera in Adelaide when John Cowley left was John V. Sanders, who subsequently went to Cambridge for a Ph.D. with Dr P. Bowden and then returned to Australia to build up a tradition of electron microscopy and diffraction within the CSIRO Division of Tribophysics in Melbourne.

Within Chemical Physics, Lloyd Rees and John Cowley, with help for a while from John A. Spink, explored the new possibilities introduced by the precision electron optics of the RCA microscope and of their first laboratory-built diffraction instrument. With this 'modern' equipment they could achieve the resolution in the diffraction patterns needed to reveal the new phenomena of refraction effects, dynamical spot splitting and shape transforms of small regularly shaped crystals such as those of MgO and ZnO smokes (Cowley and Rees 1947; Cowley, Rees and Spink, 1951).

Alex F. Moodie, a graduate of St Andrews University in Scotland, joined the electron diffraction group in 1948 and Peter Goodman from the University of Melbourne was added in 1952. Then the group remained unchanged until 1962 when John Cowley left to become a professor at the University of Melbourne and Andrew W. S. Johnson was appointed. They were an isolated group with very little contact with the rest of the world of electron diffraction until a trickle of visiting workers started, with James A. Ibers (USA) in 1956, Shigeya Kuwabara (Japan) in 1959 and Anil Goswami (India) in 1960

2. ELECTRON DIFFRACTION IN THE CHEMICAL PHYSICS DIVISION

In about 1950 the Chemical Physics group found that with the two-lens optics of their diffraction camera they could focus the incident beam on the specimen and obtain patterns from very small regions of the specimen without spreading out the diffraction spots very much. The specimen regions illuminated were probably not less than a few thousand Ångstroms in diameter, but the possibility of obtaining single crystal transmission patterns from a wide range of materials became apparent. Attempts were made at crystal structure analysis using single crystal patterns (see Cowley and Rees, 1958; Cowley, 1967). Gradually it was realized, however, that difficulties were arising, partly from disorder in the structures of those layer-lattice materials which gave good looking patterns and partly because dynamical scattering effects were thought to be present but there seemed no way of calculating them.

At that time it seemed that the Bethe dynamical theory was far too complicated for use in any but the two-beam approximation. Obviously the two-beam approximation was quite inappropriate for the diffraction patterns containing hundreds of spots which were given by thin single crystals.

The way around this difficulty came from an accidental diversion into physical optics, involving the rediscovery of the self-imaging properties of periodic objects in coherent illumination, or the formation of Fourier images. This was the Talbot Effect, found by Fox Talbot a hundred years earlier but apparently forgotten after some partial investigations at the turn of the century. Following their exploration of this effect with two-dimensional gratings, it was natural that Cowley and Moodie (1957) should try to extend it to three dimensional periodic objects and so develop the multi-slice formulation of many-beam dynamical theory.

The immediate utility of this formulation derived from the fact that it provided the basis for some useful approximations, such as the phase object approximation (Cowley and Moodie, 1962) which allowed rough estimates of dynamical effects to be made. It was not until almost 1965 that access to large digital computers allowed programs to be developed, on the basis of this formulation, for the accurate calculation of diffraction pattern intensities and electron microscope image intensities (see Goodman and Moodie, 1974). By this time (same reference) it had been well established that the multi-slice formulation of dynamical theory was completely equivalent to, and gave results consistent with, the Bethe formulation and also the formulations due to Sturkey, Fujiwara, Fujimoto, Niehrs and Tournarie. The rivalry of competing 'theories' disappeared. The choice remained, from a rich field of alternatives, of the formulation appropriate for any particular purpose.

Work on the accurate determination of structure amplitudes using dynamical diffraction effects began when Peter Goodman, on leave in Berlin, worked with Gunter Lehmpfuhl to revive the Kossel–Möllenstedt techniques of convergent beam electron diffraction (Goodman and Lehmpfuhl, 1965). From the convergent beam patterns it was possible to measure directly the variation of diffracted beam intensity with angle of incidence. The comparison with carefully computed values was further developed in Chemical Physics to provide the first example of the use of dynamical diffraction effects to give data of high accuracy. In favorable cases structure amplitudes were determined with an accuracy approaching one half per cent.

By this time John Cowley had established a group working on electron diffraction and electron microscopy in the School of Physics, University of Melbourne, where he was joined by Alan E. Spargo in 1963. Some of his students, including David Cockayne, Tony McMahon and Elizabeth Hewat (née Chidzey), worked with the Chemical Physics group. Others (Peter Turner, Ian Pollard, Ralph Holmes) made use of the variation of intensity of diffraction intensities with thickness, visible in electron micrographs of wedge-shaped crystals, to refine structure amplitudes. A large part of the effort in the School of Physics was directed towards the study of diffuse scattering, due to thermal vibrations and the positional disorder of atoms, in electron diffraction patterns when strong dynamical scattering was present. Important contributions to the theory of scattering were made by Peter Fisher, Andrew Pogany and Peter Doyle and progress on the description of the state of short-range order was made by Glenn Shirley.

A new direction of the work in Chemical Physics was initiated with the consideration of symmetry in electron diffraction patterns under dynamical scattering conditions. The key work on the theory was done by Alex Moodie in conjunction with visiting scientist, Jon K. Gjønnes of Oslo (Gjønnes and Moodie, 1965). The experimental work was led by Peter Goodman who showed how determination of absolute symmetry, including the presence or absence of a center of symmetry, and also handedness, could be made from convergent beam diffraction patterns (Goodman, 1975).

Crystal Structure Imaging

High resolution electron microscopy has recently been developed as a tool for the study of the structures and defects of thin crystals with a resolution approaching interatomic distances. The impact of this technique is being felt in an increasing range of fields of solid state science including mineralogy, solid state chemistry, metallurgy and ceramics. Since both the techniques and theoretical basis for these developments derived to an important degree from the background of electron diffraction work built up in Melbourne it is appropriate to give some account here of the early days of the subject.

David Wadsley of the CSIRO Division of Mineral Chemistry had for many years been analyzing complicated mixed oxide systems by X-ray diffraction methods, showing that in regions of the phase diagrams, previously thought to contain nonstoichiometric compounds of continuously variable composition, there were in fact series of super-lattice structures, sometimes well ordered but often disordered. Conversations between Dave Wadsley, visiting scientist Bob Roth (N.B.S., Washington) and Alex Moodie led to the realization that the resolution of available electron microscopes should be sufficient to allow the periodicities of these superlattices to be seen directly. Attempts to see these structures were made by John Allpress and John Sanders in the CSIRO Division of Tribophysics in 1967–8. Their success exceeded the expectations. Not only were the periodicities visible, but there appeared to be direct correlations between well defined features in the image intensities and particular types of atom groupings in the structures. The possibility of finding the structures of inorganic crystals and, more importantly, the form of the crystal defects present by direct observation in the electron microscope was an exciting prospect.

The resolution available to them with the tilting stage of their microscope, necessary for correctly aligning the crystals, was no better than about 7 Å and insufficient to separate the nearest neighbour metal pairs in even the most favourable orientations. However they were able to gather a great deal of new structural data on structures, defects and growth mechanisms of superlattices in the titanium–niobium, niobium–tungsten and related oxide systems (see Allpress and Sanders, 1973).

From the theoretical point of view it seemed at first unlikely that recognizable images of atom configurations could be produced in the presence of the very strong dynamical scattering effects known to occur in such materials for thicknesses greater than 10–20 Å. Gradually confidence in the image interpretation was built up as known structures were seen to be reliably imaged and as the appropriate theory was developed. The conditions under which interpretable images could be obtained were defined. Firstly, the rough approximations to the dynamical theory were used and then the accurate multislice computing methods of Goodman and Moodie (1974) were applied. In a classical series of papers under the general title 'n-Beam Lattice Images', by Alex Moodie, Peter Goodman, John Sanders, Andy Johnson, Michael A. O'Keefe and others in the period 1972–75, the imaging process was explored, the importance of parameters and approximations in the computing methods was evaluated and the foundations were laid for a practical, reliable method of structural analysis.

Experimentally the method received a considerable boost in 1971 when Sumio Iijima, working with John Cowley who was by that time at Arizona State University, showed that with one of the newer generation of high resolution electron microscopes it was possible to obtain resolution of better than 4 Å (Iijima), 1971). Then nearest neighbour atom rows in materials such as the niobium oxides could be clearly distinguished. More detailed and reliable information on structures could be obtained, while much less prior knowledge of the atom configurations was required. More recently, by the use of special high resolution microscopes in the 100–200 keV range and also of the high voltage (1 MeV) high resolution microscopes, the resolution limit has been improved to the

2–3 Å range and both the power and the range of applications of the methods have been greatly increased.

It is regrettable that no microscope capable of even 4 Å resolution was available in Melbourne until 1975 when one was installed at the School of Physics, University of Melbourne, to be used very effectively by Les Bursill and Alan Spargo. Examples of their work and an excellent general survey of the present status of the subject are given in the proceedings of the 1979 Nobel Symposium 'Direct Imaging of Atoms in Crystals and Molecules' published in *Chemica Scripta*, **14** (1978–9).

REFERENCES

Allpress, J. G. and Sanders, J. V.: 1973, *J. Appl. Cryst.* **6**, 165–190.
Cowley, J. M.: 1967, *Progress in Materials Science* **13**, 269–321.
Cowley, J. M. and Moodie, A. F.: 1957, *Acta Cryst.* **10**, 609–619.
Cowley, J. M. and Moodie, A. F.: 1962, *J. Phys. Soc. Japan* **17**, Supplement B-II, 86.
Cowley, J. M. and Rees, A. L. G.: 1947, *Proc. Phys. Soc.* **59**, 287–302.
Cowley, J. M. and Rees, A. L. G.: 1958, *Rep. Progr. Physics* **21**, 165–225.
Cowley, J. M. Rees, A. L. G. and Spink, J. A.: 1951, *Proc. Phys. Soc.* **64**, 638–644.
Gjønnes, J. K. and Moodie, A. F.: 1965, *Acta Cryst.* **19**, 65–67.
Goodman, P.: 1975, *Acta Cryst.* A**31**, 804–810.
Goodman, P. and Lehmpfuhl, G.: 1965, *Z. Naturforsch.* **20a**, 110–114.
Goodman, P. and Moodie, A. F.: 1974, *Acta Cryst.* A**30**, 280–290.
Iijima, S.: 1971, *J. Appl. Phys.* **42**, 5891–5893.

S. C. MOSS AND P. GOODMAN

II.29. Short-Range-Order Scattering: J. Cowley's School

I. THE MIT YEARS, 1947–1949: S. C. MOSS

John M. Cowley arrived at the Physics Department of MIT in late summer 1947 to do his Ph.D. with B. E. Warren. It was at that time becoming increasingly clear that X-ray diffuse scattering would quantitatively best be done on single crystals with counter methods. The past had yielded quite lovely examples of TDS (thermal diffuse scattering) photography especially by the Lonsdale group and it was actually Dame Kathleen who, on a brief visit to Cambridge, had suggested to (or urged) Warren to go over to single crystal methods. This was in particular with respect to the order–disorder transitions in metallic alloys. Z. W. Wilchinsky had done his Ph.D. research on powder studies of short-range order in Cu_3Au with only modest, however tempting, success; the time was thus ripe for a 'proper' study as Bert Warren used to say.

There were problems facing the experimenter: the weak diffuse background scattering from such systems was measured, barely jokingly, in the standard Warren unit of mpf (millicounts per fortnight); the X-ray source was a workhorse homebuilt unit with monstrous kenotrons, which among other things, generated a frightening amount of soft X-rays; the diffractometer was a jerry-built and rigged surveyor's instrument, put together in the local physics shop, on which θ and 2θ were hand set and the Geiger counter and scaler were primitive.

The thesis problem presented to John Cowley was the quantitative measurement of short-range order in single crystal Cu_3Au and measurements had to be made above 390 °C in an H_2 atmosphere. John designed a nice little furnace to hold the (borrowed) crystal which fitted on this newly designed goniometer and he collected diffuse scattering data above T_c at 405 °C, 460 °C and 550 °C. These data were taken point-by-point throughout a symmetry volume of reciprocal space and were corrected for Compton and thermal scattering. They were then Fourier inverted by hand, using an ingenious alternative to Beevers–Lipson strips (the latter aids were not around), according to a diffraction theory developed by Cowley and Warren (Bert Warren has always referred to the cofficients of this triply periodic series as the Cowley short-range order parameters, α_{ij}. John Cowley calls them the Warren parameters).

Having thus determined extensive sets of short-range order parameters, or pair correlation functions, for disordered Cu_3Au, it remained to compare them with existing theory. The problem now was that existing theory was largely insufficient to the purpose. For long-range order there was Bragg–Williams or Bethe, while for $T > T_c$ there were variations of the Bethe theory which yielded mainly the nearest neighbour order parameter. John set out to improve on this situation. Through a most formidable combination of intuition and statistical thermodynamic derring-do, he derived what is now called the Cowley Theory in order–disorder. It is one of the best closed form approximations we have for long-range order in a first order transition and provides very interesting coupled equations in the mean-field approximation for the pair correlations above T_c; it permitted the evaluation of pair-energies and the calculation of an approximate phase diagram for the Cu–Au system.

With his dissertation and the two papers that followed (experiment and theory) (see Cowley, 1950), John Cowley ushered in a new era of X-ray diffuse scattering in alloys. The entire exercise, including the usual doctoral exams etc., took 2 years and after this brief kinematic excursion – this short walk through the world of weak scattering – he returned to Australia in the summer of 1949 to electron diffraction and affairs dynamical.

University of Texas, Austin, Tex., U.S.A.

II. MELBOURNE UNIVERSITY, 1964–1969: P. GOODMAN

A. The Fine-Structure Observations

The Cowley (1950) theory was adequate to explain the appearance of smooth peaks in the diffuse scattering of electrons or X-rays. In 1952 however Raether (1952) reported that the diffuse peaks from a disordered alloy, Cu_3Au (i.e. at $T > T_c$) were split; similar phenomena were later reported for Cu_3Pd (Watanabe, 1959) and for CuAu (Sato, Watanabe and Ogawa, 1962). Since these observations were made by electron diffraction there was a general unwillingness to accept this behaviour as a property of the alloy, and rather to regard it as peculiar to either E.D. or to thin films. Cowley chose to pursue this matter further. By 1965 he could report:

Recently, with increasing understanding of the theoretical basis for electron diffraction intensities, we have become more bold in interpreting such effects and in this we are encouraged by continued success.

and in reference to Raether's results:

these seemed to indicate the apparent contradiction of a long-range correlation in a system with only short-range order. The observation was generally attributed to some peculiarity of thin films or of the electron diffraction method since the effect had not been observed with X-rays. After Dr D. Watanabe and P. M. J. Fisher in our laboratory had examined these possibilities, we concluded that the most likely source of the discrepancy lay in the limitations of the X-ray method (Watanabe and Fisher, 1965).

Den Watanabe was at that time on a visiting Fellowship from Tohoku University (Sendai). Another visitor (in 1968–1969), was Si Moss who took the investigation a stage further. He had shown, in 1965, that the doubling of the diffuse spots could be observed from bulk samples, by using X-ray diffraction at a sufficiently high resolution and with sufficiently long exposure. At that stage Moss, with his background in metal-physics, had a fresh insight into the problem that was to influence substantially the future of the S.R.O. subject.

He now sought to interpret the diffuse scattering in terms of a $V(k)$ potential in momentum space which is the Fourier transform of the interaction potential between neighbouring atoms. He had derived, with P. Clapp, a simple intensity expression, apparently at first thinking about the X-ray experiment:

$$I(k) = C[1 - (T_c/T)\{V(k)/V(k_m)\}]^{-1},$$

(Clapp and Moss, 1966), where C is a normalization constant and k_m is the position in the reciprocal space where $V(k)$ assumes an absolute minimum – usually one of the vectors of the ordered reciprocal lattice. He then went on to consider the Cu–Au data, and the electron diffraction experiments.

Essentially Moss's (1969) theory was this: the Fermi surface in a fcc alloy will depart from sphericity, and in particular will be flattened in directions of strongest nearest-neighbour interaction (i.e. [110]). This flattening, in momentum space, corresponds to an oscillating extension, in real space, of the interaction potential, over several nearest-neighbour distances. Thus he saw that it was entirely reasonable that fine-structure should arise in the S.R.O. from this potential even in the absence of actual long-range ordering, and should have a similar geometry to the super-lattice below T_c (Krivoglaz, 1969*; Wilkins, 1970).

This theory was an extension of Kohns' (1959) prediction of a scattering anomaly at $q = 2k_F$, where q is the momentum change on scattering, and k_F the Fermi wavenumber. The geometry of the planar section of reciprocal space containing the fine-structure is explained in the following way: when the 'anomaly' surfaces, with flat sections perpendicular to [110], are laid down at the reciprocal lattice points, there will occur an anomaly *enhancement* when the flat (projected straight) anomaly surfaces intersect. This development is described in Moss's (1969) own words:

The phenomenon we would like to interpret is the very weak splittings that have been observed in the diffuse scattering from Cu–Au alloys *above* T_c. These splittings were first seen in the electron diffraction patterns of Cu_3Au by Raether (1952) later by Marcinkowski and Zwell (1963) and still more recently by the present author (Moss, 1965) using X-rays. These are all single-crystal data which show the general disklike scattering distribution first seen by Cowley (1950) superimposed on which are the splittings in question. While the $CuAu_3$ scattering differs greatly from Cu_3Au, splittings have also most recently been seen there as well (Watanabe and Fisher, 1965).

These splittings [observed for a Cu_3Au sample quenched from 900 °C, about the (110) position] are so similar to the split spots one sees below T_c in an equilibrium long-period superlattice that for years they have been interpreted as direct evidence for the existence of antiphase microdomains above T_c (Moss, 1965). By invoking the ideas presented so far, however, he suggested that *these splittings are simply the reflection of strong enough anomalies in $V(k)$ to appear*, via Equation (1), *in $I(k)$.* To clarify the point, we have included in Figure 1 a section of the Cu Fermi surface in the hko plane of reciprocal space, where the flatness normal to the $\langle 110 \rangle$ direction is apparent. It is this property which is responsible

* Moss (personal communication) attributes the original theory to Krivoglaz, and regards his own work as a realization of that theory. [Added in proof, P.G.]

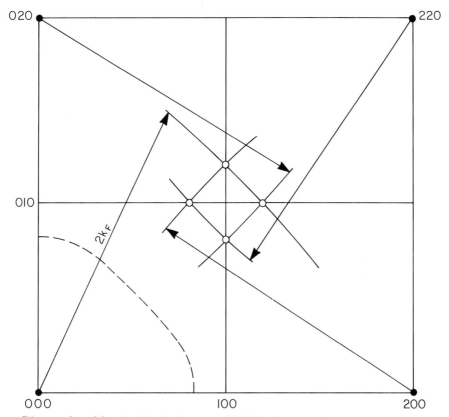

Fig. 1. Diagram from Moss (1969) showing a partial 'Kohn' construction for Cu in the *hk*o plane. Points where anomaly surfaces cross are indicated by small circles.

for the strong anomalies in the vicinity of (110). From Figure 1 we see that the result of folding the reciprocal lattice with the $2k_F$ surface is to enhance the singularity along the axes of observed splitting.

Using this Cu Fermi surface gives a separation of the split peaks from the central (110) position of $\sim \frac{1}{10}$ of the distance between (110) and (100). The experimental separation in Figure 1 is about $\frac{1}{11}$ suggesting that the Cu Fermi surface serves disordered Cu_3Au rather well. The Au Fermi surface has a $\langle 110 \rangle$ dimension ($k_F \langle 110 \rangle$) about 2 % smaller than the Cu surface – it is flatter normal to $\langle 110 \rangle$ – and this very small change means, via Pythagoras' theorem, that the splitting separation decreases to nearly $\frac{1}{20}$ of the (100)–(110) distance. The experimental value of the separation from (110) of the split diffuse peaks in $CuAu_3$, is $\sim \frac{1}{17}$ suggesting that the Fermi surface at disordered $CuAu_3$ is more like that of Au.

He further went on to give conditions for observability of the fine-structure

Such singularities will appear as minima in $V(k)$, and should be observable as images of those pieces of Fermi surface responsible for them. The factors that control how detectable they will be can be listed: (1) The flatness of the Fermi surface in the direction of interest. The flatter – the better. (2) The diffuseness of the Fermi surface at the temperature in question. If it is sharp, the singularities will be well defined.

With the completion of this work (in 1969), the subject of S.R.O. electron diffraction analysis was virtually turned around, and given a quantitative interpretation.

This incidently is a starting point for a whole series of observations of 'incommensurate' superlattice structures and the so-called 'charge-density wave' observations of a future decade.

REFERENCES

Clapp, P. C. and Moss, S. C.: 1966, *Phys. Rev.* **142**, 418.

Cowley, J. M.: 1950, *J. Appl. Phys.* **21**, 24.

Cowley, J. M.: 1965, *Electron Diffraction and the Nature of Defects in Crystals*. Aust. Acad. of Sci. publication. J-5.

Marcinkowsky, M. J. and Zwell, L.: 1963, *Acta Met.* **11**, 373.

Krivoglaz, M. A.: 1969, *Theory of X-ray and Thermal Neutron Scattering by Real Crystals*. Plenum Press, New York.

Kohn, W.: 1959, *Phys. Rev. Letters* **2**, 393.

Moss, S. C.: 1965, *Local Atomic Arrangements Studied by X-ray Diffraction*; Gordon and Breach, Inc, N.Y.

Moss, S. C.: 1969, *Phys. Rev. Letters* **22**, 1108.

Raether, H.: 1952, *Angew. Physik* **4**, 53.

Sato, K. Watanabe, D. and Ogawa, S.: 1962, *J. Phys. Soc. Japan* **17**, 1647.

Watanabe, D.: 1959, *J. Phys. Soc. Japan* **14**, 436.

Watanabe, D. and Fisher, P. M. J.: 1965 *J. Phys. Soc. Japan* **20**, 2170.

Wilkins, W.: 1970, *Phys. Rev. B* **2**, 3935.

J. V. SANDERS AND P. GOODMAN

II.30. Fourier Images and the Multi-Slice Approach

This anniversary volume seems an appropriate place to record some of the background to the multi-slice formulation of electron diffraction which led to the modern techniques of calculating accurately the intensities of up to a thousand or so interacting beams with a crystal. This problem seemed, in 1955, to be far too complex to have an easy solution. There were many influences on the developments in Melbourne, but one event seems to have had more influence than most, acting as a 'catalyst' to further theoretical progress. Prior to 1955 thin-single-crystal structure analysis had been restricted by the fact that the most reliable N-beam interpretation available was still the kinematic theory – Bethe's theory and its extensions at that time seemed most inappropriate to these extensive N-patterns. The event referred to was Alex Moodie's 'accidental' discovery of Fourier images. It happened that at home one evening, looking at a distant street light through a window which was fitted, as is normal in Australia, with a fly-wire mesh, he was able to see enlarged images of the mesh somewhere in the mid-distance. By readjusting the focus of his eye he could in fact find a series of such images. At that time he could find no reference to the effect, nor did his colleagues, skilled in optics, know of it. (As it turned out Talbot had described this phenomenon about a hundred years earlier, but this early reference was not discovered for some years, and is irrelevant in the context.) He and Cowley therefore worked out the theory of Fourier images, and showed analytically how the periodic component of a phase object could produce sets of real and virtual magnified and demagnified images if illuminated by a spherical wave front (approximated by a paraboloidal wave front for mathematical convenience). They checked the effect quantitatively in an optical microscope with a phase grating and postulated that this mechanism was a possible means of obtaining aberration-free magnification which could be used in an electron microscope (Cowley and Moodie, 1951).

The Cowley–Moodie theory predicts that the spacing of the Fourier images depends upon the spacing within the object, the order of the image, and the distance of the source. Table I shows typical values for the distance from the object to the first Fourier image plane for a distant source. It can be seen that in the optical case for a fly-wire screen the images are about 2 m apart, a change of focus easily detectable by eye within a few metres of a periodic object. For the electron case, modern instruments easily resolve the (200) fringes in gold crystals; here the separation of the Fourier images is only a 'few clicks' on the

TABLE I

	Moodie (light)	Gold (200) 100 keV	(Electrons) 1 MeV
Wavelength (d)	0.5 μm	0.037 Å	0.0087 Å
Lattice spacing (d)	1 mm	2.04 Å	2.04 Å
First Fourier image distance	2 m	110 Å	500 Å

finest focus control ($\sim$ 30 Å per step) of the electron microscope. The situation is more relaxed at 1 MeV.

Cowley and Moodie also investigated to some extent the intensity distribution on planes between those at which Fourier images appeared. An important result of this work which appears in *Fourier Images* IV (Cowley and Moodie, 1960) is that for a crystal (periodic potential $\phi(r)$) in an electron beam, the intensity distribution near (distance ε away) a Fourier image plane depends upon the value of $\varepsilon\nabla^2\phi$. In this case $\nabla^2\phi = 4\pi\,p(x,y)$, where $p(x,y)$ is the projected charge distribution (Poisson's equation), and $I_{xy}\propto\varepsilon\,p(x,y)$. Thus the image of a crystal in an electron microscope, focussed at a distance ε from a Fourier image plane should show the projected structure of the crystal.

This result led to the belief that out-of-focus images in an electron microscope could contain direct structural information. Experimental observation and calculation, carried out together at a later date, gave some feeling for the appropriate values of ε required for this approximation to be useful, and showed that this interpretation could be applied over a much greater thickness range than could the kinematic theory.

Cowley and Moodie appreciated that the just-out-of-focus image near the zero order Fourier image plane would also contain realistic, non-periodic structural information about crystals. The break-through in electron micro-scopy which this represents is best stated by saying that out-of-focus imaging was now seen in terms of holography, with the central directly-transmitted beam behaving as a reference wave. Earlier 'inventions' of devices for phase retardation of the central beam were no longer required: ε, or lens defocus, was in the control of the microscopist, the lens-transfer function was not. Knowing the latter, a value of ε could be found to optimise the structural information obtained.

Thinking in terms of spherical (paraboloidal) wave fronts within the crystal led to alternative formulations for diffraction through crystals in terms of a set of convolutions (Cowley and Moodie, 1958). In particular, the theory of overlapping crystals producing moiré patterns followed (Cowley and Moodie, 1959) and led to the multi-slice formulation which described the dynamical scattering of an electron wave through a thick crystal (Cowley and Moodie, 1957). This theory is most convenient for use in computation, and is now used routinely in calculating, with precision, the amplitudes and phases of up to 1000 beams in crystals with quite complex structures as a function of the thickness of the crystal. At a recent meeting on 'Direct Imaging of Atoms in Crystals and Molecules', (Nobel Symposium – 47, 1979), 13 of the 34 contributors or over $\frac{1}{3}$, coming from 6 different countries, made direct use of multi-slice calculation for their analysis.

Apart from the application to electron scattering there was a further development of Fourier imaging for light-optical problems, largely unknown to electron diffractionists. The technique has been applied to the restoration of images with only their periodic components (Damman, Groh and Koch, 1971), to the construction of multiple images from a single object (Bryngdahl, 1973), and more recently to the testing of Fresnel zone plates (Harburn and Williams, 1975).

In electron optics itself, highly coherent sources have only recently become available. In a development somewhat parallel to the advent of the laser in light optics, new instruments, using field-emission tips operating at 100 kV, have introduced a new field of spherical-wave electron optics (Cowley, 1980). In this way the development of diffraction theory reported here has come a complete circle. Originally described for spherical wave illumination, Fourier images led to a formulation of diffraction theory suitable for plane-wave illumination (in which of course spherical-wave treatment is given to scattering within the crystal). The new field of spherical wave illumination allows the originally described phenomenon i.e. the existence of Fourier images to very high magnification, or order, to be observed directly. With 2-beam conditions the Fourier-image remains in focus for all source-to-specimen distances (Cowley 1975), and 2-beam lattice fringes have been recorded up to 'infinite magnification'. In this form the phenomenon is equivalent to Ronchi's figures in optical testing (Cowley, 1980). With more than 2 beams spherical wave optics leads directly to structure-phase information (Spence and Cowley, 1978).

REFERENCES

Bryngdahl, O.: 1973, *J. Opt. Am.* **63**, 416.
Cowley, J. M.: 1975, *Diffraction Physics*, North Holland Press.
Cowley, J. M.: 1980, Proceedings of the 6th Aust. Conf. on Electron Microscopy, *Micron* **11**, 229.
Cowley, J. M. and Moodie, A. F.: 1951, *Proc. Phys. Soc.* **B70**, 486.
Cowley, J. M. and Moodie, A. F.: 1957, *Acta Cryst.* **10**, 609.
Cowley, J. M. and Moodie, A. F.: 1958, *Proc. Phys. Soc.* **71**, 533.
Cowley, J. M. and Moodie, A. F.: 1959, *Acta Cryst.* **12**, 423.
Cowley, J. M. and Moodie, A. F.: 1960, *Proc. Phys. Soc.* **76**, 378.
Damman, H., Groh, G. and Kock, M.: 1971, *Appl. Optics* **10**, 1575.
Harburn, G. and Williams, R. P.: 1975, *Optica Acta* **22**, 37 45.
Nobel Symposium 47: 1979, proceedings: *Chemica Scripta* 1978–79, Vol. 14.
Spence, J. E.N. and Cowley, J. M.: 1978, *Optik* **50**, 129–142.

P. B. HIRSCH

II.31. *Electron Microscopy of Defects in Crystals*

Personal recollections of developments at the
Cavendish Laboratory, Cambridge, 1946–1962.

1. INTRODUCTION

This article consists of my personal recollections of the early developments at the
Cavendish Laboratory, Cambridge, of electron microscopy of defects in crystals
by the 'diffraction contrast' technique. The history up to 1956, when the first
observations of moving dislocations were made (Hirsch, Horne and Whelan,
1956) has been published recently (Hirsch, 1980), as one of the papers in the
Royal Society Symposium on the 'Beginnings of Solid State Physics', organised
by Sir Nevill Mott. In the present article the period covered is extended to about
1962, by which time the kinematical (Hirsch, Howie and Whelan, 1960) and
dynamical theories of image contrast of defects (Whelan and Hirsch, 1957a,b;
Hashimoto, Howie and Whelan, 1960, 1962; Howie and Whelan, 1961, 1962)
relevant in the present context, had been published. Although this time limit
excludes the later contributions to, *inter alia*, inelastic scattering (due to
plasmons, single electron excitations and phonons), and their effects on image
contrast and absorption, the image contrast theories developed by 1962
provided the general framework, based on the column approximation, for image
contrast calculations of many types of defects, and in particular established the
essential ground rules for characterising stacking faults and dislocations. Thus,
by 1962, the technique had been put on a firm basis, and had been developed to
a level at which it could be and was quite generally applied by many groups both
in the U.K. and elsewhere to the study of defects in crystals. A summer school
organised by the Institute of Physics and the Physical Society was held in
Cambridge in July 1963, which concentrated on the basic electron diffraction
and image contrast theory and applications to the study of defects (Hirsch,
Howie, Nicholson, Pashley and Whelan, 1965).

2. EARLY BEGINNINGS

In the early 1940s there was a controversy concerning the origin of the
broadening of X-ray diffraction lines from cold-worked metals. Lipson (Stokes,
Pascoe and Lipson, 1943) attributed the broadening to internal strains, Wood
(1939) to small particle sizes. W. L. Bragg suggested that Wood's hypothesis

could be tested directly by limiting the size of the X-ray beam sufficiently so that spotty rather than continuous Debye–Scherrer rings were produced. The late J. N. Kellar and I, who had joined the Crystallography Department of the Cavendish Laboratory in 1946, were given the project, supervised by W. H. Taylor, to develop an X-ray microbeam technique, based on a rotating anode X-ray tube and a moderately fine focus, and a collimator system consisting of glass capillaries. This enabled us to obtain intense X-ray beams down to about 10 microns in diameter in practice. The method worked well for cold-worked aluminium, from which rings consisting of fairly clear spots were obtained; from their number a particle size of about 2 μm was deduced (Kellar, Hirsch and Thorp, 1950). The technique was also used by Gay and Kelly (1953) to determine the microstructure of a number of other deformed metals, including Fe. However, in some cases, in particular beaten gold foil, it proved impossible to obtain rings with clearly resolved spots, suggesting particle sizes (in this case) of about 2000 Å (Kelly, 1953).

Long exposure times (typically ten hours) limited the X-ray technique to particle sizes greater than about 1 μm, and since intense fine beams of electrons were available in electron microscopes, we asked J. W. Menter, who was a research student in F. P. Bowden's laboratory (Research Laboratory for the Physics and Chemistry of Rubbing Solids, Department of Physical Chemistry, later transferred to the Cavendish Laboratory and renamed Physics and Chemistry of Solids) to take electron diffraction patterns in his Metropolitan Vickers EM3 microscope. Spotty diffraction rings were obtained and their analysis is described by Hirsch, Kelly and Menter (1955) in 'The Structure of Cold-Worked Gold I: A Study by Electron Diffraction'. Menter also took some electron micrographs of the same specimens; these showed complicated structures, and their analysis was to form the subject of Part II of this pair of publications. Part II was never published, but the micrographs led to the idea that individual dislocations might be imaged directly by electron microscopy.

The micrographs of beaten gold foil showed bands of contrast parallel to the traces of (111) planes. Streaks on the diffraction patterns suggested stacking faults parallel to (111) planes; the analysis used was M. S. Paterson's (1952), who had spent some time in Orowan's Metal Physics group in the Cavendish in the early 1950s. It seemed plausible that the contrast on the micrographs was caused by faults, probably by the phase change of the electron waves due to the displacement of the crystal at the faults. Furthermore, if, as proposed by Heidenreich and Shockley (1948), dislocations in fcc lattices were dissociated into partials bounding a ribbon of stacking faults, the possibility existed of imaging individual dislocations. At that time I had no clear understanding of how to estimate the effect of the dislocation strainfield on the image.

By this time Heidenreich (1949, 1951) had published his important transmission electron microscope (TEM) study of electrolytically thinned specimens of aluminium foil in the beaten and annealed states. He interpreted 'extinction contours' in terms of the dynamical theory of perfect crystals, and found that deformed aluminium consisted of subgrains about 2 μm in diameter in good agreement with the X-ray microbeam results. (I remember being impressed by the obvious power of the TEM technique which enabled him to get direct images of substructures with exposures of 10 seconds, compared with exposures of several hours using the X-ray microbeam technique.) Although no evidence of images of individual dislocations was reported, it seemed plausible that there was a better chance of imaging dislocations in the relatively simple structures in beaten and partially annealed aluminium, than in the complex structures of beaten gold. Electron microscopists at that time were rather sceptical of getting good images from foils 1000 Å thick or more, but I hoped that an effect similar to the Borrmann effect for X-ray diffraction, on which I had done a little theoretical work (Hirsch 1952), might apply. M. J. Whelan joined me as a research student in 1954, to explore the possibility of observing dislocations by 'diffraction contrast'.

At about the same time (1954) Jim Menter had moved to Tube Investment Research Laboratories at Hinxton Hall and took delivery of one of the earliest Siemens Elmiskop I electron microscopes in 1955. He realised that the resolving power was sufficient to reveal directly the distortion of the lattice planes close to the dislocation cores in crystals with large unit cells, and this led him to produce in December 1955 and publish in 1956 (Menter, 1956) his beautiful pictures of edge dislocations in platinum phthalocyanine, by direct lattice resolution.

Meanwhile in the Cavendish Laboratory, Cosslett had also taken delivery of an Elmiskop I, operated by Bob Horne, and in October 1955 Whelan was able to get micrographs taken on both Al and Au on that instrument. Subgrain boundaries in Al were found to consist of individual dots or short lines, which we thought were dislocations because their spacing agreed well with the prediction of Frank's (1950) formula, using the measured values of misorientation across the boundaries. However, we could not be certain, since it was possible that the observed effects could be due to moiré patterns from overlapping crystals.

Our insistence on micrographs being accompanied by diffraction patterns meant that Bob Horne operated the microscope in a mode in which one could switch easily from microscopy to diffraction, but in which the double condenser system could not be used. On 3 May 1956 Bob Horne used the microscope in the 'high resolution mode' with the double condenser system, and he also pulled out the condenser aperture to increase the beam intensity. The lines were observed to move parallel to traces of (111) planes. This left no doubt that the lines were images of individual dislocations, and that their glide was observed in these experiments. Both G. I. Taylor and Mott were delighted to see moving dislocations. Ciné films were taken and cross-slip, dislocation bowing and pinning at the surface oxide film were observed. The work was reported by Whelan at an Institute of Physics Electron Microscopy Group Conference in Reading in July 1956 (Challice, 1957), and subsequently published in the *Philosophical Magazine* (Hirsch, Horne and Whelan, 1956).

The stage was now set for proceeding in two directions: firstly, to develop the image contrast theory, and secondly, to apply the technique to problems in Metal Physics and Metallurgy. The successful observations of dislocations enabled us to obtain funds from the Department of Scientific and Industrial Research for our own Elmiskop I, delivered in 1957. A period of rapid expansion followed. We continued to be part of the Crystallographic Laboratory until about 1964, when we became an independent group in the Cavendish Laboratory, the Metal Physics Group, replacing Orowan's group bearing the same title, which had disbanded some years previously on his departure to the U.S. Throughout my period in the Crystallographic Laboratory, and later when building up the Metal Physics Group we enjoyed the fullest support from W. H. Taylor, and of course from Nevill Mott after he became Cavendish Professor in 1954. Dr Cosslett's and Bob Horne's help in our initial experiments was of course also invaluable.

3. DEVELOPMENT OF IMAGE CONTRAST THEORY

The first objective was to develop the theory of image contrast from stacking faults, on which we already had some preliminary ideas dating from the experiments on gold foil. It now turned out that, by a fortunate coincidence, Walter Bollmann, at the Battelle Memorial Institute in Geneva, had been working at about the same time on a project aimed at improving austenitic stainless steels. Following a suggestion from Dr Siegfried, the head of the Metallurgy section (who had read a paper by Castaing [1955] on transmission microscope studies of aluminium alloys), that dislocations might be seen directly by this technique, Bollmann developed the very successful electropolishing technique for stainless steel, which led him to observe dislocations in that alloy (Bollmann, 1956). Bollmann also attended the 1956 Reading Conference, and there showed Mike Whelan and myself pictures of fringes in stainless steel which we interpreted tentatively as arising from stacking faults. Stainless steel proved a very useful material to study (widely dissociated dislocations, pinning by solution hardening), and Bollmann provided us with some samples of his particular alloy which produced very good specimens on electropolishing, on which we carried out detailed contrast experiments, and compared these with theoretical treatments (Whelan and Hirsch, 1957a,b). The dynamical theory was developed by Whelan, and the kinematical theory independently by Whelan and myself. The column approximation was introduced in the derivation of the kinematical theory. The dynamical theory, involving wave matching at an inclined fault, represents the first dynamical theory treatment of diffraction contrast from a defect. In 1958 Howie joined our group as a research student, and shared in the development of the kinematical theory of diffraction contrast of dislocations which was published in 1960 (Hirsch, Howie and Whelan, 1960). In this treatment we introduced the use of the amplitude phase diagram (influenced no doubt by teaching optical diffraction theory to undergraduates) and applied the column approximation to a dislocated crystal. The column approximation was an intuitive idea, justified at that time by the small scattering angles, and by the fact that for a dislocation most of the scattering was concentrated close to the reciprocal lattice points. By the time this paper had appeared Howie and Whelan had already developed the dynamical theory of diffraction from dislocations (Howie and Whelan, 1960a, 1961, 1962), and I remember Archie Howie commenting on the kinematical theory paper that it was of 'historical interest only'!

The possible importance of anomalous absorption effects had been appreciated well before the first observations of defects had been made. The theory of the effect on extinction contours and on images of stacking faults was developed and the comparison with experiment carried out during the period of a visit from H. Hashimoto in 1959–60 (Hashimoto, Howie and Whelan, 1960, 1962), and the more general theory for dislocations was developed by Howie and Whelan (1961, 1962). Although most of the detailed image calculations were performed using the 2-beam approximation, the Howie–Whelan (1960a,b; 1961) theory was a general many-beam scattering matrix formulation, including absorption, applied to imperfect crystals, and assuming the column approximation. Although most of the work on inelastic scattering and the effect on image contrast was carried out later, it was already appreciated by 1962 that, following Takagi's (1958a,b) work, the main cause of anomalous absorption was likely to be phonon scattering, and also that (on the two-beam theory) the production of X-rays should be orientation dependent, due to the different excitations of type I and II Bloch waves (Hirsch, Howie and Whelan, 1962).

At this stage then the general framework of the image contrast theory based on the column approximation had been developed to a point at which it could be applied with confidence to the characterisation of many types of defects.

4. APPLICATIONS OF THE DIFFRACTION CONTRAST TECHNIQUE TO PROBLEMS IN METAL PHYSICS AND METALLURGY

The early observations on aluminium were soon followed by a detailed study of stainless steel, in collaboration with Bollmann (Whelan, Hirsch, Horne and Bollmann, 1957). Partial dislocations were observed for the first time, pile-ups at grain boundaries, networks, slip of screws across twin boundaries, surface pinning, and nucleation of dislocations at the edge. Networks, extended nodes, interactions of dislocations on different planes forming Lomer–Cottrell and other locks were studied by Whelan (1958a), who also introduced the method of measuring stacking fault energy from the curvature of partial dislocations at nodes. Many other applications followed quickly, including the observations of dislocation loops in quenched Al (Hirsch, Silcox, Smallman and Westmacott, 1958), stacking fault tetrahedra in quenched Au (Silcox and Hirsch, 1959a), stainless steel deformed in tension and fatigue (Hirsch, Partridge and Segall, 1959), defects interpreted as prismatic loops in neutron irradiated copper (Silcox and Hirsch, 1959b), dislocation distributions, recovery and recrystallisation processes in deformed silver (Bailey and Hirsch, 1960, 1962; Bailey 1960), dislocation distribution after fatigue (Segall, Partridge and Hirsch, 1961), dynamic observations of the annealing of prismatic dislocation loops in quenched Al (Silcox and Whelan, 1960), and of precipitation in Al 4%Cu (Thomas and Whelan, 1961), stacking fault energies of Cu and Ni alloys (Howie and Swann, 1961), deformation twinning in fcc metals (Venables, 1961) and martensitic transformation in stainless steel (Venables 1962).

It was a most exciting period. The theoretical solid state physicists had developed dislocation theory to a remarkable degree of detail, and there was therefore much scope for experimental tests of the models. Many predictions were verified, but new mechanisms and types of defects were also discovered, e.g. the stacking fault tetrahedra. It also became possible to derive directly values of important parameters, e.g. the stacking fault energy, from dislocation configurations, although in many cases the appropriate elasticity calculations remained to be done. Observations on deformed samples demonstrated the importance of elastic interactions between intersecting dislocations, leading to junction interactions. Many theoretical models were two-dimensional in nature, and the direct observations showed up three-dimensional mechanisms.

On the instrumentation side, the early experiments had to be carried out without the benefit of a double tilt goniometer stage; single tilt stereo holders were available for the Elmiskop I; double tilt stages became available in the early '60s (Valdrè, 1962). A fixed grid heating stage was designed by Whelan as early as 1958 (Whelan, 1958b) and used in the experiments on annealing of prismatic loops in quenched Al (Silcox and Whelan, 1960), but the development of suitable stages with tilting, heating, cooling and straining facilities always lagged behind the demand for them.

5. REFLECTIONS

By 1962 the Metal Physics Group in W. H. Taylor's Crystallographic Laboratory consisted probably of about 20 researchers, many (but not all) of whom were or had been involved either in developing or applying electron microscope techniques. We were particularly fortunate in having so many outstanding and enthusiastic young researchers in the group around that time. The names of those in the group in 1962, with the year of their arrival (where known) and of those who had left by then, with their years of arrival and departure, are given below. There was also close collaboration and contact with the Metallurgy Department (A. H. Cottrell, J. Nutting, G. Thomas, R. B. Nicholson, M. F. Ashby), with the TI Research Laboratory at Hinxton Hall (J. W. Menter, D. W. Pashley, P. Duncumb), and with others (e.g. P. B. Price, in Dr Bowden's group). There was of course strong support from and interaction with Sir Nevill Mott. The Metal Physics Group's motivation was primarily to solve problems in metal physics and metallurgy using increasingly the TEM technique, and the contrast theories were developed to enable us to do that. This emphasis probably differentiated the group from some of the other electron diffraction and microscopy schools elsewhere. It seems somewhat strange that the development of the 'diffraction contrast' technique for the study of defects started in the Crystallographic Laboratory in the Cavendish, and not in Orowan's Metal Physics, nor in Cosslett's Electron Microscopy Groups. In the event this 'accident' probably served us well, because our basic interest and experience in diffraction and crystallography must have helped us to develop the contrast theories. We had also been very lucky in making the initial observations on moving dislocations in aluminium; the original idea that stacking faults caused contrast because of the phase change due to the displacement at the fault proved to be correct, but the contrast of dislocation images in aluminium is of course due to the strainfield, in particular the bending of the lattice planes, not to the stacking fault ribbon. The Borrmann effect also proved to be important, but this was quite irrelevant in the first experiments. Above all we had access to a new generation electron microscope at the right time. There is no doubt that if Bob Heidenreich had taken his pictures in 1949 on a microscope with the resolution of the Siemens Elmiskop I, and with a double condenser lens, he would have seen individual dislocations, and their motion, and the developments at the Cavendish might have been rather different.

Members of the Metal Physics Group up to and including 1962:

C. J. Ball (1953–58), M. J. Whelan (1954–), J. E. Bailey (1955–59), H. H. Atkinson (1955–58), E. H. Yoffe (1955–), P. G. Partridge (1956–59), P. R. Thornton, D. H. Warrington (1956–61), T. E. Mitchell (1957–), A. Howie (1957–), R. L. Segall (1957–61), J. Silcox (1957–61), J. A. Venables (1958–61), R. M. J. Cotterill (1958–61), L. M. Brown (1960–), J. S. Lally (1960–), U. Valdrè (1960–), V. Ramamurthy (1960–), J. W. Steeds (1961–), P. C. J. Gallagher (1961–), A. J. G. Metherell (1961–), J. P. Jakubovics (1961–), G. R. Booker (1962–), J. A. Eades (1962–), R. A. Foxall (1962–), M. J. Goringe (1962–), C. R. Hall (1962–), P. M. Hazzledine (1962–).

Department of Metallurgy and Science of Materials
University of Oxford,
Parks Road, Oxford, England

REFERENCES

Bailey, J. E.: 1960, *Phil. Mag.* **5**, 833.
Bailey, J. E. and Hirsch, P. B.: 1960, *Phil. Mag.* **5**, 485.
Bailey, J. E. and Hirsch, P. B.: 1962, *Proc. Roy. Soc.* **A267**, 11.
Bollmann, W.: 1956, *Phys. Rev.* **103**, 1588.
Castaing, R.: 1955, *Rev. Métall.* **52**, 669.
Challice, C. E.: 1957, Proceedings of Reading Conference on Electron Microscopy 1956; *Brit. J. Appl. Phys.* **8**, 259.
Frank, F. C.: 1950, Carnegie Inst. of Techn. Symp. on *Plastic Deformation of Crystalline Solids* (Office of Naval Research) p. 150.
Gay, P. and Kelly, A.: 1953, *Acta Cryst.* **6**, 165.
Hashimoto, H., Howie, A., and Whelan, M. J.: 1960, *Phil. Mag.* **5**, 967.
Hashimoto, H., Howie, A., and Whelan, M. J.: 1962, *Proc. Roy. Soc.* **A269**, 80.
Heidenreich, R. D.: 1949, *J. Appl. Phys.* **20**, 993.
Heidenreich, R. D.: 1951, *Bell Syst. Tech. J.* **30**, 867.
Heidenreich, R. D. and Shockley, W.: 1948, *Report of Conference on Strength of Solids*, p. 57. London: Physical Society.
Hirsch, P. B.: 1952, *Acta Cryst.* **5**, 176.

Hirsch, P. B.: 1980, *Proc. Roy. Soc.* **A371**, 160.

Hirsch, P. B., Horne, R. W., and Whelan, M. J.: 1956, *Phil. Mag.* **1**, 677.

Hirsch, P. B., Howie, A., Nicholson, R. B., Pashley, D. W., and Whelan, M. J.: 1965, *Electron Microscopy of Thin Crystals*, London: Butterworths.

Hirsch, P. B., Howie, A., and Whelan, M. J.: 1960, *Phil. Trans. Roy. Soc.* **A252**, 49.

Hirsch, P. B., Howie, A., and Whelan, M. J.: 1962, *Phil. Mag.* **7**, 2095.

Hirsch, P. B., Kelly, A., and Menter, J. W.: 1955, *Proc. Phys. Soc.* **B68**, 1132.

Hirsch, P. B., Partridge, P. G., and Segall, R. L.: 1959, *Phil. Mag.* **4**, 721.

Hirsch, P. B., Silcox, J., Smallman, R. E., and Westmacott, K. H.: 1958, *Phil. Mag.* **3**, 897.

Howie, A. and Swann, P. R.: 1961, *Phil. Mag.* **6**, 1215.

Howie, A. and Whelan, M. J.: 1960a, *Proc. Eur. Reg. Conf. on Electron Microscopy*, Delft, Vol. 1, p. 194. Delft: De Nederlandse Vereniging voor Electronenmicroscopie.

Howie, A. and Whelan, M. J.: 1960b, *Proc. Eur. Reg. Conf. on Electron Microscopy*, Delft, Vol. 1, p. 181. Delft: De Nederlandse Vereniging voor Electronenmicroscopie.

Howie, A. and Whelan, M. J.: 1961, *Proc. Roy. Soc.* **A263**, 217.

Howie, A. and Whelan, M. J.: 1962, *Proc. Roy. Soc.* **A267**, 206.

Kellar, J. N., Hirsch, P. B., and Thorp, J. S.: 1950, *Nature* (London), **165**, 554.

Kelly, A.: 1953, Ph.D. Thesis, University of Cambridge.

Menter, J. W.: 1956, *Proc. Roy. Soc.* **A236**, 119.

Paterson, M. S.: 1952, *J. Appl. Phys.* **23**, 805.

Segall, R. L., Partridge, P. G., and Hirsch, P. B.: 1961, *Phil. Mag.* **6**, 1493.

Silcox, J. and Hirsch, P. B.: 1959a, *Phil. Mag.* **4**, 72.

Silcox, J. and Hirsch, P. B.: 1959b, *Phil. Mag.* **4**, 1356.

Silcox, J. and Whelan, M. J.: 1960, *Phil. Mag.* **5**, 1.

Stokes, A. R., Pascoe, K. J., and Lipson, H.: 1943, *Nature* (London) **151**, 137.

Takagi, S.: 1958a, *J. Phys. Soc. Japan* **13**, 278.

Takagi, S.: 1958b, *J. Phys. Soc. Japan* **13**, 287.

Thomas, G. and Whelan, M. J.: 1961, *Phil. Mag.* **6**, 1103.

Valdré, U.: 1962, *J. Sci. Instrum.* **39**, 279.

Venables, J. A.: 1961, *Phil. Mag.* **6**, 379.

Venables, J. A.: 1962, *Phil. Mag.* **7**, 35.

Whelan, M. J.: 1958a, *Proc. Roy. Soc.* **A249**, 114.

Whelan, M. J.: 1958b, *Proc. 4th Intern. Congr. for Electron Microscopy*, Berlin, Vol. 1., p. 96. Berlin: Springer Verlag.

Whelan, M. J. and Hirsch, P. B.: 1957a, *Phil. Mag.* **2**, 1121.

Whelan, M. J. and Hirsch, P. B.: 1957b, *Phil. Mag.* **2**, 1303.

Whelan, M. J., Hirsch, P. B., Horne, R. W., and Bollmann, W.: 1957, *Proc. Roy. Soc.* **A240**, 524.

Wood, W. A.: 1939, *Proc. Roy. Soc.* **A172**, 231.

LORENZO STURKEY

II.32. The Development of the Scattering Matrix: A Personal Historical Memoir

Born in the Piedmont section of South Carolina, my mother was a school teacher, my father a merchant cotton broker and sometime farmer. Since I grew up and went to school in the middle of the great depression our home was richer in books than money. It certainly contained more books than the school library. Because I had inhaled all the books most of my time in school I was bored, and I was much more interested in circuses and theatre. I even became a semi-professional magician on the way. I received my B.S. degree from The Citadel in South Carolina and went on to graduate study at the University of Ky (Kentucky). From there I went directly into industry and I only left when it became impossible for me to do any real thinking.

1. AT DOW CHEMICAL

In 1939 I came to the Dow Chemical Company to work with Ludo Frevel on electron diffraction. By the end of 1940 we had designed and constructed a high resolution electron diffraction camera with much assistance and encouragement from L. O. Brockway at the University of Michigan in Ann Arbor. The camera itself had a continuously variable voltage supply from 25–80 kilovolts, and a camera length of about one metre. Very soon we had observed most of the usual ED phenomena: cross-grating patterns from thin flakes, unusual intensities and forbidden reflections, peculiar shape effects, and reflections lying only on the Ewald sphere for thicker and perfect crystals like Si. Frevel insisted that I read all the available literature on electron diffraction and compile a complete bibliography of trustworthy references. This was fortunate, for one of our first reflection patterns was of $Mg(OH)_2$ with considerably altered intensities from X-ray values. Feitknecht had already hinted at this effect in all flake-shaped crystals and explained it as due to absorption in directions near the flake surface. After this, we examined all relative intensities with caution and dashed J. D. Hanawalt's hope for a separate electron diffraction identification index for powder patterns like that for X-ray patterns.

One of our attempts to provide an easily preparable and reproducible standard for calibration purposes led to the peculiar diffraction pattern from

MgO smoke, with line splitting and groups of spots oddly arranged. To digress somewhat, Bob Heidenreich had come to the company in 1940, and in 1941 he had been assigned to operate the newly acquired RCA EMB electron microscope. He looked at our MgO samples and identified them as usually cubes. Similar preparations of CdO smoke also gave cubic shapes and the same peculiar splitting of diffraction lines. Since I had Fröhlich's book (Fröhlich, 1936), the chapter on electron diffraction completely explained this phenomenon and the double splitting at Bragg angles. (This book had already become my bible for clear, explicit interpretations of electron phenomena and I have always relied heavily on it.) Sometime during World War II Rees of CSIRO Australia came to visit us to see our diffraction apparatus. I believe this was somewhere near 1943 or '44. Later we heard that Cowley and Rees had built an instrument in Australia and were considering publication of MgO results. Hillier at RCA had by then also observed these peculiar effects but could offer no explanation. So I was convinced by Frevel to publish two short notes on MgO and CdO, especially since I had discovered I could calculate the inner potential from the double splitting of spots in cases that were obviously two-beam cases (Sturkey and Frevel, 1945; Sturkey, 1948).

2. EXPERIMENTAL OBSERVATION OF PENDELLÖSUNG–1942

In 1942 Heidenreich had confirmed that MgO crystals were perfect cubes. He also observed some crystals showing periodic thickness contours. At the time we were not working closely together – he was in another laboratory even though it was only a block from mine. He published a rather peculiar explanation of this thickness-intensity periodicity, and after I became aware of this paper, we started working closely together and decided that this phenomenon must be 'pendellösung'. We took the two-beam theory already derived by Blackman – and later by Caroline MacGillavry – for the Laue case. In these equations we inserted sensible constants and found close agreement with calculated scattering coefficients (Heidenreich and Sturkey, 1945). The particular reflection responsible for the 'pendellösung' was determined by Heidenreich by observing the motion of the reflected image as he moved back and forth from focus in the EM with a large aperture. I was prepared for such an explanation because of my work with the apparent double refraction from MgO cubes, and because I had been trying to understand Bethe's equations.

Heidenreich left for Bell Labs in 1945.

3. DEVELOPMENT OF THE SCATTERING MATRIX

The theoretical basis for the interpretation of electron diffraction patterns (and electron micrographs) was proposed by Bethe (1928). He took the one particle time-free Schrödinger equation and assumed it held for monochromatic electrons. He also assumed that the potential energy could be expressed in terms of the local potential encountered inside the medium. For a periodic medium the local potential could be expressed as a Fourier series in terms of the reciprocal lattice of the crystal. Thus the equation could be expressed as

$$\nabla^2 \psi + \frac{4\pi^2}{\lambda^2}\left(1 + \frac{P}{\varepsilon}\right)\psi$$

where

$$P = V_0 + \sum_{hkl} V_{hkl}\exp\{i2\pi \mathbf{g}_{hkl}\cdot\mathbf{r}\}.$$

(I found this a much more usable notation than Bethe's.) The various V's are expressed in volts and E is the accelerating voltage to which the electrons have been subjected. The $\mathbf{g}$'s are reciprocal lattice vectors. By some manipulation, Bethe found an infinite matrix which could be solved by making its determinant zero:

$$\mathbf{M}_{ii} = 2\,p\,k_i + p^2 \quad (p = \text{a parameter})$$

$$\mathbf{M}_{ij(i\neq j)} = \frac{4\pi^2}{\lambda^2}\frac{V_{ij}}{E}, \quad \text{etc.}$$

Bethe then neglected the term p^2. I found that I could rewrite the determinant by a little matrix manipulation as a simpler one. (This involved dividing each row by $2k_i = (4\pi/\lambda)\cos\theta_i$.)

$$\mathbf{M}'_{ii} = P$$

$$\mathbf{M}'_{ij} = \frac{\pi V_{ij}}{\lambda E \cos\theta_i}.$$

The determinant then became the auxiliary equation for a simple, first-order differential equation of the form

$$\frac{\partial}{\partial y}\Big(\psi_i(y)\Big) = i\mathbf{M}''\psi_i(y)$$

$$\mathbf{M}''_{ij} = \frac{\pi V_{ij}}{\lambda E \cos\theta_i} \qquad \mathbf{M}''_{ii} = 0.$$

This equation has a simple solution:

$$\psi_i(y) = \exp(iy\mathbf{M}''_{ij})\psi_i(0).$$

It became apparent to me that Bethe's equations could apply – in the Laue case, at least – only to reflections whose reciprocal lattice vectors lie on the 'Ewald circle'. Caroline MacGillavry (1940) had already hinted at some such reservations in her paper explaining the convergent beam experiments of Kossel and Möllenstedt (1939). After all, with only one incident wave of propagation vector $\mathbf{k}_0$, only waves can occur where $\mathbf{k}_0 + \mathbf{g}_i = |\mathbf{k}_0|$. However, the idea of the exponential of a matrix as the solution of a multiple scattering problem was distinctly appealing. For the exponential of any square matrix is always convergent and can be calculated without determining its characteristic values. I did calculate the intensities for several spots lying on the Ewald circle – explicit solutions for two, three and four spots, and numerical results for a larger number of spots.

The important conclusion I reached from this exercise was that the so-called 'anpassing' could only be a parameter, not a physically significant variable. After all, can we talk about a polydimensional 'dispersion surface'? The idea of the exponential of a matrix as the solution of the Laue case for many reflections, some not lying on the Ewald circle, was especially enticing, and I proceeded to hunt for a proper matrix which I began to call the 'Scattering Matrix'. The generality of such an $\exp(\mathbf{S}_{ij})$ solution should enable many types of problem to be solved. For instance, in a complex stack of crystals:

$$\psi(y_1 + y_2 + \ldots) = \ldots \exp\{iy_3\mathbf{S}_3\} \exp\{iy_2\mathbf{S}_2\} \exp\{iy_1\mathbf{S}_1\}\psi(0)_i.$$

I wanted to find a matrix independent of thickness for then

$$\psi(2y)_i = \exp(iy\mathbf{S}_{ij})^2\psi(0)_i, \quad \text{etc.}$$

Needless to say, I have never found such a general matrix valid for arbitrary incident angles on a thin plate, and only one valid for exactly normal incidence. I did find several approximate ones, but I have always questioned their validity for any but very, very small thickness – about 30 Å and 100 kV electrons. My working principles in approaching the theory have been these:

(a) The diffraction process in the Laue case may be described as the rotation of a single unit vector in 'diffraction space' by the crystal it penetrates into another unit vector having components along the various diffraction directions $\mathbf{k}_i$ so that

$$\psi(\mathbf{k}_0)_0\psi^*(\mathbf{k}_0)_0 = 1 = \sum_i \psi(\mathbf{k}_i)_y\psi^*(\mathbf{k}_i)_y.$$

(b) This is a unitary transformation and may be expressed as the exponential of $\mathbf{H}(y)$, where $\mathbf{H}$ is an Hermitian matrix. Thus:

$$\psi(\mathbf{k}_i)_y = \exp\{i\mathbf{H}_{(y)}\}\psi(\mathbf{k}_i)_0.$$

It is this $\mathbf{H}(y)$ that I have called the 'scattering matrix' $\mathbf{S}(y)$. (I only obtained the proof that this is *always* possible in 1963 from Gantmacher's book, but it was sufficient for my purposes that $\exp\{i\,\mathbf{S}(y)\}$ was unitary and could be computed for any value of $\mathbf{S}$ whether singular or not, as long as it was Hermitian.)

(c) The various vectors, $\mathbf{k}_i$, considered must have their directions determined at the entrance surface by the grating equations

$$n\lambda = \mathrm{d}(\sin\theta_{\text{incidence}} + \sin\theta_{\text{diffs.}})$$

The processes I went through in deriving a proper $\mathbf{S}(y)$ have been described in my lecture celebrating 'Fifty Years of Electron Diffraction' at Asilmar, California (1977); (Sturkey, 1977).

I could obtain a linear matrix with either positive or negative values of $(\mathbf{k}_y)_i$ along the diagonal, or my final non-linear matrix. This came also from the orthogonality principle for non-homogeneous equations, and even from Bethe's equations by integrating only over the volume of the crystal. All these $\mathbf{S}$ give the correct values for thin crystals, and when I had to do my calculations by hand, I chose the linear matrix with negative $(\mathbf{k}_i)$ along the diagonal since it seemed closer to Darwin's method for the Laue case. I applied the $\exp\{i\mathbf{S}\}$ to many problems, but I found only the non-linear one gave the *complete* transition from cross-grating patterns to spots lying on the Ewald circle as the crystal got thicker.

4. FURTHER PERSONAL HISTORY

I had read Zachariasen's (1945) book – he had sent Frevel a complimentary copy in 1946. He seemed to me to have the only consistent theory of dynamical diffraction, and he taught me how to tell a thick crystal from a thin one – i.e. a crystal may be considered thick at 30 Å for electrons but of the order of a micron for X-rays.

By 1950 I had made a number of numerical calculations, especially one concerning the (222) reflexion in Si. Therefore I gave a paper at the annual summer meeting of the (now) ACA at New Hampshire, August, 1950. I continued to attend most of the local meetings. But I continually found the state of the theory locally to be very disappointing. (At meetings I tried to point out errors. Because of this I acquired a reputation as a monster who devoured all people who dared to give a paper on ED or even EM at society meetings!)

In 1957, the I.U.Cr. planned its international meeting in Montreal, Canada. Bill Lipscomb, who had been a friend of mine at the University of Kentucky and who knew of all my work, called me and told me that a special symposium on ED was planned for this meeting with invited guests from all over the world. He suggested that I attend and contribute to the symposium. Since the symposium was to follow the main conference, I managed to meet all the foreign visitors first who were concerned with ED: the Russians Pinsker and Vainshtein; the Japanese Miyake, Honjo and Uyeda; Germans Molière and Raether; and Professor M. Blackman. (I had met Cowley before while he was at M.I.T. with Warren.) It appeared that new calculation methods would come from me and Cowley – a pre-symposium night was selected so that Cowley and I could talk at length about our methods. I described applications of the $\mathbf{S}$-matrix formulation – Cowley talked about convolutions and the phase-grating approximation. At this conference Professors Blackman, Miyake and Uyeda became my very close scientific friends. Professor Uyeda asked me early the next year to write out my methods in detail for him to take to a conference in Japan.

In 1961, I was invited to present a paper at the Kyoto Conference. Here Professor Ewald showed some interest in my calculations once I described diffraction as a single unitary transformation, or rotation in diffraction space.

Blackman insisted I publish my results as soon as possible. He offered to get them published in *Proc. Phys. Soc.* and to proof-read the paper himself. I came home, wrote the paper in a couple of weeks. This same year, at ASTM's request, I wrote a paper called 'Practical Considerations in the Interpretation of ED Patterns'. I went to Australia for the first time in 1965. This was an international meeting and I got to know a number of Cowley's students (University of Melbourne) and Moodie's co-workers. My most pleasant experience here however, was at the Spring School in Warburton. Here people were most interested in ED theory, and Professor Ewald and I became friends. We all discussed methods of calculation and I understood for the first time Cowley and Moodie's approach to calculation (see: Sturkey, 1962; 1963).

In 1974, I went to Australia again, but this time I gave three papers all related to the proper development of a calculation method. One of these was on the Rytov approximation. When I first used this by inserting the first Born approximation into a scattering matrix I was not sure whether it was truly proper mathematically, but some correspondence with Dr J. B. Keller of the Courant Institute at N.Y. University reassured me. I did present an easy derivation of my non-linear matrix based on the orthogonality principle, and for the first time presented some calculations comparing the non-linear results with the two prevailing linear ones.

REFERENCES

Bethe, H. A.: 1928, *Ann. d. Phys.* **87**, 55ff.
Blackman, M.: 1939, *Proc. Roy. Soc.* A**173**, 68.
Frölich, H.: 1936, *Elektronentheorie der Metall*, Julius Springer, Berlin.
Gantmacher, F. R.: 1959, *Theory of Matrices* Vol. I and II, Chelsea Pub. Co., N.Y.
Heidenreich, R.D.: 1942, *Phys. Rev.* **62**, 291.
Heidenreich, R. D. and Sturkey, L.: 1945, *J. Appl. Phys.* **16**, 97.
Heidenreich, R. D., Sturkey, L., and Woods, H.L.: 1946, *J. Appl. Phys.* **17**, 127.
Kossel, W. and Möllenstedt, G.: 1939, *Ann. d. Phys.* 113.
MacGillavry, Caroline H.: 1940, *Physica* **7**, 329.
Sturkey, L.: 1948, *Phys. Rev.* **73**, 183.
Sturkey, L.: 1950, ACA Summer Meeting, New Hampton, N.H., Abstract A-3.
Sturkey, L.: 1957, *Acta Cryst.* (Abstract) **10**, 858.
Sturkey, L.: 1962, *Proc. Phys. Soc.* **80** part 2, 321–354.
Sturkey, L.: 1963, ASTM Spec. Tech. Pub. No. 339, *Symposium on Advances in Electron Metallography* 31–45.
Sturkey, L.: 1974, *Diffraction Studies of Real Atoms and Real Crystals*, Australian Academy of Science, 273, 361, 371.
Sturkey, L.: 1977, *Trans. ACA* **13**, 1–13.
Sturkey, L. and Frevel, L. K.: 1945, *Phys. Rev.* **68**, 56 (See errata).
Zachariasen, W. H.: 1945 *X-ray Diffraction in Crystals*, John Wiley & Sons, N.Y.

F. FUJIMOTO

II.33. From Scattering Matrix to Electron Channelling

One day in 1955 I visited Professor S. Takagi and asked for instruction in physics. Before this I was a research assistant without a supervisor, with a passion for music rather than physics. When I finished a course at music school, I felt anxious about my future. Takagi's was the biggest group in our institute, and I liked his personality. The subject was not important for me.

My work in electron diffraction began with the design of a new goniometer, and with a study of papers on the dynamical theory. Professor Kohra, who was then a member of the group, kindly indicated many famous and fundamental papers on dynamical theory; they were mainly German works such as those by Bethe, Altmann and Niehrs, etc. The German language had been one of my worst subjects in high school.

When the construction of a new diffraction camera was near completion Professor Takagi showed me a paper by Germer (1942) on the intensity anomaly of the (222) reflection from CuCl crystals. I tried to measure its intensity by using a CdS detector mounted in the new diffraction camera, and at the same time to explain the crystal-size-dependence of the intensity theoretically (Takagi and Fujimoto, 1960; Fujimoto, 1961).

At that time a new computer, using parametrons, had been developed by a group of the science faculty of our university. The computer was placed in a small room without air-conditioning, and with only electric fan cooling. I planned to solve the secular equation of the dynamical theory using this computer. I was allotted three sessions on the computer, mostly at times around midnight. Unfortunately, or fortunately, I could not get correct results at all: every time some trouble developed in the computer. Therefore I gave up my computing efforts, and tried to solve the problem analytically.

In the fall of 1958 I obtained the scattering matrix (Fujimoto, 1959). Professors Takagi and Kohra congratulated me on this result. The matrix has a simple expression. However, this was inconvenient to apply to practical cases. I continued to modify the matrix in order to obtain a more convenient formula.

One day early in 1959, I attended a seminar of Professor S. Miyake's group and Mr K. Fujiwara reported on his new theory. I was surprised to find that his result obtained by the Born expansion was exactly the same as mine.

The Electron Diffraction Conference at Kyoto in 1961 was one of the most exciting events in my life and its was my great pleasure that Professor H. Raether invited me to give lectures in his department. At the end of September 1962 I visited him in Hamburg. This was my first stay abroad. Taking this opportunity,

I was in Europe for three years, and visited Professor K. Molière in Berlin and Dr P. B. Hirsch in Cambridge. A lecture by Dr M. W. Thompson from Harwell and Cambridge University opened my eyes to the new field of channelling.

After returning from Europe, I started working on the channelling of positive ions and its applications to nuclear and solid state physics. For these experiments my old goniometer was very useful. However, it was always my wish to find the relationship between channelling and diffraction. So I started an investigation using high voltage electron microscopes, with Dr Y. Uchida and Professor H. Fujita, and studied the transition from the wave mechanical region to the particle one. At this time, a work of the Munich group of professor R. Sizmann (1970) stimulated me, because they interpreted the channelling effect of electrons in an axial direction of a crystal by classical orbits of electrons, that is, the rosette motion of electrons trapped around an atomic string (Fujimoto,

Fig. 1. The varying geometrics of axial Kossel patterns. Clockwise from upper left, these are: 'firework', 'flower', 'tablecloth' and 'feather' patterns. The variations are a result of varying accelerating voltage and atomic number. They are: Si at 1200 kV, W at 700 kV, W at 1300 kV and Nb at 600 kV.

Komaki, and Uchida, 1971; Fujimoto, Takagi, Komaki *et al.*, 1972). Another group explained the channelling peak by a weaving motion of electrons instead of the rosette one (Nip, Hollis and Kelly, 1968).

I was very happy to have been able to enjoy an Australian life with my family for four months from 1971 to 1972 through the kindness of Mr P. Goodman and Dr Z. Barnea. After Melbourne, we flew directly to Munich and worked in the laboratory of Professor Sizmann for one year. During my stay in Munich, I often visited Drs Lehmpfuhl and Kambe in Berlin, and was able to establish that the weaving and rosette motions of electrons corresponded to the two-dimensional *s*-state and *p*-, *d*- states of electrons around an atomic string, respectively and, moreover, that the number of bound states increases with the electron energy and potential depth of the atomic string. What fascinated me were the very beautiful Kossel patterns around a crystal axis. These patterns should be geometrical ones at low energies and then change into flower, tablecloth, feather and firework, with increasing electron energy and crystal potential (see e.g. Fujimoto, 1977 and Figure 1).

A few years ago, I discussed with my co-worker, Dr K. Komaki, what consequences there could be from the transitions between bound states, but we came to no definite conclusions. Sometime later it was reported that photon radiation, the so-called 'channelling radiation', can be expected from the transitions and, in the last year, channelling radiation was observed at high energy.

The channelling effect of electrons, including channelling radiation, can be understood from the classical behaviour of electrons as particles, as well as from their quantum mechanical behaviour as waves (Fujimoto and Komaki, 1980). My present interest in electron diffraction is in studying phenomena in the transient region between classical and wave mechanical theories. Rather, phenomena belonging to the classical concept are attractive for me because they are understandable without calculations which need a computer.

REFERENCES

Fujimoto, F.: 1959, *J. Phys. Soc. Japan* **14**, 1558.
Fujimoto, F.: 1961, *J. Phys. Soc. Japan* **16**, 936.
Fujimoto, F.: 1977, *Proceedings 5th. International Conference on High Voltage Electron Microscopy, Kyoto.* p. 271.
Fujimoto, F. and Komaki, K.: 1980, Submitted to *Phys. Letters.*
Fujimoto, F., Komaki, K., and Uchida, Y.: 1971, *Phys. Stat. Sol.* (a) **8**, K71.
Fujimoto, F. Takagi, S. Komaki, K. Koike, H., and Uchida, Y.: 1972, *Rad. Effects* **12**, 153.
Germer, L. H.: 1942, *Phys. Rev.* **61**, 309.
Nip, H. C. H., Hollis, M. J., and Kelly, J. C.: 1968, *Phys. Letters* A**28**, 324.
Takagi, S. and Fujimoto, F.: 1960, *J. Phys. Soc. Japan* **15**, 1607.

B. B. ZVYAGIN

II.34. *Electron Diffraction Studies of Minerals in the USSR*

1. INTRODUCTION

It was clear from the very beginning that electron diffraction had some special advantages in the study of minerals. Its application in the USSR has been favoured by the works of Z. G. Pinsker (1949) which revealed the significance of the oblique texture (OT) patterns, of B. K. Vainshtein (1956), who introduced the analysis of electron diffraction intensities for crystal structure determinations, and by the industrial production of electron diffraction cameras.

Clay minerals were the first object of study, and OT patterns have been used for electron diffraction studies, initially in the Soil Institute and in the Geochemistry Institute (Moscow) and, later, to a wider extent in the All-Union Geological Institute (Leningrad). In the course of time the number of minerals investigated increased to include phyllosilicates in general, and layered and pseudolayered minerals of different kinds. The number of scientific organizations applying electron diffraction increased greatly in the '60–70s especially with the development of electron microscopy which became a source of another kind of pattern – i.e. the selected area diffraction (SAD) pattern. At the present time there are two scientific centres where both OT- and SAD- patterns are used in the study of a great variety of minerals: the Institute of Ore Geology and Mineralogy (IGEM) and the Geological Institute (GIN) of the USSR Academy of Sciences. IGEM is also the only place where a high-voltage camera (acceleration voltage variable up to 400 kV) is used, displaying new possibilities for the study of both single crystals and textured samples (Zvyagin *et al.*, 1979).

2. INTERPRETATION OF PATTERNS

The electron diffraction work on minerals has been accompanied by development of interpretation and calculation methods for both OT and single crystal patterns which are applicable in the general case of triclinic lattices (Zvyagin, 1964; Zvyagin *et al.*, 1979). Their essential features are the following:

A. Thin Single Crystal Patterns†

Let the incident beam be normal to a lattice plane having two periods a and b

---

† Sections A and B are considerably reduced from the original. For an adequate description of the pattern geometries, the reader is directed to Zvyagin (1964) as a necessary reference. [Added in proof – P.G.]

with an angle γ between. The third period c is characterised by the vector $\mathbf{c}$ which will have projective components along the axes $\mathbf{a}$ and $\mathbf{b}$, of X_n and Y_n, respectively. The reciprocal lattice is represented as a two-dimensional set of lattice rows parallel to $\mathbf{c}^*$. These lattice rows are characterised by a constant hk and variable l indices, for a very thin crystal. The single crystal (SC) patterns correspond to nearly planar sections of this lattice, normal to the beam and passing through the origin and are described as two-dimensional point nets hk through which the set of continuous lines $l = n$ ($n = 0$, ± 1, ± 2) is drawn. The net hk is determined by two periods proportional to $1/(a \sin\gamma \cos\psi_h)$ and $1/(b \sin\gamma \cos\psi_k)$ respectively with an angle γ'' between them. ψ_h, ψ_k are the angles between the lines Oh, Ok in the plane of the pattern and the ab plane and

$$\cos \gamma'' = \sin \psi_h \sin \psi_k - \cos \psi_h \cos \psi_k \cos \gamma.$$

In an arbitrary case the SC pattern may be considered as a cross-section of lattice rows parallel to $\mathbf{a}^*$ or $\mathbf{b}^*$ with a corresponding interchange of indices in the given formula as shown in Figure 1.

B. Oblique Texture Patterns

The reciprocal lattice of a textured sample is formed by rotating the reciprocal lattice of a single crystal around $\mathbf{c}^*$. The lattice rows hk form coaxial cylinders with radii $R_{hk} = (1/\sin \gamma)\{h^2/a^2 + k^2/b^2 - 2hk \cos \gamma/ab\}^{1/2}$; individual lattice points transform into rings at distances $D_{hkl}^* = ha^* \cos \beta^* + kb^* \cos \alpha^* + lc^*$ from the ab plane. (OT) patterns are then described as sets of ellipses hk having common minor axes $b_{hk} = L\lambda R_{hk}$, and reflexions hkl at distances $D_{hkl} = L\lambda D_{hkl}^*/\sin \phi = (hp + ks + lq)$ from the small axes (L being the distance between the specimen and the screen, λ the wave length, ϕ the tilting angle of the texture). The values b_{hk} are used for calculation of a, b, and determination of indices hk.

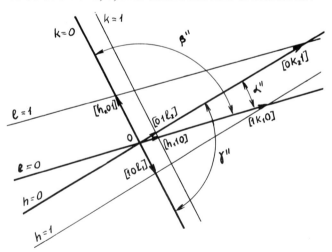

Fig. 1.   Three possibilities of presentation of an arbitrary single crystal electron diffraction pattern as plane sections of reciprocal lattice rows parallel to $\mathbf{a}^*$, $\mathbf{b}^*$, or $\mathbf{c}^*$.

General formulae which are used to determine definite plane sections of the reciprocal lattice from the OT patterns, and hence to determine unit cell parameters, may be found in Zvyagin (1964).

3. RESULTS: 1960 TO THE PRESENT TIME

OT and SC patterns (Figures 2 and 3) are original and valuable sources of structural information since they contain special and characteristic reflexion sets, and are obtained from specimens which are inaccessible to other methods. The geometry of the OT patterns yields the space features of the direct and reciprocal lattices. The *set* of ellipses characterizes one of the lattice planes which may be considered as a coordinate plane, while the sequences of reflexions *along* the ellipses is in correspondence with the length and inclination of the third period. This is especially important for groups of related structures, since these patterns make obvious features of similarity, and of difference and reveal their separate structures. Hence the OT patterns were very useful in the comprehensive study of phyllosilicates (clay minerals) and have displayed an unequalled effectiveness in application to the problems of polytypism. The SC patterns, although containing two-dimensional sets of reflexions, are superior with respect

Fig. 2.   The oblique texture electron diffraction pattern of the clay mineral nacrite, obtained at a tilt of 70° and acceleration voltage 400 kV.

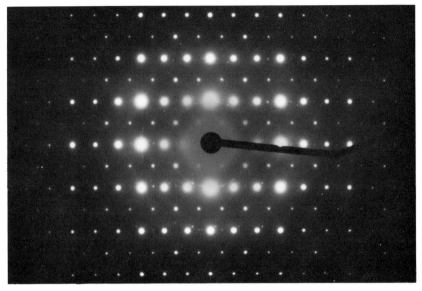

Fig. 3.   Single crystal electron diffraction pattern of the serpentine mineral lizardite, obtained by rotating the specimen about the horizontal *b*-axis by 20° from its 'normal' setting (beam perpendicular to the sample), at an acceleration voltage 400 kV.

to the extent of the diffraction field and the visibility of very weak reflexions. They may be used for construction of two-dimensional projections according to the intensity distribution or as an auxiliary means for indexing either ellipses in OT patterns or of rings in polycrystal patterns.

In cases where specimens were neither textures nor single crystals, polycrystals or aggregates of another kind electron diffraction (in particular SAD) proved to be also effective, revealing small degrees of crystallinity and microvariations of the crystal structure.

OT patterns have been utilised in several ways:

A. Crystal Structures

At the initial stage (beginning of the '60s) the OT patterns were used for the refinement of the crystal structures of the extremely fine grained clay minerals celadonite and kaolinite. Intensities of only $h0l$ and $0kl$ reflexions have been used and only structural projections on planes $Xo\mathcal{Z}$ and $0Y\mathcal{Z}$ have been considered. The same was done for the micas muscovite and phlogopite, although they are quite accessible to X-ray single crystal diffraction. It was shown nevertheless that even with a relatively primitive approach, permitting manual calculation, useful information such as the rotation angles of polyhedra bases, flattening of octahedra, elongation of tetrahedra, displacement of cations from ideal positions etc. may be obtained (Zvyagin, 1964).

At the next stage of the work, which was concerned mainly with the use of the high-voltage electron diffraction camera of IGEM (Zvyagin *et al.*, 1979), three-dimensional sets of reflexion intensities have been analysed and 9 crystal structures have been studied, representing a wide variety of dioctahedral phyllosilicates. These are composed of $1:1$ and $2:1$ layers, belonging to different polytypes, having Al or Fe^{3+} octahedra, and K or Na in interlayer sites. The structures of chapmanite and bismuthoferrite, consisting of $1:1$ layers with Fe^{3+} in octahedra and Sb or Bi in the interlayer sites, have been described for the first time. New structural data have been obtained for nacrite, the muscovites $1M$ and $2M_2$ and the paragonites $1M$ and $3T$. The structures of celadonite and paragonite $2M_1$ were additionally refined. The comparison of different structures enabled us to establish the structural significance and consequence of isomorphic substitutions of cations, and of the change of relative layer positions in different polytypes. Of particular interest are the established variation limits of the bond length Si—O, the concrete values being indicative for the stability and formation-probability of phyllosilicates. The further progress of the electron diffraction structure analysis is connected with direct registration of intensities. Being applied in the GIN of the Academy of Science for the structure refinement of celadonite and muscovites $1M$ and $2M_1$, it reduced the reliability (R) values to nearly 5%, at which value new data were obtained on the distribution of isomorphic cations, influencing the space symmetry, and, in the case of muscovite $2M_1$, the positions of the protons H were determined (Drits, Tsipursky, 1979; Tsipursky, Drits, 1977).

B. Polytypism

Since polytype structures differ only in the stacking sequence of their layers, they have two periods (a, b) in common while the third period (c) may have different values. In the OT patterns they are correspondingly characterized by a common set of ellipses but with different positions and/or intensities of reflexions along each ellipsis. Only *some* ellipses are sensitive to the changes of layer stacking which would change one polytype into another, and to stacking faults within one polytype. Polytypes of differing c values are distinguished by different reflexion positions on the ellipses which are sensitive to the layer stacking $D_{hkl} = (hp + ks + lq)$, corresponding to the values $p/q = x_n$, $s/q = -y_n$. If polytypes have equal c values they are distinguished by intensities. Under such conditions the OT patterns not only proved to be exceptionally appropriate for the study of polytypes but have stimulated the development of the theory of polytypism. In a combination of the theory and practice polytypes satisfying definite limitations may be deduced and their diffractional characteristics predicted. This makes it possible to recognize polytypes even in the presence of extensive stacking disorder, and with low quality patterns. With the use of OT patterns almost all the polytype structures of the phyllosilicates have been analyzed (Zvyagin 1964; Zvyagin et al., 1979) A comprehensive symbolic nomenclature has been proposed, which represents any polytype of a phyllosilicate as a sequence of symbols of intra-(s_i) and interlayer (t_k) displacements of adjacent tetrahedral and octahedral sheets. Notations s_i indicate also the azimuthal orientation of any pair of adjacent sheets of a layer. Indices $i,k, = 1$ indicate a displacement $(-1/6, -1/6)$

in the normal projection on the plane with the orthogonal coordinate system a, and $b = a\sqrt{3}$. Indices $i, k = n = 2, \ldots, 6$ indicate vectors obtained from the initial one by rotation $2\pi n/6$ around the normal to the layers. Since the unit cell base is c-centred all these vectors may be characterized by component multiples of $1/3$. Thus e.g. $(-1/6, -1/6)$ is equivalent to $(1/3, 1/3)$. Besides these there are three more notations t_0, t_+, t_- relating to values $(0,0)$, $(0, +1/3)$, $(0, -1/3)$.

From 32 theoretically possible kaolinite polytypes only 4 have been found in nature. One of them, halloysite, was considered previously as disordered kaolinite, and it is electron diffraction that has shown it to be an original polytype $s_3 t_+ s_3 t_- \ldots$ (Chukhrov, Zvyagin, 1966). A number of trioctahedral analogues of kaolinite polytypes (containing $3\{Mg, Fe^{2+}, Zn \text{ or } Ni\}$ instead of $2Al$) has been also identified by OT electron diffraction.

The OT patterns have revealed principal differences in the layer stacking of micas and pyrophyllite-talc. Electron diffraction may justly take the credit for the discovery of the polytype diversity of pyrophyllites and talcs. In particular the stacking sequence of the true polytypes $1TC$ and $2M$ has been established, as described by notations $s_2 s_2 t_1 \ldots$ and $s_3 s_3 t_2 s_3 s_3 t_4 \ldots$. Most striking was the discovery of quite a new Fe^{3+}-analogue of pyrophyllite, ferripyrophyllite $Fe_2^{3+} Si_4 O_{10}(OH)_2$ (Fe^{3+} replacing Al), belonging to the abovementioned modification $2M$. This modification is unambiguously identified by the dominant intensities of the reflexions (020), $(\bar{1}12)$ and (114) of the first ellipse, which are observed in spite of the low structural order of ferripyrophyllite resulting in poor quality OT patterns (Chukhrov, Zvyagin et al., 1979).

It is known that well-ordered structures are very seldom encountered among the diversity of chlorites, consisting of two kinds of layers ($2:1$ and $0:1$), which alternate. One of such structures has been found in alteration products of bauxites by means of OT patterns. A detailed analysis of these patterns has shown that it is triclinic and has the layer stacking $s_5 s_5 t_+ t_+ \ldots$; it was then recognized from some other samples giving much poorer OT patterns.

In the case of the pseudolayer Ti-silicates, or astrophyllites, the OT patterns served as a base for the deduction of 14 polytypes having different monoclinic and triclinic lattices. After that it became possible to sort out the relationships between the X-ray results obtained by different authors.

Electron diffraction has also been successfully applied to the polytypism problems of molybdenites and graphite.

C. Identification of Minerals

Under conditions when complicated problems of structure analysis and polytypism are successfully solved, the use of OT patterns for an ordinary identification of minerals may be simple and effective. Thus the main features of all phyllosilicates: the type of layers, the number of layers per repeat, the shape of the unit cell etc., may be recognized from even a visual inspection of the patterns, without measurements and calculations. It is a characteristic feature of the electron diffraction identification that it does not limit itself to indication of mineral names and polytype symbols of single phases or mixtures, but there is a tendency to reveal individual features (degree of order–disorder, kind of lattice distortions etc.) which are peculiar to single samples. in the course of electron diffraction identification work the structural peculiarity of halloysite, chapmanite, ferripyrophyllite has been established as a first stage requirement for a further structural refinement.

A number of problems of structural mineralogy have been solved with the use of SAD. In combination with OT patterns the SAD patterns helped to identify polytypes and to construct structural models. Thus, reflexion rows $02l$ with l uneven in SAD patterns of single rod-shaped crystallites were important testimonies in favour of the above mentioned monoclinic two-layer structure of halloysite (Chukhrov, Zvyagin, 1966; Gritsaenko *et al.*, 1969). The SAD patterns of fiber bundles have not only revealed the meaning of diffuse bands in the patterns of macro-specimens, but played a decisive role in the proposal of a structural model of chrysocolla, consisting of sheets of Cu-octahedra interconnected by intermediate sheets of Si-tetrahedra facing in opposite directions relative their common basal plane (Chukhrov, Zvyagin *et al.*, 1969; Gritsaenko, Zvyagin *et al.*, 1969).

Both OT and SAD patterns of polycrystalline aggregates have been highly effective in the study of iron hydroxides which had been considered earlier as amorphous or pseudo-amorphous. Not only was their crystalline state proved but new minerals were discovered among them and structural models proposed. The structures of ferrihydrite and feroxihite are thus presented as frameworks of Fe-octahedra arranged with respectively 4 and 2 layer repeats in a hexagonal close packing of O, OH, H_2O (Chukhrov, Zvyagin *et al.*, 1972; Chukhrov, 1975; Chukhrov, Zvyagin *et al.*, 1976).

Applied in an independent way SAD has given striking results in the study of remarkable hybrid minerals consisting of different kinds of layers, which separately form quite different and independent structures (e.g. of the brucite and sulfide type). In the hybrid structures these layers alternate in a unique sequence as if forming two structures, one inserted into another. These structures are characterized by pairs of different a, b periods which may be rationally related, and by a common 'c' period. As a result of intensity analysis of point patterns, structural projections on the plane parallel to the layers have been successfully obtained which revealed the distribution of cations of both layers (Organova, Drits, Dmitric, 1974).

Such an important scientific event as the discovery of a new type chain silicate with the $[Si_6O_{16}]$ radical should be put down to the credit of SAD (Drits, Goncharov, 1974). Its structure has been determined as a result of intensity analysis of point patterns and the construction of structural projections. The structural units $[Si_6O_{16}]$ have been found to represent a condensation result of 3 pyroxene chains $[SiO_3]$ and may be considered as a next step after amphibole units $[Si_4O_{11}]$ towards silicate sheets $[Si_2O_5]$.

By means of SAD, different superlattices and new structural varieties have been found in the complex system of oxides and hydroxides of Mn including such minerals as todorokite, birnessite etc. (Chukhrov, Gorshkov *et al.*, 1978; 1979).

It may be concluded that electron diffraction methods have demonstrated that they indeed are effective in the study of minerals, and that further achievements are to be expected in future.

Institute of Ore Geology and Mineralogy (IGEM)
Academy of Sciences
Moscow, USSR

REFERENCES

Chukhrov, F. V. (Ed.): 1975, *Hypergene Iron Oxides.* Moscow, Nauka Press (in Russian).
Chukhrov, F. V., Gorshkov, A. I., Sivtsov, A. V., and Berezovskaya, V. V.: 1978, *Izvestiya AN SSSR* (Geological Series) **12**, 86–95.
Chukhrov, F. V., Gorshkov, A. I., Sivtsov, A. V., and Berezovskaya, V. V.: 1979, *Izvestiya AN SSSR* (Geological Series) **1**, 83–90.
Chukhrov, F. V. and Zvyagin, B. B.: 1966, *Proc. Intern. Clay Conf. Jerusalem* Vol. 1, 11–26.
Chukhrov, F. V., Zvyagin, B. B., Drits, V. A., Gorshkov, A. I., Ermilova, L. P., Goilo, E. A., and Rudnitskaya, E. S.: 1979, *Proc. Intern. Clay Conf. Oxford* 1978, 55–64.
Chukhrov, F. V., Zvyagin, B. B., Ermilova, L. P., and Gorshkov, A. I.: 1972, *Proc. Intern. Clay Conf., Madrid* 333–341.
Chukhrov, F. V., Zvyagin, B. B., Ermilova, L. P., Gorshkov, A. I., and Rudnitskaya, E. S.: 1969, *Proc. Intern. Clay Conf., Tokyo* 141–150.
Chukhrov, F. V., Zvyagin, B. B., Gorshkov, A. I., Ermilova, L. P., Korovushkin, V. V., Rudnitskaya, E. S., and Yakubovskaya, N. Yu.: 1976, *Izvestiya AN SSSR* (Geological Series) 5, 5–24.
Drits, V. A., Goncharov, Yu. I.: 1974, *Collected Abstracts of IX IMA Meeting* W. Berlin, 78.
Drits, V. A., Tsipursky, S. I.: 1979, *Collected Abstracts*, 5th *ECM*, Copenhagen, 73.
Gritsaenko, G. S., Zvyagin, B. B., Boyarskaya, R. V., Gorshkov, A. I., Samotoin, N. D., and Frolova, K. E.: 1969, *Methods of Electron Microscopy of Minerals* Moscow, Nauka Press.
Organova, N. I., Drits, V. A., and Dmitric, A. L.: 1974, *Amer. Miner.* **59**, 199–200.
Pinsker, Z. G.: 1949, *Electron Diffraction* Moscow, Ac. Sc. USSR Press, (in Russian); English translation: Butterworths Scientific Publications, London, 1953.
Tsipursky, S. I., and Drits, V. A.: 1977, *Izvestiya AN SSSR* (Physics series), **41**, 2263–2271.
Vainshtein, B.K.: 1956, *Structure Analysis by Electron Diffraction* Ac.Sc.USSR Press, Moscow (in Russian); English translation: Pergamon Press Oxford, New York, 1964.
Zvyagin, B. B.: 1964, *Electronography and Structural Crystallography of Clay Minerals* Moscow, Nauka Press (in Russian); English translation: *Electron-Diffraction Analysis of Clay Mineral Structures*, Plenum press, New York, 1967.
Zvyagin, B. B., Vrublevskaya, Z. V., Zhukhlistov, A. P., Sidorenko, O. V., Soboleva, S. V., and Fedotov, A. F.: 1979, *High-Voltage Electron Diffraction in the Study of Layered Minerals* Nauka Press, Moscow.

S. A. SEMILETOV AND R. M. IMAMOV

II.35. *Electron Diffraction Investigation into Thin Film Structures*

1. INTRODUCTION

Thin films of metals, semiconductors and dielectrics are widely used in modern science and technology both in fundamental investigations and in the manufacture of various devices.

Radio and electrical engineering, acoustics and optics, metallurgy and chemistry, microelectronics and computational engineering are just some of the fields in which thin films are used. In science the use of thin films gave new fundamental results and a series of discoveries. Thin films can be studied by methods inapplicable to macrospecimens. Such methods include high-resolution electron microscopy, tunnelling of electrons, the use of electrical fields with field strengths up to 10^7–10^8 V cm^{-1}, optical investigations in the far IR- and UV-regions of the spectrum etc.

In the context of scientific and technological revolution and rapid development of microelectronics and computational engineering the significance of thin film materials is ever increasing. Along with the traditional fields of their use there appear new branches such as atomic engineering (radiation-proof coatings of the first wall of reactors), machinery (improvement of mechanical and corrosion-resultant properties of steels and other structural materials with the use of thin film coatings), the solar energy transformation (photoelements of large surface area), and cryogenics (thin film superconductive elements and devices) etc.

At the same time the stability of thin film elements and their reliability still fall short of the requirements of modern technology, which is due to thin film sensitivity to ambient atmosphere and to higher speeds of phase transformations, diffusion and electromigration of atoms, and to accompanying processes of aging and degradation faster than those in macrospecimens. To prevent these processes and understand their nature, in particular the mechanism of phase transformations and causes of thin film devices failure and parameter changes, one should study in detail the structure of thin films and their phase composition and transformations at all stages of the preparation and exploitation of thin films. And no doubt, one of the main methods for it must be HEED.

The Electron Diffraction Laboratory at the Institute of Crystallography of the USSR Academy of Science has accumulated a lot of experimental data on the structural investigation of thin films by HEED. We shall present here in brief the main results of these investigations.

Heed began to be used immediately after Thomson's and Tartakovsky's experiments on electron transmission in thin films (Thomson and Cochrane, 1939; Tartakovsky, 1932). Yet in the initial stages of HEED development it was used only as an auxiliary tool for phase analysis of thin films. It was not until the establishment of fundamentals of electron diffraction structure analysis that HEED found its application for deciphering unknown structures. Systematic studies of the structures of semiconducting thin films were initiated by Pinsker in 1950.

All these investigations made wide use of the methods of electron diffraction structure analysis developed by Pinsker (1949) and Vainshtein (1956). The preparation of texturized thin films for complete structure determination has been greatly aided by a technique worked out by Semiletov (1954) which consists in evaporation of pure components or synthesized alloys in high vacuum and condensation of the vapours onto substrates. By varying substrate temperature of subsequent annealing it is possible to obtain specimens of desired orientation, which provide excellent oblique texture electron diffraction patterns. A pattern of oblique texture type is presented in Zvyagin's article (this volume).

2. BINARY COMPOUNDS

It should be noted that in the 'fifties the technique of single crystal growing was still fairly primitive and X-ray investigation of semiconductor structures (most of them being sulphides, selenides and tellurides) was a rather complicated procedure. Thus the preparation of texturized thin films was very important. It should also be mentioned that, recently, thin film textures have been used as electromechanical transducers in generating acoustical phonons (Semiletov and Magomedov, 1967). The first objects for the electron diffraction structure analysis of semiconductors were thin films of Se, Te (Semiletov, 1955a), CdTe (Semiletov, 1955b), bismuth selenide and telluride (Semiletov, 1954). During the investigation into Se and Te thin films no anomalies have been revealed, the parameter obtained for Te-atoms being the same (0.266) as in the X-ray study. Interesting results were obtained during the investigation of the binary system Cd–Te (Semiletov, 1955b; Semiletov, 1956a). The specimens were prepared by evaporation of Cd and Te from two independent sources placed 60 mm apart. The film obtained on the substrate had a composition varying from pure Cd to pure Te, thus representing the whole phase diagram.

In the system Cd–Te, according to metallographic data, there exists only one compound (CdTe) forming no solid solutions with the components. The electron diffraction patterns showed no changes in Cd, Te or CdTe lattice parameters, i.e. there were no indications of the formation of solid solutions in thin films. Yet, in addition to the cubic phase CdTe, known for macrospecimens (sphalerite structure), in films it was discovered that there was also a hexagonal modification, CdTe with the wurtzite structure.

In most thin films there were observed crystals of both phases; it was also established that the quantitative ratio of both modifications depends on condensation rate, substrate temperature and film thickness. The indexing of spot electron diffraction patterns obtained from epitaxial CdTe films grown on mica permitted the determination of the mutual orientation of cubic and hexagonal phases. It was also suggested that there is a continuous structural transition from one phase to the other and the formation of crystals with variable structure. The detailed structure investigation also helped to explain the appearance of an anomalous high-voltage electromotive force in CdTe films exposed to light (Semiletov, 1962).

Later it was shown that the formation of two phases and crystals with variable structure is typical not only for CdTe films but also for those of indium antimonide and some other compounds of the $A^{II}B^{VI}$ and $A^{III}B^{V}$ types (CdS, CdSe, ZnS, ZnSe, ZnTe, InAs, GaAs and GaSb). During the investigation of the Bi–Se and Bi–Te binary systems (Semiletov, 1955c; Semiletov and Pinsker, 1955) specimens were prepared both by evaporation of the components from independent sources and by that of alloys of the composition Bi_2Se_3 and Bi_2Te_3. The existence has also been established of new phases and other considerable variations of the unit cell parameters for Bi_2Se_3 and Bi_2Te_3 due to the formation of solid solutions. Later X-ray investigation has confirmed the existence of wide ranges of homogeneity for Bi_2Se_3 and Bi_2Te_3 and also revealed in these systems many new phases, which represent a new type of ordered solid solution (Imamov and Semiletov, 1970). Using HEED, it has also been established that the films of Bi_2Se_3 and Bi_2Te_3 possess tetradimite structure. The same structure has also been found for Sb_2Te_3 (Semiletov, 1955b). Thus, the identity of structures has been proved for thin films and macrospecimens of the above considered compounds.

The study of the systems In–Se,, In–Te and Ga–Te has yielded rich experimental data. The indexing of electron diffraction patterns from InSe gave a hexagonal unit cell with parameters $a = 4.041$ $c = 16.90$ Å. The structure is built by packets along the z-axis, each packet consisting of 2In and 2Se atoms, forming the sequence . . . SeInInSe . . . (Semiletov, 1958). The binding between the packets is obviously very weak which results in the perfect cleavage along the bases. It was shown that the hexagonal modification of GaTe also has the same structure (Semiletov and Vlasov, 1963), yet it is metastable and formed only in thin films in highly nonequilibrium conditions (fast condensation of vapours onto cold substrate). Brief annealing of the phase at 100–$120°C$ was accompanied by transformation into a monoclinic modification which had been known for macrospecimens.

During the investigation of In_2Se_3 thin films (Semiletov, 1960; Semiletov, 1961a,b,c,) it was established that the compound had at least two hexagonal modifications. The α-modification, stable at room temperature, is formed by two-layer hexagonal packing of Se-atoms, with most In atoms occupying tetrahedral voids while the remaining 1/16 of them are situated in octahedral sites. Two layers of occupied voids alternate with two empty ones (Semiletov, 1961b,c).

The β-modification, stable above 300°C, is also formed by the two-layer hexagon packing of Se-atoms. But in contrast to the α-form all the In atoms occupy tetrahedral voids (Semiletov, 1960; Semiletov, 1961a). The structures of the orthorhombic phases in the systems In–Se and In–Te were studied independently by X-ray and electron diffraction methods, the results obtained being in good agreement (Man, Karakhanian and Imamov, 1974; Man, Imamov and Semiletov, 1976; Hogg, Sutherland and Williams, 1973). These investigations resulted also in refining of the chemical composition of the orthorhombic phases which proved to be In_4Se_3 and In_4Te_3 instead of In_2Se and In_2Te correspondingly.

These structures are characterized by infinite chains built by five-membered rings of metal and halogen atoms and connected by strong metal–metal bonds (In—In = 2.77 and 2.78 Å for In_4Se_3 and In_4Te_3, respectively). The presence of such bonds in (ab)-planes explains the strong anisotropy of physical properties observed for these compounds.

During electron diffraction investigation into thin films in the systems Ga–Te, In–Se and In–Te (Semiletov and Vlasov, 1963; Semiletov, 1961c) cubic phases were also found having primitive lattice Ga_3Te_{4-x}, In_3Se_{4-x}, and In_3Te_{4-x}. Such compositions were attributed to the phases, as in the phase diagrams they were situated between the phases of composition $A^{III}B^{VI}$ and $A_2^{III}B_3^{VI}$. The analysis of the lattice parameters and intensity distributions for the reflections indicated the same structure type for all the phases which was confirmed by structure determination. Structure determination also resulted in a refinement of the composition for the phases ($x = 0.16$). A distinctive feature of these phases is the presence of two type tetrahedra: those consisting of halogen atoms Se (Te) only (as in β-In_2Se_3) and those consisting of 3 metal 3In (3Ga) and one halogen atom Se (Te) (as in InSe).

The data on the γ-phase in the system Tl–Te were inconsistent. The composition was given either as Tl_5Te_3 or Tl_2Te though the unit cell was considered to be cubic for both compositions. The complete structure determination carried out independently by X-ray and HEED (Man, Imamov and Pinsker, 1971; Bhan and Schubert, 1970) showed that the composition of the phase was, in fact, Tl_5TE_3 while the structure was tetragonal. Atoms in this structure form layers perpendicular to the c-axis, interatomic distances In—In, In—Te and Te—Te do not vary significantly, and the interaction takes place not only between the Tl- and Te-layers but also between Tl-layers, which makes this structure similar to intermetallic compounds of the Mn_5Si_3-type.

The same structure has also been established by the electron diffraction method for the compound Tl_5SE_3 (Man, Parmon, Avilov and Imamov, 1980). Very interesting results have been obtained during the electron diffraction study of halogenides of Ag and Cu. For the system Ag–Se only one high-temperature-disordered phase β-Ag_2Se had been known. In HEED studies there were also found cubic phases enriched by silver in comparison with the β-Ag_2Se: these were fcc cubic phase ($a = 6.90$ Å) of approximate composition $Ag_{13.5}Se_4$ and a cubic phase representing solid solution on the basis of Ag ($a = 4.10$–4.11 Å) (Chou Tsin-lian and Pinsker, 1962). A low-temperature orthorhombic phase α-Ag_2Se has also been deciphered in our laboratory (Pinsker, Chou Tsin-lian, Imamov and Lapidus, 1965). This ordered structure can be described as slightly distorted (in respect to the orthorhombic symmetry) diamond-like structure of Ag-atoms while Se-atoms form flat zig-zag chains parallel to the c-axis. On the basis of these structural data one could expect metallic conductivity, which, in fact, was registered in later investigations of electrical and physical properties.

The structure of the hexagonal phase in the Ag–Te system determined by HEED is rather complicated (Imamov and Pinsker, 1966). Hexagonal phases of composition close to Ag_2Te were repeatedly reported in the literature, there were some X-ray data on their parameters and symmetry but no complete structure determination had been performed. Electron diffraction structure determination of this phase in the system Ag–Te has been performed on the basis of 450 observed structure factors. The composition of the phase had been established as Ag_7Te_4. Only part of all the atoms in the structure form Ag_2Te groupings and therefore it cannot be classified either as pure coordination or as molecular structure. Electron diffraction study of the system Cu–Te showed that all the known phases exist in the concentration range between 50 and 32 at% Te. The compound CuTe of stoichiometric composition has orthorhombic structure, X-ray and HEED data being in coincidence (Baranova and Pinsker, 1964). Moving through phase diagram towards pure Cu, one reaches tetragonal ricardite. According to various X-ray data, the composition of the phase (both natural and synthetic) lies in the range $Cu_{1.33}Te$–$Cu_{1.50}Te$. Electron diffraction study has revealed in this range two tetragonal modifications with different lattice parameters and compositions: γ^I-phase of composition $Cu_{1.36}Te$ ($a = 3.98$ $c = 6.12$ Å) and γ^{II}-phase of composition $Cu_{1.50}Te$ ($a = 3.98$ $c = 6.55$ Å) (Baranova and Pinsker, 1969). In both phases Cu-atoms are situated statistically.

While studying thin films in the system Cu–Te, there have also been established some hexagonal phases: β^I-$Cu_{1.75}Te$; β^{II}-$Cu_{1.75\ 0.4}Te$ and β^{III}-$Cu_{1.84}Te$ (Baranova, 1967; Baranova, 1968; Baranova, Avilov and Pinsker, 1973). The lattice parameters of these phases are in multiple proportions, i.e. for β^I: $a = 2a_0$ $c = c_0$; for β^{II} $a = a_0$ $c = 3c_0$ and for β^{III} $a \approx 2a_0$ $c \approx 3c_0$, where $a_0 = 4.15$ and $c_0 = 7.20$ Å. Despite the multiplicity of the parameters and overlapping of the concentration fields the phases belong to different and rather complicated structure types. But all of them are characterized by significant scatter in interatomic distances and somewhat shortened bonds in comparison with the sum of atomic or covalent radii.

3. TERNARY COMPOUNDS

The stoichiometric mixed telluride of the composition $CuAgTe_2$ has also been studied by HEED (Avilov and Baranova, 1972) and proved to be isotypic to orthorhombic telluride CuTe. For another mixed telluride $CU_{0.80}Ag_{0.96}Te$ there has been established a hexagonal structure (Avilov, Baranova and Pinsker, 1976). Though the phases $CU_{0.80}Ag_{0.90}Te$ and $\beta^{II}Cu_{1.84}Te$ have the same symmetry $(P3m1)$ and close lattice parameters they possess different, though related, structures. Two-thirds of Cu-atoms and 2Ag-atoms occupy their positions statistically but with different probability. Along with the study of binary semiconductors our laboratory has for many years been carrying out investigations of ternary semiconductors. A great contribution has been made by Imamov, who developed the technique of preparation of ternary thin films, and also by Man, Baranova and Avilov. Many of the ternary phases were discovered while studying pseudobinary systems formed by compounds of the types Bi_2Te_3 (Bi_2Se_3 and Sb_2Te_3) and PbSe (SnTe, PbTe and GeTe). They have compositions $A^{IV}B_2^VC_4^{VI}$, $A_2^{IV}B_2^VC_5^{VI}$, $A^{IV}B_4^VC_7^{VI}$ and $A_3^{IV}B_2^VC_6^{VI}$ (Talybov and Vainshtein, 1961; Agajev and Semiletov, 1965; Agajev, Talybov and Semiletov, 1968; Petrov and Imamov, 1970; Vainshtein, Imamov and Talybov, 1969; Petrov and Imamov, 1969; Imamov, Semiletov and Pinsker, 1970).

These phases crystallize in hexagonal or orthorhombic systems and usually possess a pseudoperiod along the c-axis. Disregarding kinds of atoms, the structures can be described by cubic close packing, in which interlayer distances do not exceed 1.80–2.00 Å. These peculiarities are clearly seen on electron diffraction patterns from textures. Preliminary structure models have been deduced from the considerations of electron diffraction pattern geometry and intensity distribution of reflections. These models have been confirmed by the Fourier method. The structures determined can be classified according to the number of layers along the c-axis, as it is correlated with the composition and depends on the relative quantity of binary phases participating in the formation of ternary compounds. The coordination of atoms is octahedral both for initial binary and resulting ternary compounds, while the structures of the latter are closely related to the Bi_2Te_3-type. For $AgBiTe_2$ (Pinsker and Imamov, 1964), $TlBiSe_2$ (Semiletov and Man, 1959; Man and Semiletov, 1962) two types of structure have been established: NaCl with disordered distribution of atoms (cations) in the unit cell and $NaInO_2$ with ordered distribution of Ag and Bi, Tl and Bi or Tl and Sb atoms respectively. In the ordered (rhombohedral) structures Te(Se)-atoms build layers ... ABCABC ... along the c-axis (hexagonal setting) while Ag(Tl) and Bi-atoms occupy all the octahedral sites in an orderly way.

If one denotes disordered layers of Ag, Tl or Bi by small letters (no matter which) and ordered layers of Ag, Tl or Bi by small letters with indices, the sequence of layers for the disordered NaCl-type structure is:

$$\dots b\overset{|}{A}cBaC\overset{|}{b}AcBaC\dots,$$

and for ordered $NaInO_2$-type structures:

$$\dots b_{Ag}\overset{|}{A}c_{Bi}Ba_{Ag}Cb_{Bi}Ac_{Ag}Ba_{Bi}Cb_{Ag}\overset{|}{A}c_{Bi}\dots,$$

where vertical strokes show the identity period along the c-axis.

TABLE 1

Compound	Comment	Reference
$CuAsSe_2$	Sphalerite type structure either (a) ordered rhombohedral or (b) random Cu As distribution	Imamov and Petrov, 1968
$TlSbSe_2$	Orthorhombic. Zig-zag chains of Se (Se-Se = 2.15 Å) held between nets of Sb.	Pinsker, Semiletov and Belova, 1956
$CuSbSe_2$	$CuSbS_2$ structure type.	Imamov, Pinsker and Ivchenk, 1964
$AgTlTe_2$	Tetragonal phase: corner linked $AgTe_4$ tetrahedra with Tl atoms at center of slightly distorted Te atom cube.	Imamov and Pinsker, 1964
$GaTlTe_2$ $GaInTe_2$ $InTlTe_2$	TlSe structure type Different cations (Ga, In and Tl) correspond to different voids in the TlSe structure.	Avilov, Agajev, Guseinov and Imamov, 1969
$AgTlSe_2$	Six membered rings of Se atoms between flat, centered hexagonal nets of Ag/Tl.	Imamov and Pinsker, 1965

Thus, in this case, ordering consists in the occupancy by the atoms of their sites in specific planes which results in the formation of a superperiod $c_{hex} = 2\,a_c\sqrt{3}$ and transition from cubic $m3m$ into rhombohedral $\bar{3}m$ symmetry.

Other compounds with an analogous formula were found to have different structures, as documented in Table I.

For rather a long time the electron diffraction laboratory has also been studying atomic structures of oxides of transitional metals (Klechkovskaya, Troitskaya and Pinsker, 1965; Khitrova, 1966; Klechkovskaya and Khitrova, 1968; Khitrova and Pinsker, 1970), those of Bi and Sb (Zavialova, Imamov and Pinsker, 1964; Zavialova and Imamov, 1968; Zavialova and Imamov, 1972), and some other compounds. Without going into particulars of their structures, we should like to underline that many disordered structures have been found among oxides, which has enriched our knowledge of these compounds (works of Khitrova, Zavialova and Klechkovskaya).

Considering the perspective of EDSA development, it should be noted that some important classes of compounds have been insufficiently studied. These are, in particular, oxides, sulphides and hydrides of transitional and rare-earth metals and their analogues. Here many problems must be solved concerning investigations into specific phases and systems, and a better understanding and generalization of common features of their structure obtained.

4. CONCLUSION

In structure determinations of the phases considered above, the kinematical theory of electron scattering has been applied and, in some cases, corrections for extinction have been made. The comparison of the atomic coordinates received by HEED with those obtained in the X-ray determinations has proved that in the case of texture electron diffraction patterns the use of the kinematical theory is justified. The precise measurements of reflection intensities on the electron diffraction patterns of PbSe and Bi_2Se_3 has also revealed that in the presence of heavy atoms extinction effects are rather significant. It has been established, in particular, that in photographing electron diffraction pictures one should carefully choose beam inclination angle, as dynamic effects are stronger for some specific angles. As to the effects of many-beam scattering, we believe that they do not significantly affect the results of our structure investigations.

Institute of Crystallography
Academy of Sciences
Moscow, USSR

REFERENCES

Agajev, K. A. and Semiletov, S. A.: 1965, *Kristallografiya* **10**, 109.
Agajev, K. A., Talybov, A. G., and Semiletov, S. A.: 1968, *Kristallografiya* **13**, 59.
Avilov, A. S., Agajev, K. A. Guseinov, G. G. and Imamov, R.M.: 1969, *Kristallografiya* **14**, 443.
Avilov, A. S. and Baranova, R. V.: 1972 *Kristallografiya* **17**, 219.
Avilov, A. S., Baranova, R. V., and Pinsker, Z. G.: 1976, *Kristallografiya* **21**, 89.
Baranova, R. V.: 1967, *Kristallografiya* **12**, 266.
Baranova, R. V.: 1968, *Kristallografiya* **13**, 803.
Baranova, R. V., Avilov, A. S., and Pinsker, Z. G.: 1973, *Kristallografiya* **18**, 1169.
Baranova, R. V. and Pinsker, Z. G..: 1964, *Kristallografiya* **9**, 104.
Baranova, R. V. and Pinsker, Z. G.: 1969, *Kristallografiya* **14**, 274.
Bhan, S. and Schubert, K.: 1970, *J. Less-Common Metals* **20**, 229.
Chou Tsin-lian and Pinsker, Z. G.: 1962 *Kristallografiya* **7**, 66.
Imamov, R. M. and Petrov, I. I.: 1968, *Kristallografiya* **13**, 412.
Hogg, J. H. C., Sutherland H. H., and Williams, D. J.: 1973, *Acta Cryst.* 1973 B**29**, 1590.
Imamov, R. M. and Pinsker, Z. G.: 1964, *Kristallografiya* **9**, 743.
Imamov, R. M. and Pinsker, Z. G.: 1965, *Kristallografiya* **10**, 199.
Imamov, R. M. and Pinsker, Z. G.: 1966, *Kristallografiya* **11**, 17.
Imamov, R. M., Pinsker, Z. G., and Ivchenko, A. I.: 1964, *Kristallografiya* **9**, 853.
Imamov, R. M. and Semiletov, S. A.: 1970, *Kristallografiya* **15**, 972.
Imamov, R. M., Semiletov S. A., and Pinsker, Z. G.: 1970, *Kristallografiya* **15**, 278.
Khitrova, V. I.: 1966, *Kristallografiya* **11**, 204.
Khitrova, V. I. and Pinsker, Z. G.: 1970, *Kristallografiya* **15**, 540.
Klechkovskaya, V. V. and Khitrova, V. I.: 1968, *Kristallografiya* **13**, 523.
Klechkovskaya, V. V., Troitskaya, N. V., and Pinsker, Z. G.: 1965, *Kristallografiya* **10**, 37.
Man, L. I., Imamov, R. M., and Semiletov, S. A.: 1976, *Kristallografiya* **21**, 628.
Man, L. I., Imamov, R. M., and Pinsker, Z. G.: 1971, *Kristallografiya* **16**, 122.
Man, L. I., Karakhanian, R. K. and Imamov, R. M.: 1974, *Kristallografiya* **19**, 1166.
Man, L. I., Parmon, V. S., Avilov, A. S., and Imamov, R. M.: 1980, *Kristallografiya* **25**, 1070.
Man, L. I. and Semiletov, S. A.: 1962, *Kristallografiya* **7**, 844.
Petrov, I. I. and Imamov, R. M. : 1969, *Kristallografiya* **14**, 699.

Petrov, I. I. and Imamov, R. M.: 1970, *Kristallografiya* **15**, 168.

Pinsker, Z.G.: 1949, *Diffraktsia Electronov*. Moscow-Leningrad: Nauka.

Pinsker, Z. G., Chou Tsin-lian, Imamov, R. M., and Lapidus, E. L.: 1965, *Kristallografiya* **10**, 276.

Pinsker, Z. G. and Imamov, R. M.: 1964, *Kristallografiya* **9**, 347.

Pinsker, Z. G., Semiletov, S. A., and Belova, E. L.; 1956, *DAN SSSR* **6**, 1003.

Semiletov, S. A.: 1954, *Trudy Instituta kristallografii AN SSSR* **10**, 76.

Semiletov, S. A.: 1955a, *Trudy Instituta kristallografii AN SSSR* **11**, 115.

Semiletov, S. A.: 1955b, *Trudy Instituta kristallografii AN SSSR* **11**, 121.

Semiletov, S. A.: 1955c, *Zh. technicheskoi fisiki* **25**, 2336.

Semiletov, S. A.: 1956a, *Kristallografiya* **1**, 306.

Semiletov, S. A.: 1956b, *Kristallografiya* **1**, 403.

Semiletov, S. A.: 1962, *Fizika tverdogo tela* **4**, 1241.

Semiletov, S. A.: 1958, *Kristallografiya* **3**, 288.

Semiletov, S. A.: 1960, *Kristallografiya* **5**, 704.

Semiletov, S. A.: 1961a, *DAN SSSR* **137**, 594.

Semiletov, S. A.: 1961b, *Fizika tverdogo tela* **3**, 746.

Semiletov, S. A.: 1961c, *Kristallografiya* **6**, 200.

Semiletov, S. A. and Magomedov, Z. A.: 1967, *Kristallografiya* **12**, 376.

Semiletov, S. A. and Man, L. I.: 1959, *Kristallografiya* **4**, 414.

Semiletov, S. A. and Vlasov, V. A.: 1963, *Kristallografiya* **8**, 877.

Semiletov, S. A. and Pinsker, Z. G.: 1955, *DAN SSSR* **100**, 1079.

Talybov, A. G. and Vainshtein, B. K.: 1961, *Kristallografiya* **6**, 514.

Tartakovsky, P. S.: 1932, *Experimentalnyje osnovaniya volnovoi teorii* Moscow-Leningrad:GTTIh.

Thomson, G. P. and Cochrane, W.: 1939, *Theory and Practice of Electron Diffraction*. London: McMillan.

Vainshtein, B. K.: 1956, *Strukturnaya elektronografiya*. Moscow: Izdatelstvo Akademii Nauk SSSR.

Vainshtein, B. K., Imamov, R. M., and Talybov, A. G.: 1969, *Kriatallografiya* **14**, 703.

Zavialova, A. A., Imamov, R. M., and Pinsker, Z. G.: 1964, *Kristallografiya* **9**, 857.

Zavialova, A. A. and Imamov, R. M.: 1968, *Kristallografiya* **13**, 49.

Zavialova, A. A. and Imamov, R. M.: 1972, *Zh.Strukt.Khim.* **13**, 869.

L. V. VILKOV

II.36. *The Development of Gas Phase Electron Diffraction in the USSR*

1. BEGINNING OF GAS ELECTRON DIFFRACTION IN THE USSR

The development of gas phase electron diffraction in the USSR had firm foundations laid by extensive electron diffraction studies on condensed phases carried out in this country since the early 1930s. It was not, however, until the 1950's that actual gas phase work began. The first gas phase electron diffraction unit was constructed specially for the Department of Chemistry of Moscow State University on the initiative of A. V. Frost and P. A. Akishin in 1950 (Frost, Akishin, Gurvich *et al.* 1953). This instrument, of horizontal type, was made of brass and had no sector. Diffraction patterns were registered on five small (6 × 6 cm) photographic plates. The unit was equipped with a glass evaporator used at low and moderate temperatures and a metal ampoule heated by radiation which, however, could only be used at temperatures not higher than 1000°C.

In 1954 when the Department of Chemistry moved into the new University building, P. A. Akishin instituted the laboratory for molecular studies by electron diffraction. Two more electron diffraction units of horizontal type were built which had better vacuum systems than their predecessor and were equipped with electron guns of the screened type. They were reserved for studies of inorganic substances. Electronic bombardment was applied to heat samples to temperatures about 2500°C. The ampoules were made of molybdenum, tantalum, tungsten, and platinum. Photographic emulsions were protected from radiation by coating the photoplates with black India ink (Akishin, Vinogradov *et al.* 1958). Later, the original brass instrument was reconstructed for the purpose of studying organic substances. New pumps were installed to increase pumping velocity. A new metallic evaporator that could be heated up to 300°C was made. The diffraction patterns were registered on four larger (6 × 9 cm) photographic plates (Vilkov, Mastryukov and Akishin, 1963).

2. NEW STAGE OF DEVELOPMENT

Since the late 1950s, the sector-microphotometer method and computer treatment of electron diffraction data have been used. Optical densities are

318

Fig. 1.   A. Ivanov (left) and A. Demidov with the EG–100 M instrument, and the evaporator for high-temperature studies (being held by Demidov).

measured by oscilating the plate about the pattern centre during radial scanning.

The appearance of commercial electron diffraction units EG-100 (Models A and M) whose production was started in the Ukraine (Sumy) in the early 1960s (Levkin and Alekseev, 1963) was a great stimulus to the development of gas phase electron diffraction studies in the USSR. These instruments of the vertical type, however, were intended primarily for crystal diffraction and needed certain alterations and additions for gas work such as an evaporator, a sector assembly, and a cold trap. The vacuum and high voltage systems were reasonably good: a vacuum of about 10^{-5} torr was maintained in the diffraction chamber, the accelerating voltage could be varied stepwise by 20 kV from 40 to 100 kV, the high voltage drift was only about 0.01% per hour, and the electron beam was about 0.1 mm in diameter and had a current of about 0.2 μA. Fifteen plates (9×12 or 13×13 cm) could be loaded at one time and patterns could be taken at three camera distances of about 200, 400 and 700 mm.

The production of these EG-100 units greatly increased the scope of gas phase electron diffraction studies in the USSR and in Hungary. In the USSR the number of groups engaged in high-temperature electron diffraction studies increased to five and those studying organic compounds to three. The high-temperature groups are:

1. V. P. Spiridonov, E. Z. Zasorin, A. A. Ivanov, A.A. Ischenko, G.V. Romanov, E.V. Erokhin and A. V. Demidov (Department of Chemistry, Moscow State University).

2. N. G. Rambidi and V. V. Ugarov (All-Union Institute of Meteorological Service, Moscow).

3. Yu. S. Ezhov (High-Temperature Institute of the USSR Academy of Sciences, Moscow).

4. K. S. Krasnov, G. V. Girichev (Institute of Chemical Technology, Ivanovo).

5. G. I. Novikov, I. M. Zharskii, G. E. Slepnev, V. N. Kupreev (Byelorussian Technological Institute, Minsk).

Organic compounds are studied by

1. L. V. Vilkov, V. S. Mastryukov, A. V. Golubinskii, N. I. Sadova, L. S. Khaikin, M. V. Popik (Department of Chemistry, Moscow State University).

2. V. A. Naumov, N. M. Zaripov (Arbusov Institute of Organic and Physical Chemistry, Kazan).

3. N. V. Alekseev (first in the Institute of Organo-Element Compounds of the USSR Academy of Sciences, now in the State Scientific Research Institute of Chemistry and Technology of Organo-Element Compounds, Moscow). One of the original units for high-temperature studies was transferred from Moscow to Ivanovo in the mid 1960s. The Moscow unit was reconstructed: a new sector assembly was made, a box for two 18×24 cm photoplates was attached, and an electron gun of an EG-100 instrument was installed (Akishin, Rambidi and Spiridonov, 1967). Similar changes were made in the Ivanovo instrument. Also various modifications of the EG-100 unit were introduced (Ivanov, Spiridonov, Demidov and Zasorin, 1974; Mastryukov, Dorofeeva, Golubinskii, Vilkov et al., 1975). A. A. Ivanov designed a special radiation-heated evaporator to attain temperatures of up to 2000°C and A. A. Ivanov and E. Z Zasorin used a split-shaped nozzle and electron beam of a large (ca. 2 mm) diameter to register scattering at very low pressures (Spiridonov and Zasorin, 1979). Recently, A. V. Golubinskii and M. V. Popik have described a new evaporator for organic compounds (Popik, Vilkov, Mastryukov and Golubinskii, 1979). The use of metallic seals makes it possible to work at nozzle temperatures up to 900°C. The gas flow through the nozzle, is regulated by a needle valve. During the experiment, scattering patterns from standard substances (e.g. benzene or CS_2) can be taken to determine the electron beam wavelength. A cold trap containing zeolite has proved to be very effective for removing sample gases from the diffraction chamber.

Rambidi and co-workers have designed a vertical-type electron diffraction instrument, termed ELPRIM, having a volume of about 230 l. Electron diffraction patterns are registered on a single 18×24 photoplate (Rambidi, Ezhov et al., 1974). The large volume allows easy installation of extra apparatus inside the diffraction chamber. Over the years the refinement procedure underwent still more radical changes than the experimental technique. First, visual intensities were treated by the method of trial and error. All the theoretical curve calculations were performed manually with the help of hand calculators. Only a few parameters were determined. Radial distribution curves were used extensively. The least-squares method in its up-to-date form was introduced only in the early 1970's. Earlier, each parameter was refined separately by a step-by-step procedure. We owe this progress to the kindness of our Norwegian collegues who sent us their programme for treatment of electron diffraction data (Anderson, Seip, Strand and Stolevik, 1969). Similar programmes were written by Zaripov (Kazan). Novikov (1979) and Spiridonov, Prikhod'ko and Butaev (1979) suggested the use of some numerical methods to avoid manual operations and accelerate treatment of the experimental data.

3. STUDIES OF INORGANIC SUBSTANCES

Thanks to helpful cooperation from synthetic chemists, we have been able to study stereochemical patterns for many classes of compounds. Among inorganic compounds, much work has been done on various metal and non-metal halides, oxyhalides, oxides, sulphides, metal salts of oxygen-containing acids, halo-complexes.

Vapours of alkali metal halides contain monomeric MX, and dimeric, $(MX)_2$, species (Vilkov, Rambidi and Spiridonov, 1967). Measurements of dimers of lithium halides, Li_2X_2 (where X is F,Cl,Br), have given values for the geometrical parameters $r(Li—X)$ and $r(X...X)$. The electron diffraction patterns of all Group II metal halides, $M^{II}X_2$, were found to agree with linear structures for their molecules. Studies of electrical deflection of molecular beams, however, proved barium halides and some strontium and calcium halides to be non-linear. The discrepancy still remains unexplained (Vilkov, Rambidi and Spiridonov, 1967). Of the group III metal halides, $AlCl_3$ has a practically planar configuration at 1150 K (Spiridonov, Zasorin, 1979) whereas $M^{III}Cl_3$ molecules with M = La, Pr, Gd, Tb, Ho, Lu have pyramidal configurations at 1000 at 1200 K (Krasnov, 1979). In recent work by Ischenko, Spiridonov and Strand, energy characteristics of metal halides were estimated. The vibronic coupling constant for VBr_4 and barriers to pseudorotation in $NbCl_5$ and $TaCl_5$ were determined (Spiridonov, Ischenko and Zasorin, 1978).

Much work has been done on the B_2O_3 molecule. The most recent results obtained at 1620 K are indicative of a planar structure with a bent BOB fragment ($\angle BOB$ 137.5°) containing nonequivalent B—O and B=O bonds and linear O—B=O groups (Spiridonov and Zasorin, 1979). The monomeric oxides, Al_2O, Ga_2O, In_2O, Tl_2O feature wide (of about 150°) MOM bond angles (Vilkov, Rambidi and Spiridonov), 1967).

Chromium, molybdenum and tungsten trioxides form trimers and tetramers in the gas phase and have cyclic structures. The results of joint work with Hungarian colleagues, M. Hargittai and I. Hargittai, show that the six-membered ring of $(WO_3)_3$ has a chair conformation at 1700 K; a similar ring of $(MoO_3)_3$ is nearly planar (Spiridonov, and Zasorin, 1979).

The alkali metal atoms M are two-coordinate in a number of molecules of their salts and coordination compounds such as M_2SO_4, M_2CrO_4, M_2MoO_4, M_2WO_4, $MReO_4$, MNO_3, $M^IM^{II}X_3$, and $M^IM^{III}X_4$ where X stands for a halogen; that is they form four-membered rings of the type

$$M \langle \substack{O \\ O} \rangle A \quad or \quad M \langle \substack{X \\ X} \rangle M^{II(III)}.$$

Ugarov, Petrov and Rambidi have shown that the caesium atom in $CsAlCl_4$ is three-coordinate,

$$Cs—Cl—\overset{\displaystyle Cl}{\underset{\displaystyle Cl}{Al}}—Cl$$

The molecules of alkali metal salts and coordination compounds are not rigid, as follows from the large M^I—O and M^I—X amplitudes (as large as 0.2 Å against usual values of about 0.050Å) and still larger amplitudes for nonbonded distances (Rambidi, 1975; Spiridonov, Ishchenko and Zasorin, 1978).

Quite recently, Spiridonov and Gershikov developed the method of the so-called equilibrium r_e^h structure and of force field determination from electron diffraction data for small molecules such as BF_3, $AlCl_3$, $AsBr_3$, $SbBr_3$, $SNCl_4, B_2O_3$, Their method is based on the harmonic vibration approximation (Spiridonov and Zasorin, 1979).

4. STUDIES OF ORGANIC SUBSTANCES

The studies of organic compounds include halogen-, nitrogen-, oxygen-substituted derivatives; cyclic, bicyclic and polyhedral (carborane) compounds, many organoelement compounds, and metal carbonyls. The main results are summarized in the book by Vilkov, Mastryukov and Sadova (1978).

Some trends in the carbon-halogen bond lengths were observed in 1957 using only a very rough visual method of scattering pattern treatment. Thus, the C—Cl and C—Br bonds were shown to undergo elongation in acetyl halides CH_3COX (X = Cl,Br) and allyl halides CH_2=$CHCH_2X$. The CH_2=$CHCH_2X$ and CH_3OCH_2X molecules were found to have the *gauche* conformation (Vilkov, Rambidi and Spiridonov, 1967).

The main purpose of the studies of nitrogen and oxygen derivatives was to find the dependence of bond configurations at N and O on the substituents. The first work in that direction showed that the presence of C=O, C=C, N=O double bonds as in dimethylformamide, *N*-methylpyrrole, tetramethylurea, *N*-nitroso-dimethylamine and *N*-dimethylaminoformyl chloride, resulted in the planar or near-planar configuration of the three-coordinate nitrogen atom, whereas in *N*-dimethylaniline and *N*-cyanodimethylaniline, the bond configuration at the amine nitrogen atom is flattened only slightly. The flattening is accompanied by shortening of the central N—C and N—N bonds.

The presence of Si, P, and Sn substituents on N leads to almost planar bond configurations at nitrogen, e.g. in $[(CH_3)_2N]_3SiCl$, $(CH_3)_2NPCl_2$, $[(CH_3)_2N]_3P$, $(CH_3)_2NPF_4$, $[(CH_3)_2N]_2PCl$, $[(CH_3)_2N]_4Sn$, and $[(CH_3)_3SN]_3N$. The bond angles at nitrogen in $(CH_3)_2NPOCl_2$ and

$[(CH_3)_2N]_2SO$ are, however increased only slightly from $(CH_3)_3N$ ($111°$). On the other hand, the amino group causes elongation of neighbouring Si—Cl and, particularly, P—Cl bonds. The nitrogen atom remains pyramidal in aziridine derivatives, N-acetyl- and N-phenylaziridine, and tris(aziridinyl)phosphine. The bond angle at the two-coordinate oxygen atom was found to increase to $140°$ in $[(CH_3)_3Ge]_2O$ and $[(CH_3)_3SN]_2O$. The POC bond angle also increases somewhat (to $120°$) in $P(OCH_3)_3$, $P(OC_2H_5)_3$, and $P(OC_2H_5)_3$.

Since 1968 Sadova has carried out systematic studies of nitro-compounds. The C—Cl and C—Br bonds in $XC(NO_3)_2$ (X = Cl, Br) are found to be shortened markedly from XCH_3. The bond configuration at the amino nitrogen atom in nitroamines was shown to change from planar in $CH_3(CH_2Cl)NNO_2$ to pyramidal in $CH_3N(NO_2)_2$ and $CH_3(Cl)NNO_2$.

Naumov, Mastryukov and Alekseev have carried out numerous studies of cyclic compounds, beginning with cyclopropane derivatives and a new class of compounds, diaziridines, containing the CN_2 ring. Four-membered rings in di-X-silacyclobutane ($X=$ H, Cl), 4-silaspiro[3.3]heptane, azacyclobutane and P-oxo-P-chloro-1-thia-4-phosphacyclobutane were found to be nonplanar. In five- and six-membered rings, involving P-Cl groups, the preferred conformations are those with the P—Cl bond in the axial positions. Bicyclo [3.1.0]hexane was shown to have a boat conformation. The results obtained for bicyclo[3.3.1]nonane at 65 and 400°C demonstrated that the chair–chair conformation was more stable than the chair–boat one, the energy difference between the two conformers being of about 2.5 kcal mol^{-1}.

In the early '70s Alekseev's group proved that the cyclopentadienyl derivatives $C_5H_5A(CH_3)_3$ where A is Si, Ge, Sn did not have half-sandwich structures and contained no π- bonds but only 'normal' element-to-ring σ-bonds.

Mastryukov has carried out numerous studies of carborane structures, such as $1,6-C_2B_4H_6$, $1,10-C_2B_8H_{10}$, 1,2- and $1,12-(CH_3)_2C_2B_{10}H_{10}$, 1,7-and $1,12-I_2C_2B_{10}H_{10}$, $1,12-CHB_{10}H_{10}X$ ($X=$ P,As).

As with many other branches of science, the development of gas phase electron diffraction studies in the USSR was promoted by contacts and collaboration with foreign colleagues, especially those of Hungary, Norway, the USA, Japan, Belgium and the Netherlands.

The three books (M. Hargittai and I. Hargittai, 1976; Vilkov, Anashkin *et al.*, 1979; Vilkov, Mastryukov and Sadova, 1978) recently published in the USSR have been very useful for students and research workers in the field of gas phase electron diffraction.

Department of Chemistry
Moscow State University
Moscow, USSR

REFERENCES

Akishin, P. A., Rambidi, N. G. and Spiridonov, V. P.: 1967, *Characterisation of High Temperature Vapours*. J. L. Margrave, (Ed) New York: J. Wiley. pp. 300-358.

Akishin, P. A., Vinogradov, M. I., Danilov, K. D., Levkin, N. P., Martinson, E. N., Rambidi, N. G., and Spiridonov, V. P.: 1958, *Prib. Tekhn. Eksperim.* **2**, 70–75.

Andersen, B., Seip, H. M., Strand, T. G., and Stolevik, R., 1969, *Acta Chem. Scand.* **23**, 3224-3234.

Frost, A. V., Akishin, P. A., Gurvich, L. V., Kurkchi, G. A., and Konstantinov, A. A. 1953, *Vest. Mosk. Univ.* No. 2, 85–95.

Hargittai, M. and Hargittai, I.: 1976, *Molecular Geometries of Coordination Compounds in the Vapour Phase.* Mir Publ., Moscow. (In Russian)

Ivanov, A. A., Spiridonov, V. P., Demidov, A. V., and Zasorin, E. Z.: 1974, *Prib. Tekhn. Eksperim.* 270–272.

Levkin, N. P. and Alekseev, N. V. 1963, *Zh. Strukt. Khim.* **4**, 327–333.

Krasnov, K. S.: (Ed.) 1979, *Molcular Constants of Inorganic Compounds.* Khimiya Publ. (In Russian).

Mastryukov, V. S., Dorofeeva, O. V., Golubinskii, A. V., Vilkov, L. V., Zhigach, A. F., Laptev, V. T. and Petrunin, A. B.: 1975, *Zh. Strukt. Khim.* **16**, 171–176.

Novikov, V.P.: 1979, *J. Mol. Strukt.* **55**, 215–221.

Popik, M. V., Vilkov, L. V., Mastryukov, V. S., and Golubinskii, A. V.: 1979, *Prib. Tekhn, Eksperim.* 274–276.

Rambidi, N.G.: 1975, *J. Mol. Struct.* **28**, 77–88; 89–96.

Rambidi, N.G., Ezhov, Yu. S., Leont'ev, K. L., Grashis, F. I. Kurkov, A. A., Khor'kov, N. A., Ul'yanova, E. L., Mikhailov, Yu. S., Ostapenko, R. G., Komarov, S.A., and Ugarov, V.V.: 1974, *Special Report of the Institute of High-Temperature: Apparatus for Study of High-Energy Electron Scattering by Gases.* Acad. Sci. USSR,Moscow. pp. 3–24.

Spiridonov, V. P., Ischenko, A. A., and Zasorin, E.Z.: 1978, *Uspekhi Khim.* **47**, 101–126.

Spiridonov, V.P., Prikhod'ko, A. Ya., and Butayev, B.S.: 1979, *Chem.Phys.Letters* **65**, 605–609.

Spiridonov, V. P., and Zasorin, E.Z.: 1979, *Report of the 10th Materials Research Symposium on Characterisation of High Temperature Vapors and Gases:* NBS Special Publ. **561,1**, 711–755.

Vilkov, L. V., Mastryukov, V. S., and Sadova, N.I.: 1978, *Determination of Geometrical Structures of Free Molecules.* Khimiya Publ., Leningrad. (In Russian).

Vilkov, L. V., Mastryukov, V. S., and Akishin, P. A.: 1963, *Zh. Strukt, Khim.* **4**, 323–326.

Vilkov, L. V., Anashkin, M.G., Zasorin, E.Z., Mastryukov V. S., Spiridonov, V. P., and Sadova, N. I.: 1979, *Introduction to Gas Phase Electron Diffraction Theory.* Moscow State University Publication.

Vilkov, L. V., Rambidi, N. G., and Spiridonov, V. P.: 1967, *Zh. Strukt. Khim.* **8**, 786–812.

PART THREE

The Present Subject

A. F. MOODIE

III.1. Notes on the Theory of Forward Elastic Scattering

The theory of the elastic scattering of electrons in transmission through a slab of perfect crystal has a long and involved, even devious history. Yet, in essence, Bethe provided a solution in 1928, and, in a current notation, an answer can be written down compactly. The Fourier transform of the wave function is given by,

$$| u \rangle = \sum_1 \exp\{iMz\} | 0 \rangle , \qquad (1)$$

where M is a constant matrix describing the structure of the crystal and its orientation, z is the thickness of the crystal, $|0\rangle$ is the boundary *ket* vector, and the summation extends over the exceptional lattice index.

It would therefore appear that, apart from detail, development in this topic might be summarised in two references, one to Bethe (1928), and the other to any of the introductory texts on the theory of the scattering matrix (say, Mattuck, 1967). Yet the quarter of a century that saw some extension of Bethe's analysis was succeeded by a period of extensive reformulation that continues to the present day.

The reason for this rather surprising state of affairs would appear to lie in purpose. It was Bethe's concern to answer a question, central in physics, namely whether Schrödinger's theory of mechanics provides a foundation, adequate for the description of the first order wave effects, observed when a beam of electrons is scattered by a crystal of known structure. In contrast, subsequent formulations, starting from a quantum mechanics now known to be sufficient for the purpose, have been designed to extract crystallographic information from the known distribution of scattered intensity. In the notation of Equation (1), Bethe demonstrated that the intensity distribution $U_h U_h^*$ is correctly predicted, at least least to first order, for a known M, whereas subsequent formulations have, in one way or another, been designed to extract the structural information from M, for a known $U_h U_h^*$. This latter problem remains unsolved in general, and continues to provide a strong incentive for further theoretical work.

The theoretical background to Bethe's analysis had been set by two major works, Ewald's account of the dynamical scattering of electromagnetic waves, and Born's theory of quantum scattering. Born, in his lectures in Scotland, described the difficulties he met in attempting to set up a scattering theory within the framework of Heisenberg's matrix mechanics, and how, on hearing of

Schrödinger's theory, he saw that a wave equation provided just the starting
point that he required for an analysis of the general problem. The result of this
analysis, while profoundly influencing virtually every branch of physics, of
course offers no immediate prescriptions for the solution of specific problems
and, in particular, for the problem of scattering by a crystal.

Ewald's theory, on the other hand, while describing propagation in a crystal,
is concerned with a different radiation, and with field equations which are
Lorentz invariant, as distinct from the Schrödinger equation which is Gallilean
invariant. One of the several complicating factors which then emerges in the
X-ray problem is polarisation and, elsewhere in this volume, Bethe remarks on
the technical simplifications which accrue from dealing with a scalar wave
function. But electrons are charged, and consequently interact much more
strongly with the crystal than does the electromagnetic field, so that the
inevitable approximations are more difficult of access. Further, and particularly
at this early stage, the interpretation of electron interaction in terms of a wave
motion raised fundamental questions in interpretation.

Bethe starts his analysis, in a notation, modified for typographical convenience, by considering the Schrödinger equation,

$$\nabla^2\psi + \frac{8\pi^2 me}{h^2}(E+V)\psi = 0. \tag{2}$$

The primary concern is with elastic scattering, so that E is a constant, and
Bethe writes $(8\pi me/h^2)\,E = K^2$, so that $K/2\pi$ is the wave number for the electron
outside the crystal.

But the central feature of a crystal is its periodicity, so Bethe expands the
potential inside the crystal in a Fourier series, writing $(8\pi^2 me/h^2)\ V =
\Sigma_h v_h \exp\{2\pi i(\mathbf{h}\cdot\mathbf{r})\}$. Here $\mathbf{h}$ is a vector of the dual, or reciprocal lattice, and $\mathbf{r}$ is
the current vector. Contact is made with a tradition extending back through
Ewald, Debye and Bravais to Fraunhofer and Young. Distances that are large in
real space are inversely small in diffraction space.

The differential equation has now to be solved, and the technique adopted is
to substitute an admissable form for ψ, thereby deriving equations of
consistency. It is apparent to Bethe that one plane wave cannot subsist in the
crystal, but that a particular set can, namely those for which

$$\psi = \sum_h \psi_h \exp\{i(\mathbf{k}_0 + 2\pi\mathbf{h})\cdot\mathbf{r}\} \tag{3}$$

These waves have, as wave numbers, vectors which are the vector sums of a wave
number $k_0/2\pi$, as yet undetermined, and the vectors $\mathbf{h}$, of the reciprocal lattice.
This is a crucial insight. In retrospect the connection with the ubiquitous
Floquet's theorem and band theory can be seen. Bloch, in fact, in describing his
approach to the conduction problem during the same period, has recently
(1980) remarked, "By straight Fourier analysis I found to my delight that the
solution to the Schrödinger equation differed from the de Broglie wave of a free
particle only by a modulation with the period of the potential".

The whole problem now devolves on the calculation of $\mathbf{k}_0$.

Writing $\mathbf{k}_h = \mathbf{k}_0 + 2\pi\mathbf{h}$, the trial wave function is substituted in the wave equation, and the coefficients of individual exponentials are equated to zero to give,

$$\psi_h(K^2 + v_0 - k^2{}_h) + \sum_g{}' v_g \psi_{h-g} = 0, \tag{4}$$

where the prime in the summation indicates that the term $g = 0$ is omitted.

If this triply infinite set of dispersion equations could be solved, a complete solution to the non-relativistic elastic scattering problem would follow, on imposing standard boundary conditions, which are shown to be equivalent to the requirement that the tangential components of the wave vectors on the two sides of the boundary should be equal. Specifically for the solution number i, there would result a set of wave vectors $\mathbf{k}_h^i$, and a set of amplitudes ψ_h^i, one for each reciprocal lattice vector, and this set would represent one of the solutions for a wave in the crystal, the 'Bloch wave' $\psi^i(\mathbf{r}) = \Sigma_h \psi_h^i \exp\{i\mathbf{k}_h^i \cdot \mathbf{r}\}$. This description, in terms of dispersion surfaces and Bloch waves, has retained its power and utility throughout the entire history of the subject.

The algebraic problems posed by Equation (4) are, however, formidable. The approximation is therefore made that experimental conditions can be chosen such that, apart from the origin, only one point of the reciprocal lattice lies in the neighbourhood of the Ewald sphere. Under those conditions only two wave fields will be strongly excited. The perturbing effect of the weak beams is estimated in terms of the strong beams, and is found to act as a pseudopotential. With a final approximation that effectively excludes scattering in the neighbourhood of ninety degrees, the problem is reduced to the solution of a quadratic.

Explicit expressions are given for the Bragg case, that appropriate to the experiments of Davisson and Germer. The paper ends with that analysis of the atomic scattering factor for electrons, based essentially on the Fourier transform of Poisson's equation, which is in current use.

The whole dispersion model, familiar in classical physics, is now firmly based in quantum mechanics.

It was during this early period that von Laue (1935) discovered the most fundamental of all the symmetries in electron diffraction, that of reciprocity. This refers to the invariance of the wave function of the diffracted beam when the centre of the Ewald sphere is reflected across the diffraction vector.

The experimentalists now, however, found a wealth of application in the geometry of diffraction and microscopy, and it was not until 1939 that Blackman derived an explicit expression for the two-beam approximation, in the Laue case. The result, neglecting weak-beam interactions, and in Blackman's notation, is

$$I = I_0 \sin^2\{A(W^2 + 1)^{\frac{1}{2}}\}/(W^2 + 1), \tag{5}$$

where $A = vH(\cos \theta_1{}^{1/2}/\{2k \cos \theta_2 (\cos \theta_2)^{1/2}\}$, $W = k\zeta/v$, $k^2 = K^2 + v_0$, ζ is the excitation error, θ_1, θ_2, are the angles made with the normal to the surface, and H is the crystal thickness.

When compared with the single scattering, or kinematical result, namely that the scattered wave field is described by the intersection of the Ewald sphere with the Fourier transform of the finite crystal, the essentially new phenomenon for the Laue case is seen to be the coupling of the two beams; in Ewald's vivid model, there is an analogy with the motion of coupled pendulums. As Sturkey (1948), Kato (1949, 1952a, b), Kato and Uyeda (1951), and Molière and Niehrs (1954) showed, it is this coupling which generates the fine structure observed in the scattering from polyhedral crystals; or, equivalently (Heidenreich and Sturkey, 1945), the thickness fringes observed in the electron microscope.

Wagenfeld (1958) extended the treatment to three and four beams.

The implication for structure analysis, in Blackman's (1939) words, is that, " ... the integrated intensity is proportional to v, which is one of the characteristic features of the dynamical theory for thick films". This matter is discussed by Gjønnes in the present volume.

Since, in the two-beam approximation, the forward scattered beam is coupled to the diffracted beam through the inverse vector, the structure amplitudes appear in the product $v_h v_{\bar{h}}$, and the intensities of the diffracted beams are independent of the phase of v_h. From the very structure of the dispersion equations it is evident that this cannot hold in the three-beam approximation, so that, on this ground alone, it would be anticipated that the higher approximation would contain qualitatively new effects, and, in fact, this proved to be the case. Heidenreich (1950), showed that special space-group-forbidden reflections have finite intensity; and Kambe (1957a,b) demonstrated that the sign of the product $V_h V_g V_{g-h}$ could be determined directly from the diffracted intensities generated by a centro-symmetric crystal. This key result represents the first direct determination of phase in crystallography.

A number of authors had found it technically convenient to adapt the methods of band theory, put the dispersion approach into matrix notation and, with approximations appropriate to the forward scattering of fast electrons, appeal to the highly developed theory associated with the eigenvalues and eigenvectors of a Hermitian matrix. With the amplitudes of the diffracted beams labelled $\ldots j \ldots$, the off diagonal elements, a_{ij}, of this matrix are proportional to v_{i-j}, while the diagonal elements, a_{ii}, are proportional to ζ_i, the excitation errors, substantially measures of the deviations from the Bragg condition. The N-beam approximation is thus seen to involve the solution of the Nth order secular equation.

In this context, the rich physical content of the three-beam approximation can be seen reflected in the complexities of the cubic equation, and some of the consequences of van der Waal's equation are brought to mind. In fact, no compact and explicit solution for the general three-beam approximation, analogous to that of Equation (5) for the two-beam has yet appeared in the literature; a direct approach leading to expressions of quite surprising complexity. The possibility of extending this analysis (as distinct from numerical evaluation) to a large number of beams seems limited and, further, it is difficult to overlook Abel's theorem on the quintic.

Yet the experimentalists, in the early work on convergent-beam diffraction, had shown that many-beam effects were significant and now, with accelerating voltages which flattened the Ewald sphere, thereby exciting hundreds of simultaneous reflections, they launched into detailed structure analysis (for instance, Cowley 1953).

The need for alternative formulations was acute, and these emerged in abundance, each with its own particular strength. Happily, priority, that most sterile of phenomena, cannot be assigned, since preliminary results were freely presented at a number of conferences.

As early as 1950, Sturkey, at the summer meeting of the American Crystallographic Association, had reported on the scattering matrix representation. The essence of this idea is best conveyed by a direct quotation from his paper of 1962. "If there is no absorption, we may consider the emergent beams $\phi_0(t)$ and $\phi_i(t)$ as the components of an n-dimensional vector of unit amplitude (length) and we may say that the diffraction process is a unitary transformation that rotates the initial vector of one component in this n-dimensional diffraction space to the final position with n components." (t, here, is thickness.) The connection with the theory of Lie groups is established, since every unitary matrix can be written as $S = \exp\{iR\}$, where R is Hermitian. It might be noted that Sturkey refers to R as the scattering matrix, whereas a conventional usage defines S as that operator. Sturkey (1962) shows that the matrix R is just that which emerges in the eigenvalue treatment, so that (ignoring upper layer lines) the solution may be written as

$$| u \rangle = \exp\{iM\mathcal{Z}\} | 0 \rangle . \tag{6}$$

This bears the hallmark of the new formulations; no reference is made to the roots of an algebraic equation, which are seen as quantities which, if derived, are to be, and can be, eliminated.

But, if correct, equation (6) must be compatible with a dispersion model. In a key paper, Fujimoto (1959) showed this to be true. Building on the work of Niehrs and Wagner (1955), and starting from the dispersion equations, he derived Equation (6) by, in effect, treating the eigenvalues as dummy variables and exploiting the orthogonality of the eigenvectors. This result is now so well known that its central importance in the subsequent development of the whole theory is sometimes overlooked.

In the course of his derivation, again building on the work of Niehrs and Wagner, he wrote down the relation,

$$\psi_h = \sum_i {}^i t_0^* {}^i t_h \exp\left\{i\frac{\mathcal{Z}}{2k} {}^i\lambda\right\} \exp\{i\mathbf{k}_h \cdot \mathbf{r}\}, \tag{7}$$

where ${}^i t_h$ is the hth component of the eigenvector with eigenvalue ${}^i\lambda$. Once more, this relation is so well known as to be taken for granted, but its implications in, for instance, the determination of handedness, and inversion of diffracted intensity, were not to be appreciated for some fifteen years.

By an explicit expansion of the scattering matrix he obtained two expressions for the wave function of a diffracted beam; the first, a series in powers of $(z/2k)$, the thickness expansion; and the second, a series in powers of the electron-crystal interaction, the Born series. This latter proved to be the same as that derived by Cowley and Moodie (1957), and by Fujiwara (1959), both of whom had started from entirely different models.

Cowley and Moodie, stimulated by Cowley's interest in structure analysis (see the article by Gjønnes), envisaged a crystal cut into slices. A projection of the potential in each slice modified the phase of the electron wave by multiplication, while propagation between slices further modified the phase, in accordance with Huygen's principle, by convolution. The alternation of those operators, dual under the Fourier transform, greatly simplifies the analysis.

On taking the limit, with the thickness of the slices going to zero as the number goes to infinity, the product remaining constant, and equal to the thickness of the crystal, Cowley obtained an expression for the wave amplitude in the form of the Born series,

$$u(h) \ = \ \sum_n E_n(h) Z_n(h) \ , \tag{8}$$

where E_n (h) is an operator which convolutes the structure amplitudes n times and sums over the exceptional index l. This operator acts on Z_n, a known, purely geometric function of the excitation errors and the thickness.

Expressions for single (kinematical) scattering, and for the two-beam approximation, emerge as special cases of Equation (8). The first is obvious, but the derivation of the second proved surprisingly lengthy, and the significance of this was not appreciated until several years later, when it emerged as an aspect of the parametrisation of SU (2) (Hurley, Johnson, Moodie, Rez and Sellar, 1978).

On taking the relativistic limit, as the wavelength goes to zero, a high voltage limit is obtained, and this leads to the phase-grating, and eventually to the charge-density approximations.

Since the whole slice approach can be shown to be a special case of Feynman (1948) quantum mechanics, and since Schrödinger's equation without back scattering can be obtained from it by a Taylor expansion of the wave function (Goodman and Moodie, 1974), it is natural that Fujimoto's interaction expansion leads to the same result.

With the work of Fujiwara (1959), a solution is at last obtained in the spirit of Born's analysis; for Fujiwara starts with the solution to Schrödinger's equation in the form of the integral equation,

$$\psi(\mathbf{r}) \ = \ \psi_{(0)}(\mathbf{r}) - (2me/\hbar^2) \int G^0{}_K(\mathbf{r},\mathbf{r}_i) \ V(\mathbf{r}_i) \psi(\mathbf{r}_i) \ d\mathbf{r}_i \ , \tag{9}$$

where $G^0{}_K$ $(\mathbf{r},\mathbf{r}_i)$ is the Fourier integral form of the Green's function.

He iterates the equation, using contour integration on the complex k_z plane to obtain explicit expressions, and finally derives the Born series. The methods exhibit, in particularly clear form, the multiple coherent scattering aspects of dynamical interaction.

In addition, his approach, in the present writer's experience, is particularly accessible to those working in other fields.

In a particularly compact and clear analysis, Tournarie (1960, 1961, 1962) Fourier-transformed Schrödinger's equation, not in the three spatial coordinates, which leads to Bethe's dispersion equations, but in two, the exceptional coordinate being z. This semi-reciprocal treatment generated matrix equations of the form

$$\frac{d^2u}{dz^2} = -Gu ,$$

and the persuasive analogy with simple harmonic motion is underlined. But this equation describes low energy electron diffraction, with the attendant discussions on the meaning to be attached to $G^{\frac{1}{2}}$ (Kambe, 1967; Dederichs, 1971). These papers, along with Tournarie's, are central to the theory of low energy diffraction, and give precision to the nature of the approximations made in neglecting back scattering. Before the publication of these papers, this latter point had been the subject of a good deal of not always profitable discussion. Broadly, the problem is complicated by a branch point with the attendant tunneling phenomena. Tournarie proceeds by defining a matrix Green's function, and, in effect, invoking a chronological operator. For present purposes, it may be acceptable to assume that $G^{\frac{1}{2}}$ (with generalised spring constants in mind) can be expanded as a Taylor's series and truncated after the second term (which accounts for the ubiquitous $\frac{1}{2}$ in all of these formulations) to give the equation for the Laue case as,

$$\frac{du}{dz} = iMu . \tag{10}$$

Sturkey's and Fujimoto's scattering matrix solution is now recovered, and another link between the formulations is established. Equation (10) is consistent with the Schrödinger equation for no back-scattering.

In entirely independent investigations, Hirsch, Howie and Whelan (for references see the article by Hirsch in this volume) in their analysis of the intricate problem of diffraction from a distorted crystal, reformulated Darwin's (1914) treatment of X-ray scattering, and obtained a set of coupled differential equations.

When the crystal is distorted these lead to the celebrated Howie–Whelan equations, extensively discussed in a number of articles in this volume. When the crystal is undistorted, the equations are shown to be isomorphic with Equation (10). Howie and Whelan establish the relationship between their formulation, and the eigenvalue method in a number of ways, but, perhaps, most strikingly, by solving the single Nth order differential equation, which can be obtained from the N first order equations.

All of these formulations were published within a few years of each other and, at the time, many of the interconnections were either not known or thought to be obscure. There was even a feeling that some of the agreements might prove fortuitous. But these problems were substantially resolved at the conference in Kyoto in 1961. It was a remarkable meeting, most aptly summed up by Professor Ewald, who was heard to remark that for the first time in his life he had entered a conference room where most of the participants were aware of the implications of the word 'dynamical'.

It was at this conference that Gjønnes (1962) (apart from presenting his accounts of inelastic scattering) resolved the problem (Miyake, 1959) of dynamical potentials; and that Fujiwara (1962) presented his results on relativistic scattering. Carrying through a Green's function analysis on the Dirac equation, Fujiwara showed that, to adequate approximation, the results for the Schrödinger equation could be used, provided that relativistic corrections were made for the wavelength and mass of the electron.

All of the new formulations showed that Friedel's Law fails, so that the presence of a centre of symmetry should be qualitatively detectable. This point had been established much earlier for reflection diffraction in the three-beam approximation (Miyake and Uyeda, 1955; Kohra, 1954) but, curiously, no definitive treatment had been given for the Laue case, the most widely used in structural investigations.

But, again, it was the experimentalists who provided the stimulus for application and development. The importance of Kossel and Möllenstedt's convergent beam technique has already been mentioned in this context, and now, with Goodman and Lehmpfuhl's extension and development of the method, a new range of observations demanded quantitative interpretation. Further, electron microscopy had been extended into the MeV range, and the resolution of more conventional instruments dramatically increased. Fast and accurate computers had also become generally available, efficient numerical methods were developed, and the N-beam formulations were subjected to stringent tests as detailed structural work was undertaken.

Further problems in symmetry were raised. Cowley and Moodie (1959) had shown that certain reflections would remain forbidden if suitable restrictions were placed on the angle of incidence. It was now shown (Gjønnes and Moodie, 1965) that screw axes and glide planes would generate loci of zero intensity which pass through the Bragg position for the appropriate beams. These results are, of course, compatible with those of Heidenreich, who considered special space-group forbidden reflections.

Pogany and Turner (1968), using a Green's function technique, analysed von Laue's theorem on reciprocity, now set in a specifically N-beam context, and Buxton, Eades, Steeds and Rackham (1976) classified the point symmetries of zone-axis patterns. The results of this exhaustive analysis, derived by explicit appeal to group theory, can best be summarised by quotation: "We have obtained a set of simple rules which can often be applied quickly and easily to enable the crystal point group of the specimen to be determined".

Ohtsuki and Yanagawa (1966) now brought the techniques of diagram analysis to the field, on recasting the theory in second quantised form. These techniques have had most extensive application in the problems associated with inelastic interactions, where an N-body treatment is required, and hence are discussed elsewhere in this volume.

In the midst of this N-beam activity further subtleties were uncovered in the three-beam approximation, and yet again Bethe's theory was invoked. It was found (Nagata and Fukahara, 1967; Watanabe, Uyeda and Kogiso, 1968) that, at a specific orientation, a confluence in the secular equation developed at a sharply defined voltage, the critical voltage, and that the experimental requirements could frequently be met in high voltage instruments. In fact, two of the dispersion surfaces not only touch, but also exchange symmetry (Metherell and Fisher, 1969). Hewat and Humphries (1974) later carried the technique to the level of high precision. In extending the analysis, Gjønnes and Høier (1971) showed that, for a centro-symmetric crystal, the secular equation is confluent at any wavelength for a unique angle of incidence, which they calculated in terms of the structure amplitudes v_h, v_g, v_{g-h}, a result which they then exploited in structure analysis.

Towards the close of this fifty-year period, the ramifications of the subject, already considerable, expanded to those unmanageable proportions which require detailed articles on any one development. Fortunately, a most lucid account has recently been given by Howie (1978), and further, many of these topics are discussed in other articles in this volume. A few matters may, however, be mentioned, albeit briefly.

The theory associated with high-voltage investigations has now developed into a field in its own right (see Fujimoto, this volume). In the broadest sense, the classical problem, in the form, for instance, of the Hamilton-Jacobi equations, presents formidable difficulties. Nevertheless, those concerned with channelling have been able to find connections between certain, classical trajectories, and quantities which, previously, had been defined exclusively in terms of wave propagation (for instance, Howie, 1968; Chadderton, 1968). Alternatively, classical trajectories can be quantized (see Fujimoto, this volume). The whole can be set in the context of classical, semi-classical, and wave descriptions (Berry, 1971; Berry, Buxton and Ozorio de Almeida, 1973), and developed into a graphic description of, at least, the qualitative aspects of the distribution of scattered intensity. The approach to the classical limit remains, however, as a delicate problem.

The scattering-matrix formulation provides a link with intermediate energies, but the group theoretic and algebraic opportunities which it offers have scarcely been explored, in particular, in inversion. As a simple, but concrete example, the two-beam result can, of course, be recovered immediately on expanding the matrix on a Pauli basis, thereby establishing contact with the highly developed theory of two-level systems, but the corresponding development in SU (3) remains to be exploited.

Finally, in an attempt to reduce computational complexities and, at the same time, achieve a more direct understanding of convergent beam patterns, cellular methods have been introduced (Kambe, Lehmpfuhl and Fujimoto, 1974; Ozorio de Almeida, 1975).

But these concepts from band theory stem from the theories of Bethe and Bloch, and we can close with the beginning.

It must be clear to those who have seen a zone-axis convergent-beam pattern, or the lattice image of a moderately thick crystal, that the interplay of theory and experiment has never been more productive. There can be no doubt that the young theorists are responding.

CSIRO
Melbourne, Australia

REFERENCES

Berry, M. V.: 1971, *J. Phys. C: Solid St. Phys.* **4**, 697.
Berry, M. V., Buxton, B. F., and Ozorio de Almeida, A. M.: 1973, *Radiat. Effects* **20**, 1.
Bethe, H. A.: 1928, *Ann. Phys.* **87**, 55.
Blackman, M.: 1939, *Proc. R. Soc. Lond. A.* **173**, 68.
Bloch, F.: 1980, *Proc. R. Soc. Lond. A* **371**, 24.
Buxton, B. F., Eades, J. A., Steeds, J. W., and Rackham, G.M.: 1976, *Phil. Tr.A. Proc. R. Soc. Lond.* **281**, 171.
Chadderton, L. T.: 1968, *Phil. Mag.* **18**, 1017.
Cowley, J. M.: 1953, *Acta Cryst.* **6**, 522.
Cowley, J. M. and Moodie, A. F.: 1957, *Acta Cryst.* **10**, 609.
Cowley, J. M. and Moodie, A. F.: 1959, *Acta Cryst.* **12**, 360.
Darwin, C. G.: 1914, *Phil. Mag.* **27**, 315.
Dederichs, P. H.: 1971, KFA-JÜL report JÜL-797-FF.
Feynman, R. F.: 1948, *Rev. Mod. Phys.* **20**, 367.
Fujimoto, F.: 1959, *J. Phys. Soc. Japan* **14**, 1558.
Fujiwara, K.: 1959, *J. Phys. Soc. Japan* **14**, 1513.
Fujiwara, K.: 1962, *J. Phys. Soc. Japan* **17**, Supplement BII, 118.
Gjønnes, J.: 1962, *Acta Cryst.* **15**, 703.
Gjønnes, J. and Høier, R.: 1971, *Acta Cryst.* **A27**, 313.
Gjønnes, J. and Moodie, A. F.: 1965, *Acta Cryst.* **19**, 65.
Goodman, P. and Moodie, A. F.: 1974, *Acta Cryst.* **A30**, 280.
Heidenreich, R. D.: 1950, *Phys. Rev.* **77**, 271.
Heidenreich, R. D. and Sturkey, L.: 1945, *J. App. Phys.* **16**, 97.
Hewat, E. A. and Humphreys, C. J.: 1974, *Proc. Int. Conf. on Diffraction, Melb.* (Canberra: Australian Academy of Sciences), 76.
Howie, A: 1968, *Brookhaven National Laboratory Report* No. 50083, 15.
Howie, A.: 1978, *Electron Diffraction 1927-1977* (ed. P. J. Dobson, J. B. Pendry and C. J. Humphreys) 1. Inst. Phys. Conf. Ser. No. 41, London.
Hurley, A. C., Johnson, A. W. A., Moodie, A. F., Rez, P. and Sellar, J. R.: 1978, *Electron Diffraction 1927-1977*, (ed. P. J. Dobson, J. B. Pendry and C. J. Humphreys) 34. Inst. Phys. Conf. Ser. No. 41, London.
Kambe, K.: 1957a, *J. Phys. Soc. Japan* **12**, 13.
Kambe, K.: 1957b, *J. Phys. Soc. Japan* **12**, 25.
Kambe, K.: 1967, *Z. Naturf.* **22a**, 422.
Kambe, K., Lehmpfuhl, G. and Fujimoto, F.: 1974, *Z. Naturforsch.* **A29**, 1034.
Kato, N.:1949, *Proc. Japan. Acad.* **25**, 41.
Kato, N.: 1952a, *J. Phys. Soc. Japan* **7**, 397.
Kato, N.: 1952b, *J. Phys. Soc. Japan* **7**, 406.
Kato, N. and Uyeda, R.: 1951, *Acta Cryst.* **4**, 227.
Kohra, K.: 1954, *J. Phys. Soc. Japan* **9**, 690.
von Laue, M.: 1935, *Ann. Phys.* **23**, 705.
Mattuck, R. D.: 1967, *A Guide to Feynman Diagrams in the Many-Body Problem.* McGraw-Hill, New York.
Metherall, A. J. F. and Fisher, R. M.: 1969, *Phys. Stat. Solidi* **32**, 551.
Miyake, S.: 1959, *J. Phys. Soc. Japan* **14**, 83.
Miyake, S. and Uyeda, R.: 1955, *Acta Cryst.* **8**, 335.

Molière, K. and Niehrs, H.: 1954, *Z. Phys.* **137**, 445.
Nagata, F. and Fukahara, A.: 1967 *Jap. J. Appl. Phys.* **6**, 1233.
Niehrs, H. and Wagner, E. H.: 1955 *Z. Phys.* **143**, 285.
Ohtsuki, Y. H. and Yanagawa, S.: 1966, *J. Phys. Soc. Japan* **21**, 326.
Ozorio de Almeida, A. M.: 1975, *Acta Cryst.* A**31**, 435.
Pogany, A. P. and Turner, P. S.: 1968, *Acta Cryst.* A**24**, 103.
Sturkey, L.: 1948, *Phys. Rev.* **73**, 183.
Sturkey, L.: 1962, *Proc. Phys. Soc.* **80**, 321.
Tournarie, M.: 1960, *Bull. Soc. Franc. Miner. Cryst.* **83**,179.
Tournarie, M.: 1961, *C. R. Acad. Sci.* **252**, 1961.
Tournarie, M.: 1962, *J. Phys. Soc. Japan* **17**, Suppl. B-II, 98.
Wagenfeld, H.: 1958, *Doctoral Thesis*, Berlin Free University.
Watanabe, S., Uyeda, R. and Kogiso, M.: 1968, *Acta Cryst.* A**24**, 249.

M. A. VAN HOVE

III.2. Surface Crystallography by Low Energy Electron Diffraction

1. INTRODUCTION

The past decade has seen an impressive growth in the use of low energy electron diffraction (LEED) for surface crystallography. It is now becoming a routine matter to determine the three-dimensional position of surface atoms at the simpler clean surfaces and in the simpler layers of adsorbed atoms and diatomic molecules. Over 100 such surfaces have so far been subjected to detailed surface crystallography to yield bond lengths and bond angles. (Pendry, 1974; Van Hove and Tong 1974; Somorjai and Van Hove, 1979.) Many more ($\sim$ 1000) surfaces have been analyzed in terms of their two-dimensional structure (e.g. Castner and Somorjai, 1979). The large variety of unit cells formed at single crystal surfaces testifies to the richness of surface phenomena that can be studied. This variety is characteristic of interfaces in general, and is enhanced at surfaces because one often deals with *two* interfaces, namely the bulk–surface interface (which might be a bulk–adsorbate interface) and the surface–vacuum interface; these two interfaces are usually only a few Ångstöms apart and so can strongly interact with each other.

Thus one finds that single crystal semiconductor surfaces, which often reconstruct to non-bulk atomic arrangements, exhibit a variety of unit cells, depending on the heat treatment or chemical treatment or ion bombardment treatment to which the surface is subjected. While different impurities may play a role in stabilizing different surface structures, it is also thought that many structures are metastable phases characteristic of the impurity-free surface. In some cases a clean non-reconstructed surface cannot be obtained at all, indicating that it is energetically unfavorable. Similarly, one finds sequences of different unit cells when atoms or molecules are adsorbed onto a single crystal substrate, depending on the density of the adsorbate and on the temperature, as well as on the treatment. These are often stable structures. In addition, disordered arrangements of surface atoms and molecules are obviously possible and are often observed, giving for example access to the study of two-dimensional phase transitions.

The success of LEED in surface crystallography is due to the small mean free path of the electrons at the energies used, typically 5–10 Å for energies between 10 and 400 eV, in which range the wavelength varies between about 3 and 0.5 Å.

This amount of penetration into the surface is just right for studying both the topmost atoms and their relationship to the underlying atoms. The reflection mode rather than the transmission mode is obviously required under these conditions. At higher energies, where the electron mean free path is much larger, a similar surface sensitivity can be obtained by using a glancing angle of incidence of the electrons (as in reflection high energy electron diffraction or RHEED), but this energy regime has not been used nearly as much as that of LEED in surface studies.

The small mean free path in LEED is due primarily to the large inelastic cross section of the electron–surface collision. Inelastic losses to single electron excitations predominate, but losses to plasmons, phonons, and other collective excitations contribute as well. The elastically backscattered electrons are further limited in their penetration depth by the large elastic collision cross-section. Both the elastic and inelastic cross-sections are often comparable to the geometrical cross-section of the surface atoms (Pendry, 1974). This implies that each reflected electron most likely has undergone more than one atomic scattering. As a result LEED is 'dynamical', i.e. multiple scattering is the rule rather than the exception. This explains the need for a relatively complicated theory and the proliferation of different forms and approximations of the theory as a result of various successful attempts to reduce the required computational effort.

The presence of multiple scattering is responsible for the following subdivision of this review. First, we discuss in Section 2 the two-dimensional structural information available from diffraction patterns as obtained in diffraction photographs, because these are largely interpretable with kinematical arguments. Then we describe in Section 3 the main theoretical methods that have been developed to calculate the intensities of the diffraction spots, as well as the way in which structural determination is carried out. Detailed structures determined so far will be presented in Section 4 and prospects for future developments will be addressed in Section 5. The large amount of work performed with LEED to date can only be described through a few representative examples. More complete listings and discussions will be found in the literature: (e.g. Cunningham, Chan & Weinberg, 1978).

2. TWO-DIMENSIONAL INFORMATION AVAILABLE FROM LEED

2.1. Surface Lattices and Notation

Most single crystal surface studies start with surfaces that are obtained by cutting the crystal along a well-defined plane. This plane is denoted by its Miller indices (hkl). Thus a Ni(100) surface should exhibit a square array of close-packed surface nickel atoms, as is indeed observed experimentally, showing that no 'reconstruction' occurs on this clean surface. Usually low Miller index surfaces are chosen because of their relatively simple geometry involving small two-dimensional unit cells and flat atomic arrangements. But high Miller

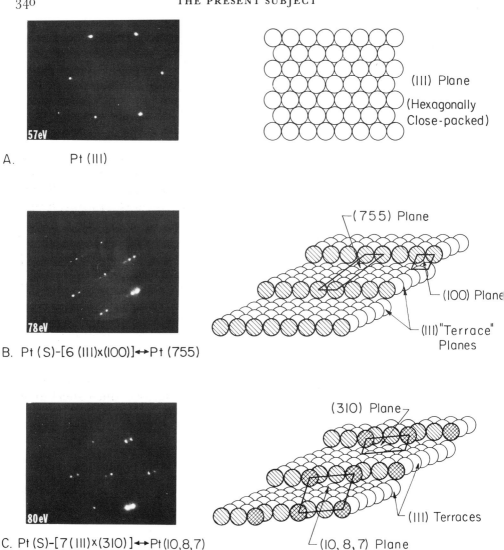

A.   Pt (III)

B. Pt (S)-[6 (III)x(I00)]↔Pt (755)

C. Pt (S)-[7 (III)×(3I0)]↔Pt (I0,8,7)

Fig. 1.   Electron diffraction patterns and atomic structures of a step-free (A), a stepped (B), and a kinked (C) surface of platinum. Steps are seen to produce splitting of the spots corresponding to the step-free surface. Only the first-order hexagonal ring of spots is clearly visible in these photographs.

index surfaces are also studied because of their special structure that includes terraces and steps or also kinks in the steps; such surfaces are called stepped or kinked surfaces. Examples of a few surfaces and the corresponding diffraction patterns for the face-centered cubic metal platinum are shown in Figure 1.

Whenever a surface exhibits a two-dimensional unit cell different from that expected from simple truncation of the single crystal, one talks of superlattices. This situation may be due to a surface reconstruction or to a layer of adsorbates that adopts a larger unit cell. Then a two-dimensional notation is used that relates the superlattice unit cell to the initially expected unit cell. In the Wood (1964) notation one speaks of a $(m \times n)R\alpha°$ superlattice to designate that it is obtained from the initially expected unit cell by expanding its two basis vectors m-fold and n-fold, respectively, and then rotating the resulting cell by an angle α (when the two basis vectors must be rotated by different angles a more general matrix notation is used instead of the Wood notation). Centered lattices, in which the center of the $(m \times n)R\alpha°$ cell has the same symmetry as its four corners, are designated $c(m \times n)R\alpha°$. By contrast, non-centered lattices (i.e. primitive lattices) are often designated $p(m \times n)R\alpha°$. When $\alpha = 0$ the $R\alpha°$ is normally dropped. Examples of commonly observed superlattices are shown in Figure 2.

2.2. *The Two-Dimensional Reciprocal Space*

In kinematic diffraction from three dimensional arrays of atoms one obtains three Bragg conditions, one for each dimension of the array. In LEED the array participating in the diffraction is reduced to a two-dimensional array with a

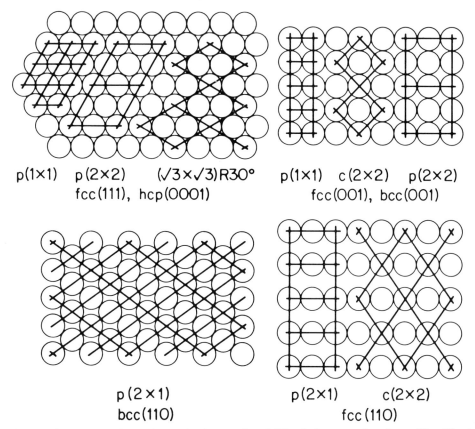

p(1×1)   p(2×2)   (√3×√3)R30°   p(1×1)   c(2×2)   p(2×2)
fcc(111), hcp(0001)   fcc(001), bcc(001)

p(2×1)   p(2×1)   c(2×2)
bcc(110)   fcc(110)

Fig. 2.   Commonly observed superlattices on low Miller index crystal surfaces. The Wood notation is used.

thickness in the third dimension of only about 5 to 10 Å. The third dimension, therefore, does not give rise to a strict Bragg condition and the diffracted beams spread out and join up in this dimension of reciprocal space into what are commonly called 'rods' perpendicular to the surface. Whenever the Ewald sphere intersects such a rod, a diffraction spot can be observed that only in exceptional cases has negligible intensity. Thus a diffraction picture taken at any given electron energy shows, in general, the entire set of diffraction spots corresponding to all the rods intersected at that energy. Varying the energy causes the spots to move gradually in the plane of the picture while a continuous change in intensity occurs. Since each spot and the associated diffracted beam correspond to a particular reciprocal-lattice rod, they are labelled by the two-dimensional reciprocal lattice vector defining that rod. The reciprocal lattice is that obtained from the two dimensional surface lattice. Thus the specularly reflected beam is labelled (00), while (10), $(\bar{1}0)$, (01), and $(0\bar{1})$ label first-order diffraction beams, etc. Superlattices induce additional 'fractional-order' spots that are accordingly labelled $(\frac{1}{2}, 0)$, $(\frac{1}{2}, \frac{1}{2})$, etc. if we take the example of a (2×2) superlattice.

The sharpness of the diffraction spots is of particular importance as it is a measure of the degree of ordering of a surface. (Houston and Park, 1970). A variable quality of ordering is particularly often encountered with layers of adsorbed atoms or molecules and is directly displayed on the diffraction screen as spot broadening and/or diffuse background between the spots. For example, an adsorbed layer that manages only to order in patches (called domains or islands, depending on the case) produces spots broadened in inverse proportion to the patch diameter, as long as this diameter is less than the coherence width of the electron beam (typically of the order of 100 Å).

Efforts to Fourier-transform the two-dimensional LEED diffraction pattern in the hope of generating a picture of the actual surface structure have largely failed because of the dynamical character of the diffracted intensities. However, the patterns themselves are often sufficient to convey a great deal about the surface structure, as we discuss next.

2.3. Overlayer Superlattices

Nearly every imaginable superlattice has been observed with overlayers of atoms or molecules on various substrates. The simpler ones are shown in Figure 2. Often there is a simple relationship between the density of adsorbates and the superlattice. For example, with oxygen deposited on Ni(100) at a density corresponding to one oxygen atom for every four surface nickel atoms, which is usually described as $\frac{1}{4}$ of a 'monolayer', a (2×2) pattern appears; when the density is doubled, a $c(2 \times 2)$ pattern is formed because the half monolayer also fills in the centers of the (2×2) cells. Of course, the exact substrate–adsorbate bonding arrangement cannot be predicted from the observed pattern, but the oxygen–oxygen distances are determined, giving much insight into the interatomic forces acting at surfaces. The mere fact that such ordering occurs so frequently is also significant in this respect as it shows that the substrate–adsorbate bonding forces can often dominate over the adsorbate–adsorbate forces. Furthermore, order–disorder transition temperatures can be measured and lead to quantitative information about these interatomic forces (Wang, Lu and Lagally, 1978).

With larger molecules the superlattice unit cell may at times be directly related to the size of the molecules and indicate how the molecule is oriented with respect to the surface. Thus the straight-chain saturated hydrocarbon molecules from propane (C_3H_8) to octane (C_8H_{18}) deposited on Ag(111) at temperatures of 100 to 200 K produce a series of unit cells that grow in proportion to the molecular size and suggest that the molecular chain axes lie parallel to the surface (Firment and Somorjai, 1977).

Many different structures also appear when metal monolayers are deposited on substrates of another metal (see for example: Biberian and Somorjai, 1979). They are often interpreted as the close packing of a layer of adsorbate metal, mostly in hexagonal arrays, but sometimes in square arrays if such an arrangement fits better on the substrate as far as relative atomic radii are concerned. Some such adsorbate structures can also be interpreted as regular arrays of small anti-phase domains with a simple basic arrangement such as $c2 \times 2$) (Huber and Oudar, 1975; Biberian and Huber, 1976). The adsorbed metal, instead of aggregating in a monolayer, can also grow into pyramidal microcrystallites with an orientation such that the fit between their base and the substrate is optimal. These results are clearly important for understanding the mechanism of epitaxial crystal growth.

Close packed hexagonal structures are also often observed with adsorbates that bind weakly to the substrate or that have a mismatch in atomic radii. Weak binding occurs especially with rare gas adsorbates. Argon deposited on the basal plane of graphite presents a particularly interesting situation. A single layer of argon has a lattice that almost fits as a $(\sqrt{3} \times \sqrt{3})R30°$ overlayer on graphite; the slight misfit in lattice constants gives rise to a small rotation of the overlayer about the surface normal by about 1° away from parallelism of the two hexagonal lattices. This angle of rotation depends on the state of compression of the argon layer which, in turn, depends on the amount of deposited argon (Novaco and McTague, 1977; Shaw, Fain and Chinn, 1978).

3. THEORETICAL METHODS OF SURFACE CRYSTALLOGRAPHY BY LEED

3.1. Single and Multiple Scattering

To obtain more than two-dimensional information about surface structures, the intensities of the LEED beams are measured. Since the electron energy is easily varied experimentally, the data are usually presented as intensity versus (accelerating) voltage or $I-V$ curves. Rotation curves are also sometimes used, where for fixed electron energy the incident polar or azimuthal angle is varied. Examples of $I-V$ curves, both experimental and theoretical, are given in Figure

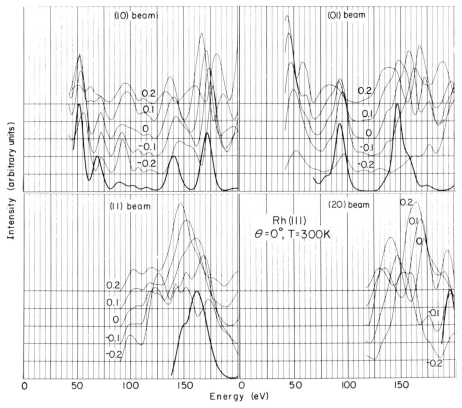

Fig. 3.   Experimental (thick lines) and theoretical (thin lines) *I–V* curves for four different LEED beams diffracted from a clean Rh(111) surface. The incident beam arrives normal to the surface which is held at room temperature. The theoretical curves, dynamically calculated and shifted to different baselines for clarity, correspond to various changes in the top interlayer spacing (-0.2, -0.1, 0, $+0.1$, $+0.2$ Å) relative to the bulk value of that layer spacing. The bulk value gives the best agreement with experiment here.

3 for a Rh(111) surface. The structure of this particular surface turns out to be an ideal truncation of the bulk structure and in the kinematical limit one would then expect the *I–V* curves to have a simple appearance: a succession of relatively widely and regularly spaced near-Lorentzian Bragg peaks of width ~ 10 eV due to the small penetration depth. (These wide Bragg peaks would be the remnants of the strict Bragg condition in the dimension perpendicular to the surface.) Instead, the *I–V* curves are much more complicated and consist of a high density of peaks at irregular positions and with irregular heights. This is the effect of multiple scattering. Low density materials show less multiple scattering than rhodium. Unfortunately few materials can be treated kinematically, a notable example being xenon surfaces, which have very large interatomic distances. Beryllium, graphite, silicon, and aluminum show significant multiple scattering effects, although a rough structural determination of surfaces of these materials should be possible with a kinematic theory. Attempts have been made

to average out multiple scattering effects (Ngoc, Lagally and Webb, 1973; Webb and Lagally, 1973), or to Fourier-transform the I–V curves and deconvolute out the kinematic structure factor (Adams and Landman, 1977), or to fit a structure by convolution (Cunningham, Chan and Weinberg, 1978). These kinematic approaches have had some success for the simplest surfaces and they should still provide approximate but efficient methods for rough structural determinations of more complicated surfaces.

Two physical ingredients can be included in the kinematic methods described above. First, the mean free path is needed to provide the proper surface sensitivity. It is simply represented by an exponential decay of all waves, using either the mean free path itself or an imaginary component of the energy (the optical potential model). Second, atomic scattering factors are sometimes included, but they cannot be calculated in the Born approximation and thus cannot be represented by the Fourier components of the electron–crystal interaction potential. Instead, phase shifts for the scattering of partial waves must be computed from first principles, because multiple scattering also occurs within each atom. This is done, after assuming spherical atoms and a constant interstitial potential (the 'muffin-tin model'), by integration of the radial Schrödinger equation for the scattering electron subject to the atomic potential composed of an electrostatic part and an exchange-correlation part (this latter part is usually calculated either in the Hartree–Fock self-consistent approximation or by adopting the $X\alpha$ approximation.) Typically 5 to 8 phase shifts are needed to describe the scattering by an atom. A conventional Debye–Waller factor correction is often applied to the resulting scattering amplitude.

3.2. Formalisms that Include Multiple Scattering

The theoretical formalisms that have been developed to treat the multiple scattering of LEED electrons in a surface stem mainly from solid-state band structure calculation methods. (See, also, Moodie, this volume.)

As a compromise between accuracy and computational efficiency, the 'muffin-tin' potential approximation mentioned above is usually used. This choice leads to the convenience of using free-space waves, such as simple spherical waves or plane waves, in the interstitial regions between the atoms.

The problem of determining the scattered electron wave is solved by first considering the scattering by individual atomic layers parallel to the surface and then combining these to obtain the scattering by the stack of layers composing the surface. One of the first successful formalisms, developed by Beeby (Beeby, 1968; Laramore and Duke, 1972), is based on a spherical wave expansion throughout. The incident electron beam is represented by a plane wave that is expanded in terms of spherical waves incident on the surface atoms arranged in layers. A set of self-consistent equations is solved to obtain the scattering of these spherical waves by each layer and another set of self-consistent equations leads to the overall scattered spherical waves. A final transformation to plane waves gives the amplitude of each scattered plane wave and thereby the intensity in each scattered beam. The self-consistency mentioned above is equivalent to including all multiple scattering to infinite order and therefore provides an exact result (exact within the assumed model). This method is, however, rather demanding of computation effort, involving in particular the repeated inversion of a matrix of dimension usually far exceeding 100. A perturbation expansion of the self-consistent equations in terms of number of scatterings has been proposed by Zimmer and Holland, (1975) which converges in most materials, but not in tungsten or platinum, for example. It saves considerable computation effort.

Spherical-wave methods are inherently inefficient computationally when compared to plane-wave methods, but the latter break down much more readily. It has been found very profitable to use plane waves to represent the scattering wave between atomic layers whenever possible. It is necessary to include in this representation two plane waves for each two-dimensional reciprocal-lattice vector (one wave moving into the surface, the other moving back out of the surface). The number of plane waves remains finite, however, because most of them correspond to exponentially decaying waves that do not reach from one atomic layer to the next. Herein lies the limitation of the use of plane waves. When two atomic layers are very close together, too many exponentially decaying plane waves contribute and have to be included. Convergence problems arise and the computation effort rises rapidly. In practice this happens for spacings below about 0.5 Å.

As a result of the limitation on the use of plane waves, a mixed representation is often chosen (the Combined Space Method) in which plane waves represent the wavefield between widely spaced layers, while spherical waves represent the wavefield between closely spaced layers and also within simple layers that have one atom per unit cell. Thus within regions of close spacings the methods described previously (due to Beeby and to Zimmer and Holland) are used to produce reflexion and transmission matrices describing the scattering properties of plane waves. It then remains to stack up these regions, whose reflexion and transmission matrices are now known, on top of each other to form the complete surface. Several methods are available for this purpose. One of the earliest ones, which is still in use, first finds the Bloch waves in the substrate (McRae, 1966, 1968; Boudreaux and Heine, 1967). Since the substrate has a periodicity perpendicular to the surface, the Bloch theorem can be used to find the eigenfunctions (the Bloch waves) in that region. The Bloch waves can then be matched through the surface interface(s) to the plane waves in the vacuum, yielding the amplitudes of the reflected beams. The only approximation in this method comes from the truncation of the number of plane waves.

A different method, proposed by Pendry (1974), first stacks one pair of layers and then iterates the process by adding layers until a slab is obtained that is thick enough compared to the mean free path to represent the surface. The finite thickness is the only additional approximation compared to the Bloch-wave method. Multiple scattering is included to infinite order between the layers by matrix inversions in the layer-pairing method as compared to a diagonalization in the case of the Bloch-wave method.

More efficiency can be gained with a perturbation method called Renormalized Forward Scattering due to Pendry (1974). This method makes a perturbation expansion in terms of the number of multiple scatterings that an electron undergoes. However, it recognizes the fact that forward scattering is usually so strong that it cannot be treated as a perturbation, but must be included exactly. Thus in this method a wave that backscatters from the nth layer below the surface has undergone $(n-1)$ forward scatterings on the way in and $(n-1)$ forward scatterings on the way out. All these scatterings are included in the lowest order of perturbation. This idea, transplanted to the spherical wave representation, gave rise to the method of Zimmer and Holland mentioned earlier. The Renormalized Forward Scattering method is at present the most efficient for calculating LEED intensities, when it converges. But it fails to converge with layer spacings less than about 1.0 Å and when very strong multiple scattering occurs.

A few words should now be devoted to the physical parameters required for input into a LEED calculation. The way atomic scattering properties are obtained has been described in the previous section. The muffin-tin constant, i.e. the interstitial potential value, is fitted to the experiment, as it is difficult to calculate or measure. The mean free path is best estimated from experimental I–V curves as it directly determines the peak widths. Atomic vibration amplitudes are poorly known at surfaces, mainly because the usual intensity-versus-energy dependence that is used in X-ray diffraction and at high electron energies produces unreliable Debye temperatures in LEED as a result of multiple scattering. However, a roughly 50% increased vibration amplitude for surface atoms as compared with bulk atoms seems to be indicated both by experiment and by theory. The mean free path and the vibration amplitudes turn out not to be critical in structural determinations.

3.3. *Application to Surface Crystallography*

Since the theoretical multiple scattering formalisms are too complex for inversion, a trial-and-error approach to the determination of atomic position is inevitable, requiring as many sets of calculations as there are plausible surface structures that one wishes to test. Therefore, computational efficiency is very important.

Each of the calculational methods described in the previous section has a limited range of applicability insofar as convergence, computer time, and core size requirements are concerned. The present approach is, therefore, to use that method which best suits the particular problem at hand. With today's increasing size of computer, the complexity of surface structure that can be handled is rapidly growing. Whereas calculations with two atoms per unit cell per layer were considered impressive only a few years ago, several calculations have now been performed with six atoms per unit cell per layer and a few with up to 19 atoms per unit cell. The limiting factors are now becoming the availability of enough computer time to enable the computation of the variation of a sufficient number of atomic coordinates and, at the same time, the ability to deduce improved trial positions from previously tested positions. This feedback loop requires the use of a reliability factor to quantify the level of agreement between experiment and theory. Several such R-factors suitable for handling LEED data have been proposed and some have been extensively used already, but further development is required here.

It is important to mention at this point that techniques other than LEED are always used to complement the information gathered about the surface under study. For example, Auger electron spectroscopy and thermal desorption spectroscopy usually provide identification of the atomic and molecular species present at a surface. Photoemission and high resolution electron energy loss spectroscopy also provide identification but can, in addition, give qualitative information about the position and orientation of adsorbed species. Many other techniques are available and used, including some that compete directly with LEED in providing atomic positions; we should mention especially surface extended X-ray absorption fine structure, angle-resolved photoemission and ion scattering spectroscopy (more detailed descriptions and references are given in Biberian and Huber (1976)). However, to this day LEED remains the most productive method in surface crystallography.

The accuracy of LEED structural determination can be roughly given as being at present of the order of 0.05 Å for bond lengths. With simple surfaces and with certain geometries this can be reduced to about 0.01 Å. For more complicated surfaces, an uncertainty of 0.1 Å is normal.

4. RESULTS OF DETAILED SURFACE CRYSTALLOGRAPHY BY LEED

At the time of writing, about 110 surface structures have been submitted to a structural determination. These fall into distinct categories, some of which are discussed next. More complete listings and analyses are published elsewhere (Van Hove & Tong, 1979; Somorjai & Van Hove, 1979).

4.1. Clean Metal Surfaces

As mentioned previously, many clean metal surfaces are found to have essentially the structure that one would expect from an ideal truncation of the substrate (see Figure 1, and the substrates shown in Figure 4). For these the only noticeable deviation from the bulk geometry occurs in the topmost interlayer spacing, which shows a clear tendency to contract by typically 0.1 Å on those surfaces where the most bonds have been cut in making the surface, such as fcc(110), bcc(100), and bcc(111). These contractions correspond to bond length reductions by 2 to 3%, i.e. by about 0.05 Å. An interesting result is that these contracted bond lengths are systematically increased by the presence of adsorbates, including contaminants.

More exciting has been the discovery that a number of clean metal surfaces

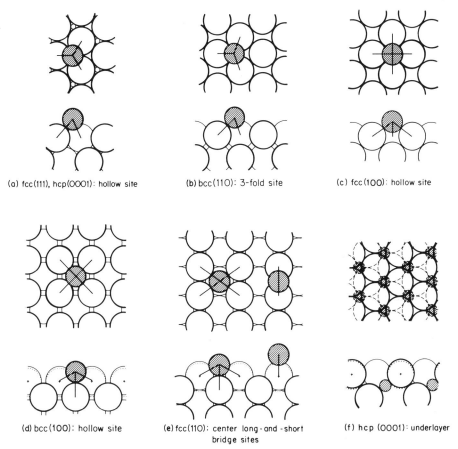

(a) fcc(111), hcp(0001): hollow site

(b) bcc(110): 3-fold site

(c) fcc(100): hollow site

(d) bcc(100): hollow site

(e) fcc(110): center long-and-short bridge sites

(f) hcp (0001): underlayer

Fig. 4. Top and side views (in top and bottom sketches of each panel) of various metal surfaces with commonly observed atomic adsorbate sites. Adsorbates are drawn shaded. Dotted circles represent clean-surface (relaxed) atomic positions.

reconstruct. Perhaps the best known example of this is the (1×5) reconstruction of Ir(100) and the related but slightly more complicated reconstructions of Pt(100) and Au(100). LEED studies indicate that the topmost atomic layer of these surfaces adopts a hexagonal close-packed arrangement on top of the square-lattice substrate. Different contractions and orientations of this hexagonal layer would explain the different patterns observed for the three metals. A warping of the hexagonal layer due to its misfit on the substrate has also been detected during our recent studies.

The (100) surfaces of the bcc metals Cr, Mo, and W provide another class of reconstruction patterns that are basically $c(2 \times 2)$. A LEED intensity analysis of W(100)$c(2 \times 2)$ indicates that the surface may have a structure compatible with a periodic displacement wave along the surface which may, therefore, be coupled to a two dimensional surface charge density wave.

Finally, we mention the Ir(110)(2×1) reconstruction which is found by LEED to probably have the missing row structure shown in Figure 5. The

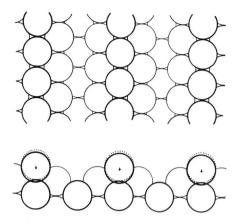

fcc (110) (2×1) missing row

Fig. 5.   Atomic arrangement in the missing-row model of the Ir(110)(2 × 1) surface. Conventions as in Figure 4.

absence of alternate rows of atoms on this fcc(110) surface exposes long, narrow facets of (111) orientation, which are known to be more stable thermodynamically than the ideal (110) surface.

4.2. Clean Semiconductor Surfaces

Most clean semiconductor surfaces exhibit superlattices, the most famous of which is undoubtedly the as-yet-unsolved (7 × 7) structure of Si(111). Fortunately many smaller unit cells occur as well. For example, Si(111)p(2 × 1) is believed to have been solved and involves the raising and lowering of alternate rows of top-layer atoms by about 0.2 Å as well as smaller adjustments in the second layer of atoms. A different rearrangement is found on Si(100)(2 × 1); namely, top layer atoms bond pairwise together to reduce the number of cut bonds by half. The subsurface atoms move somewhat under the strain of this top layer reconstruction, as shown in Figure 6.

Another interesting example of reconstruction occurs on GaAs(110). It takes place within the (1 × 1) unit cell without formation of superlattice, as shown in Figure 7. Surface As atoms are pushed slightly outward, while surface Ga atoms are pulled inward.

4.3. Atomic Adsorbates

Figure 2 shows some frequently observed superlattices for atomic adsorption on low Miller index metal surfaces. Detailed LEED analyses of many such adsorbate-substrate combinations have been carried out, with mainly the adsorbates H, N, O, Na, S, Cl, Se, Te, and I and the substrates Al, Ti, Fe, Co, Ni, Cu, Mo, Rh, Ag, W, and Ir, using various single crystal faces. The results for the location of adsorption are given, with only a few exceptions, by Figure 4 and hold for adsorbate densities of $\frac{1}{4}, \frac{1}{3}, \frac{1}{2}$ or 1 monolayer, depending on the case. A

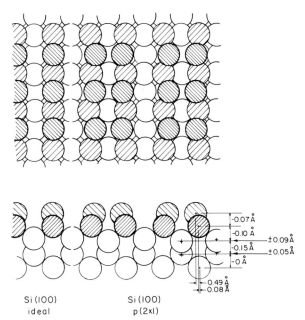

Fig. 6. Top and side views (in top and bottom sketches) of ideal bulk-like Si(100) at left and of Si(100)$p(2 \times 1)$ in the modified Schlier–Farnsworth model at right. Layer spacing contractions and intra-layer atomic displacements relative to the bulk structure are given. Shading differentiates surface layers.

general rule emerges that atoms choose those adsorption sites that provide the largest number of nearest metal neighbors. This usually is a site of high symmetry as well. Exceptions occur on fcc(110) faces, where oxygen prefers bridged sites of 2-fold coordination, while sulfur prefers 'center' sites, cf. Figure 4e. In Figure 4f is shown an 'underlayer': N on Ti(0001) in one monolayer chooses an interstitial site below the first Ti layer, closely reproducing the structure of the bulk compound TiN. Bond lengths between adsorbate and

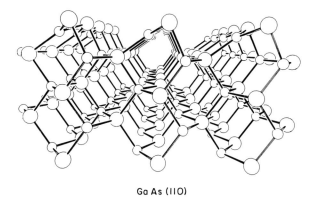

Ga As (110)

Fig. 7. ORTEP-II view, looking along surface of clean reconstructed GaAs(110). Small and large circles represent Ga and As atoms, respectively.

metal atoms are found to generally agree well with bond lengths known in compounds and in molecules. The precision of the LEED results seems sufficient to start interpreting the observed differences in terms of bond ionicity, bond order, etc.

Under the heading of atomic adsorption we can also mention the dissociative adsorption of molecules. This has been studied for CO adsorption on both Ti(0001) and on Fe(100). With the Ti(0001) substrate it seems that the C and O atoms occupy 3-fold hollow sites (Figure 4a), the C atoms forming a (2×2) array, and the O atoms forming a similar but shifted (2×2) array intermeshed with the C array. With the Fe(100) substrate the LEED analysis finds that the C and O atoms appear individually and randomly to occupy 4-fold hollow sites (Figure 4c) in a $c(2 \times 2)$ array, so that a $c(2 \times 2)$ array of unoccupied sites is present, all other sites being randomly occupied by either C or O atoms.

4.4. Molecular Adsorbates

The main molecule studied so far by LEED is carbon monoxide, which does not decompose on a number of surfaces. From non-LEED information this molecule is thought to usually stand roughly perpendicular to the surface, to which it bonds by its C end. This has been confirmed with LEED, while certain hypotheses concerning the adsorption site could also be verified and the bond lengths could now be measured. Thus CO forms at a half monolayer a $c(2 \times 2)$ structure both on Ni(100) and on Cu(100), with 'terminal' bonding, i.e. with each carbon atom bonding to one substrate atom only. The Ni—C, Cu—C, and C—O bond lengths of 1.72 Å, 1.9 Å, and 1.15 Å, respectively, are in good agreement with equivalent terminal bonding values in corresponding metal carbonyl clusters. Similarly, CO on Rh(111) at $\frac{1}{3}$ monolayer has terminal bonding with Rh—C and C—O bond lengths of 1.95 Å and 1.07 Å, respectively. (All bond lengths quoted here have uncertainties of up to ± 0.1 Å.) This last structure is illustrated in Figure 8. Non-terminal bonding of CO is also encountered, for example, on Pd(100) at a half monolayer, where bridge bonding to two Pd atoms occurs, with Pd—C and C—O bond lengths of 1.93 Å and 1.15 Å, respectively.

The adsorption geometry of small hydrocarbon molecules is also under investigation. Two structures have been analyzed with LEED, namely those of

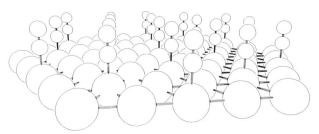

Rh(111) + $(\sqrt{3} \times \sqrt{3})$R30°CO

Fig. 8. ORTEP-II view of Rh(111) + $(\sqrt{3} \times \sqrt{3})R30°$CO.

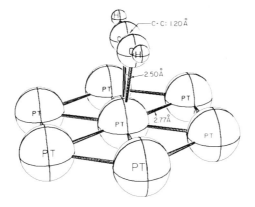

Pt (III) + metastable C_2H_2

Fig. 9. ORTEP-II view of Pt(111) + (2 × 2) C_2H_2 in its metastable structure.

acetylene (C_2H_2) and ethylene (C_2H_4) on Pt(111). Two ordered structures can be obtained, illustrated in Figures 9 and 10. A weakly-bound metastable acetylene structure, in which the molecules lie parallel to the surface, is followed in the presence of more hydrogen by a more strongly bound stable ethylidyne structure bonded approximately perpendicularly to the surface. This structure is also obtained directly with ethylene. In fact, LEED is only weakly sensitive to hydrogen and, therefore, the distribution of hydrogen atoms among the two carbon atoms in the stable structure is unknown at present. It appears that other methods than LEED wil be needed to locate hydrogen atoms at surfaces. But in any case, the ability to determine molecular structures at surfaces has obvious consequences for the understanding of the mechanisms of catalysis.

5. PROSPECTS

After a decade devoted to establishing its viability in the determination of

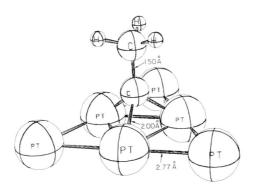

Pt (III) + ethylidyne

Fig. 10. ORTEP-II view of stable Pt(111) + (2 × 2) ethylidyne.

important and non-trivial surface structures, LEED has reached a stage of consolidation and further expansion. On the experimental side the focus is now on speeding-up and automating the time-consuming data gathering process. An increased speed, at the same time, reduces the damage done to the surface by the electron beam, a serious problem with many adsorbed molecules. On the theoretical side, the trend is towards enabling experimentalists to perform the required computations without the intervention of a theoretician. Also, new calculational methods are being developed to extend LEED to more and more complicated surface structures. One can, therefore, look forward to the further rapid development of surface crystallography by LEED in various directions. In particular, adsorbed molecules of the size of benzene, surface reconstructions that penetrate several layers deep, oxidation of metal surfaces, and some forms of disorder or defect structures are already within grasp. But serious attention to the theoretical reproducibility of the surface–vacuum interface and to the pitfalls of secondary minima in R-factors will remain essential. Finally, reliable spin-polarized low energy electron diffraction measurements are now available and enable the study of spin-orbit effects (Wang, Dunlop, Celotta and Pierce, 1980; Kirschner and Feder, 1979) as well as magnetic surface structures (Celotta, Pierce, Wang, Bader and Felcher, 1979; Wang, Kirby and Garwin, 1979).

Materials and Molecular Research Division
Lawrence Berkeley Laboratory, and
Department of Chemistry,
Berkeley, California, U.S.A.

REFERENCES

Adams, D. L. and Landman, U.: 1977, *Phys. Rev.* **B15**, 3775.
Beeby, J. L.: 1968, *J. Phys.* **C1**, 82.
Bibérian, J. P. and Huber, M.: 1976, *Surf. Sci.* **55**, 259.
Bibérian, J. P. and Somorjai, G. A.: 1979, *J. Vac. Sci. Technol.* **16**, 2073.
Boudreaux, D. and Heine, V.: 1967, *Surf. Sci.* **8**, 426.
Castner, D. G. and Somorjai, G. A.: 1979, *Chem. Rev.* **79**, 233.
Celotta, R. J., Pierce, D. T., Wang, G.-C., Bader, S. D., and Felcher, G. P.: 1979, *Phys. Rev. Lett.* **43**, 728.
Cunningham, S. L., Chan, C.-M., and Weinberg, W. H.: 1978, *Phys. Rev.* **B18**, 1537.
Firment, L. E., and Somorjai, G. A.: 1977, *J. Chem. Phys.* **66**, 2901.
Houston, J. E., and Park, R. L.: 1970, *Surf. Sci.* **21**, 209.
Huber, M. and Oudar, J.: 1975, *Surf. Sci.* **47**, 605.
Kirschner, J. and Feder, R.: 1979, *Phys. Rev. Lett.* **42**, 1008.
Laramore, G. E. and Duke, C. B.: 1972, *Phys. Rev.* **B5**, 267.
McRae, E. G.: 1966, *J. Chem. Phys.* **45**, 3258.
McRae, E. G.: 1968, *Surf. Sci.* **11**, 479.
Ngoc, T. C., Lagally, M. G., and Webb, M. B.: 1973, *Surf. Sci.* **35**, 117.
Novaco, A. D. and McTague, J. P.: 1977, *Phys. Rev. Lett.* **38**, 1286.
Pendry, J. B.: 1974, *Low Energy Electron Diffraction.* Academic Press, London.
Shaw, C. G., Fain, S. C., Jr., and Chinn, M. D.: 1978, *Phys. Rev. Lett.* **41**, 955.
Somorjai, G. A. and Van Hove, M. A.: 1979, *Structure and Bonding.* Springer, Heidelberg, Vol. **38**, p. 1.

Van Hove, M. A. and Tong, S. Y.: 1979, *Surface Crystallography by LEED*, Springer, Heidelberg.
Van Hove, M. A., Tong, S. Y., and Elconin, M. H.: 1977, *Surf. Sci.* **64**, 85.
Wang, G.-C., Dunlap, B. I., Celotta, R. J. and Pierce, D. T.: 1980, *Phys. Rev. Lett.* **42**, 1349.
Wang, S.-W., Kirby, R. E., and Garwin, E. L.: 1979, *Sol. St. Commun.* **32**, 993.
Wang, G.-C., Lu, T.-M., and Lagally, M. G.: 1978, *J. Chem. Phys.* **69**, 479.
Webb, M. B. and Lagally, M. G.: 1973, *Solid State Phys.* **28**, 301.
Wood, E. A.: 1964, *J. Appl. Phys.* **35**, 1306.
Zanazzi, E. and Jona, F.: 1977, *Surf. Sci.* **62**, 61.
Zimmer, R. S. and Holland, B. W.: 1975, *J. Phys.* **C8**, 2395.

KOZO KUCHITSU

III.3. Gas Electron Diffraction

The need for solving many problems in structural chemistry
really led to the development of this field since it afforded direct
answers to some questions of long standing.
L. O. Brockway (1936)

1. INTRODUCTION

Gas electron diffraction is probably the simplest application of the wave nature of electrons, from both experimental and theoretical standpoints. In almost all cases, electrons of about 40 keV (with wavelength of about 0.06 Å) are used, and a diffraction pattern obtained from a target gas of typically 20 torr is recorded on a photographic plate in a very short time. In the very first report, Mark and Wierl (1930a) remarked: "Unsere Versuche, einen Elektronenstrahl an Tetrachlorkohlenstoff zu beugen, ergaben bei einer Belichtungszeit von 1–3 Sekunden zwei deutliche Ringe, ein dritter ist angedeutet." In early days, it was regarded as a remarkable advantage over gas X-ray diffraction, which took tens of hours to obtain a comparable diffraction pattern but with a much narrower s range (Pirenne, 1946).

From a theoretical point of view, the target molecules can be regarded as independent (no intermolecular interference) under the conditions mentioned above, and they are oriented randomly in space. This makes the theory very straightforward and simple (see Section 3). Furthermore, a simple molecular model such as the 'independent atom model' works well in most cases. In this model, the molecule is regarded as an independent assembly of (vibrating) spherical atoms, and the dynamical effect is taken into account only within each atom (intra-atomic interference). The effect of interatomic multiple scattering becomes significant only in special cases, such as rhenium hexafluoride (Miller and Bartell, 1980) or clusters (Bartell, Raoult and Torchet, 1977; Yokozeki, 1978). The effects of chemical bonding (distortion from a spherical electron distribution in atoms) and polarization of electron clouds by incident electrons appear only in small scattering angles, and they are either negligible or approximately correctable in most applications.

On the other hand, there is a serious limitation of gas electron diffraction in comparison with diffraction in the crystal phase: in contrast to the three dimensional data from the latter, experimental information on molecular geometry obtainable from gas electron diffraction is limited to only one

356

dimension because of the random orientation of molecules in space (Debye, 1915, 1941); thus only the probability distributions of bonded or nonbonded internuclear distances, not their relative directions in regard to the molecule-fixed axes, can be obtained from gas electron diffraction. Therefore, even with an accurate measurement of diffraction intensity and careful analysis it is not always easy to determine a complete three-dimensional structure when a complicated molecule is studied. In such a case, other experimental information (such as spectroscopic) often provides valuable assistance (Hilderbrandt and Bonham, 1971; Kuchitsu, 1972a; Robiette, 1973, 1976).

2. BRIEF HISTORICAL SURVEY

2.1. Start

The concept of 'diffraction by gas molecules' was initiated in 1915. Being stimulated by the great success of X-ray diffraction by crystals, Debye (1915) calculated the interference of X-rays by free molecules and predicted a halo pattern fluctuating with frequencies determined by the internuclear distances. Debye's manuscript, which contained his famous $\sin x/x$ formula and dated 25 February 1915, was received in the editorial office of *Annalen der Physik* in Leipzig on the 27th. On the latter day, an independent paper by Ehrenfest (1915) on a similar theory was presented by Lorentz and Kammerlingh-Onnes to the Academy of Amsterdam (Debye, 1930).

It was only after 14 years that the first experimental study of gas X-ray diffraction was made in Leipzig (Debye, Bewilogua, and Ehrhardt, 1929). They were able to observe a diffraction pattern from CCl_4 vapor. Immediately afterwards, a remarkable technical improvement was made in Debye's laboratory (Bewilogua, 1931). This development was probably a real trigger of the start of gas electron diffraction which came in less than a year (January, 1930) (see Mark's article). The following footnotes refer to the suggestions and discussions on the possibility and potentiality of gas electron diffraction (unfortunately without explicit dates of the events) and show the background of the pioneer work of Mark and Wierl (1930): "Wir möchten bemerken, daß Herr W. Bothe nach einem Vortrag von Debye auf der Röntgentagung in Zürich ebenfalls die Möglichkeit von Elektronenbeugungsversuchen an Gasen zur Diskussion stellte." (Mark and Wierl, 1930a). "Herr Mark bemerkt, daß Herr Bothe gelegentlich eines Vortrages von mir in Zürich auf die Möglichkeit hinwies, Elektronenstrahlen an Stelle von Röntgenstrahlen zu verwenden. Ich darf hier zusätzlich bemerken, daß nach einem früheren Vortrage von mir in der Berliner Physikalischen Gesellschaft Herr Pringsheim eine ähnliche Bemerkung machte." (Debye, 1930).

The structural problems taken up by Mark and Wierl (1930b) were the following:

(a) differences in the C—C bond lengths caused by chemical environment (aliphatic, alicyclic, aromatic, and multiple bonding),

(b) molecular symmetries and bond lengths in simple inorganic molecules (CCl_4, $SiCl_4$, and $TiCl_4$),

(c) regularities of covalent atomic radii,

(d) comparison between the bond lengths in gas molecules and those in crystals,

(e) geometrical isomerism (*cis*- and *trans*-dichloroethylene),

(f) linearity and planarity of molecules (CS_2, benzene),

(g) internal rotation (1,2-dichloroethane),

(h) conformations of long saturated hydrocarbons (1,5-dichloropentane).

They also suggested a study of less volatile molecules with a high-temperature oven so as to investigate the distortion of benzene ring by halogen substitutions. They had a special interest in comparing their data with those determined by other experimental methods.

A remarkable improvement was made in a year or two by Wierl (1931, 1932) in the quality of the diffraction patterns, and fairly accurate geometrical parameters of many molecules were determined. For example, Wierl reported the C—C bond lengths in benzene and cyclopentane to be 1.42 ± 0.03 and 1.52 ± 0.03 Å, while the best values known at present are 1.399 ± 0.001 (Tamagawa, Iijima and Kimura, 1976) and 1.5460 ± 0.0012 Å (Adams, Geise and Bartell, 1970), respectively. He often referred to the various other structural data, which had just been determined by molecular spectroscopy. All the items mentioned above have indeed been the most interesting and important problems of structural chemistry, and they have since been investigated extensively by gas electron diffraction and other methods.

2.2. Visual Method

Early workers of gas X-ray diffraction measured the intensity recorded on a photographic plate with a microphotometer; it was easy to analyze its trace, because it exhibited clear maxima and minima. On the other hand, the corresponding trace from a photographic plate of gas electron diffraction showed only faint shoulders (Figure 1a) because of the steeply falling atomic background (see Section 3). However, experienced human eyes could subtract the background by inspection, so that the maximum and minimum positions of the 'molecular fluctuations' could be located on the plate with remarkable accuracy. The diameters of these maxima and minima were measured and compared with the corresponding values calculated from model structures. The geometrical structure which resulted in the best fit was obtained by trial and error. This method, started by Wierl (1930, 1931), was called the visual method. Several groups started to use this method, and with the application of Fourier transform to gas electron diffraction (the radial distribution function) (Pauling and Brockway, 1935), it soon became one of the most powerful methods for the determination of molecular structure in the gas phase. As early as July 1936, Brockway (1936) listed in his review the structures of 44 inorganic and 103 organic molecules of fundamental importance determined by this method and published in 50 papers.

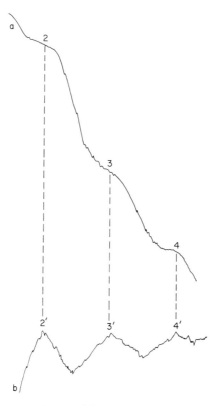

Fig. 1.   Electron diffraction intensities for CCl_4 gas (Debye, 1939). (a) Microphotometer trace of a non-sectored photographic plate. (b) Microphotometer trace of a photographic plate taken with an r^3-sector.

These structural data, together with those determined by molecular spectroscopy and X-ray crystallography, made great contributions to the establishment of the theory of chemical bonding (Pauling, 1939) (see Bauer's article). The visual method had been used in many laboratories until it was superseded after 1950 by the sector-microphotometer method and by high resolution (particularly microwave) spectroscopy. Among the various subjects studied in the visual period, the following pioneer works may be mentioned: (a) Maxwell, Hendricks and Moseley (1937) used an electric furnace, with which they evaporated alkali halides at about 1200 °C, and studied their structures. Structure studies at high temperature have since been carried out actively in several laboratories, particularly in USSR (Spiridonov and Zasorin, 1979) (see Vilkov's article). (b) The failure of the first Born approximation was demonstrated by Schomaker and Glauber (1952), and Glauber and Schomaker (1953) in their studies of metal hexafluorides; a splitting of the radial distribution peak of the metal-fluorine distance was explained by the difference in the phase shifts for the metal and fluorine atoms (see Schomaker's article).

2.3. Sector-Microphotometer Method

The use of a rotating sector, which compensates for the atomic background by a partial screening of scattered electrons, was first proposed by Trendelenburg (1933). Figure 1b was obtained by P. P. Debye (1939) as one of the first applications of this technique. A microphotometer trace of a photographic plate taken with such a sector gave a quantitative density distribution, which was convertible to electron scattering intensity. Thus the accuracy of the structure derived from gas electron diffraction was greatly improved (see article by Karle and Karle). The sector method was first applied extensively by Hassel, Finbak, and their coworkers (Finbak, Hassel and Ottar, 1941) to their studies of the structures of many organic molecules (particularly, ring molecules such as cyclohexane) (see Bastiansen's article). Karle and Karle (1949, 1950) were also among the most active pioneers of this method, by which they made quantitative analyses of intramolecular thermal motions. The corresponding theoretical analysis, which was initiated by James (1932), was extended by Morino and his coworkers (Morino, Kuchitsu and Shimanouchi, 1952; Morino and Hirota, 1955) and by Bartell (1955) (see Morino's and Bartell's articles). The sector method was spread over the world in the 1950s. When electronic computers became available, a statistical analysis of the molecular intensity using the method of least squares (Bastiansen, Hedberg and Hedberg, 1957) became a standard technique (see Hedberg's article). In favorable cases, it was possible to make precise investigations of mean-square vibrational amplitudes and shrinkage effects (summarized by Cyvin, 1968), rotational isomerism (Ainsworth and Karle, 1952), and the effect of vibrational anharmonicity (Bartell, Kuchitsu and deNeui, 1961).

The application of the sector method not only enabled studies of nuclear arrangements in a molecule but also those of electron distributions. The measurement of electron distribution in the argon atom (Bartell and Brockway, 1953) was the first successful case (see Bartell's article). Recent measurements of small angle scattering by simple molecules give information on their electronic wavefunctions (Bonham and Fink, 1974).

3. THEORY

Determination of the geometrical structure of a free molecule in the gas phase is usually based on the theory outlined below (Davis, 1971; Seip, 1973; Bonham and Fink, 1974). Deviations from this standard theory are described briefly in Section 7.

3.1. Scattering from Atoms

The scattering of a plane wave of electrons with kinetic energy E by a spherical potential, $V(r)$, is described by the following equation:

$$\nabla^2\psi + [k^2 - U(r)]\psi = 0 \tag{1}$$

where

$$k = \sqrt{2mE}/\hbar \tag{2}$$

and

$$U(r) = 2mV(r)/\hbar. \tag{3}$$

The elastic scattering amplitude, $f(\theta)$, at angle θ is given by the method of partial waves as

$$f(\theta) = (2ik)^{-1} \sum_{l=0}^{\infty} (2l+1) \left[\exp(2i\delta_l) - 1\right] P_l(\cos\theta) \tag{4}$$

$$= |f(\theta)| \exp[i\eta(\theta)]. \tag{5}$$

The phase shift, δ_l, can be calculated by solving the radial equation,

$$d^2 R_l/dr^2 + [k^2 - l(l+1)/r^2 - U(r)] R_l = 0 \tag{6}$$

and taking the phase of its asymptotic form,

$$R_l(r) \sim k^{-1} \sin(kr - \tfrac{1}{2}l\pi + \delta_l). \tag{7}$$

Numerical values of the $|f|$ and η for any atom as functions of s and k have been tabulated with sufficient accuracy (Sellers, Schäfer and Bonham, 1978). In addition, a sum of all inelastic scattering is given, using the first Born approximation, by (Morse, 1932)

$$I_{\mathrm{inel}}(s) = (4m^2 e^4/\hbar^4 s^4) S(s) \tag{8}$$

where $s = 2k \sin(\theta/2)$, and $S(s)$ is the inelastic scattering factor for X-ray diffraction (Heisenberg, 1932).

3.2. Scattering from Diatomic Molecules:
Independent Atom Model

The following assumptions are made: (1) The molecule is composed of two independent atoms with spherical charge distributions, their nuclei being separated by a distance r. (2) Their charge distributions are not influenced by vibration, i.e., they are independent of the internuclear distance r. (3) There is no interatomic and intermolecular multiple scattering; i.e., electrons encounter only single collisions with atoms.

Under the above conditions, the scattering cross-section is given by

$$I_T(s) = I_a(s) + I_{\mathrm{inel}}(s) + I_m(s) \tag{9}$$

where the atomic term, $I_a(s)$, and the inelastic term, $I_{inel}(s)$, are sums of contributions from each atom,

$$I_a(s) = |f_1|^2 + |f_2|^2 \tag{10}$$

and

$$I_{\mathrm{inel}}(s) = (4m^2 e^4/\hbar^4 s^4) [S_1(s) + S_2(s)]. \tag{11}$$

These terms constitute a uniform background.

The molecular term, $I_m(s)$, results from two averages: an average over the random orientation of the molecular axis in space leads to

$$\langle \exp(is \cdot r) \rangle = \sin sr/sr \tag{12}$$

and an average over the probability distribution of r, $P(r)dr$, results in

$$I_m(s) = 2|f_1|\, |f_2| \cos(\eta_1 - \eta_2) \int P(r) \sin sr/sr \, dr. \tag{13}$$

The probability function can be calculated from the vibrational wavefunction. If the system is assumed to be in thermal equilibrium at temperature T, $P(r)$ is given as an average over the vibrational state v,

$$P(r;T) = \sum_{v=0}^{\infty} |\psi_v(r)|^2 \, w_v(T) \tag{14}$$

where

$$w_v(T) = \exp[-E_{\mathrm{vib}}(v)/kT] \left/ \sum_{v=0}^{\infty} \exp[-E_{\mathrm{vib}}(v)/kT]. \right. \tag{15}$$

This $P(r)$ function is generally a slightly distorted Gaussian. By a polynomial expansion of the distortion, (13) can be integrated and the molecular term can be obtained as a damped sine curve with only a slight frequency modulation (Kuchitsu, 1967)

$$I_m(s) = 2|f_1|\, |f_2| \cos(\eta_1 - \eta_2) \exp(-\tfrac{1}{2}l^2 s^2) \sin s(r_a - \kappa s^2)/sr_a \tag{16}$$

where l^2 is the mean-square amplitude of vibration, and r_a is an effective internuclear distance representing the center of gravity of the $P(r)/r$ function. The phase parameter κ is nearly constant with s and is related to the anharmonicity of the potential function. The influence of κ is small unless one of the atoms is hydrogen or deuterium, or uness the potential function is very shallow and anharmonic.

3.3. Scattering from Semirigid Polyatomic Molecules

The probability distribution function for each (bonded or nonbonded) atom pair, $i-j$, can be calculated by the theory of normal vibrations. By averaging over random orientations in space, the scattering cross section can be expressed as a superposition of individual atom pairs,

$$I_T(s) = I_a(s) + I_{\mathrm{inel}}(s) + I_m(s) \tag{17}$$

where

$$I_a(s) = \sum_i |f_i|^2 \tag{18}$$

$$I_{\mathrm{inel}}(s) = (4m^2 e^4/\hbar^4 s^4) \sum_i S_i(s) \tag{19}$$

and

$$I_m(s) = 2 \sum_{i}\sum_{j>i} |f_i|\, |f_j| \cos(\eta_i - \eta_j) \exp(-\tfrac{1}{2}l_{ij}^2 s^2) \sin s\,(r_{aij} - \kappa_{ij} s^2)/sr_{aij}. \tag{20}$$

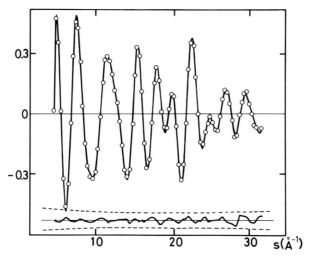

Fig. 2. Molecular intensities, defined as $sI_m(s)/[I_a(s) + I_{inel}(s)]$, for Cl_2CO (Nakata *et al.*, 1980). Experimental values are shown as open circles, and the best fit theoretical values are shown as a solid curve. The differences are shown in the lower solid curve. The broken curve represents the estimated limits of random and systematic error in the intensity measurement.

The calculation of mean-square amplitudes, l_{ij}^2, plays an important role in this theory. If the molecule has no large-amplitude motion, and if a reasonably accurate intramolecular force field is known, it is relatively easy to calculate the mean-square amplitudes by use of the theory of normal vibrations (James, 1932; Cyvin, 1968).

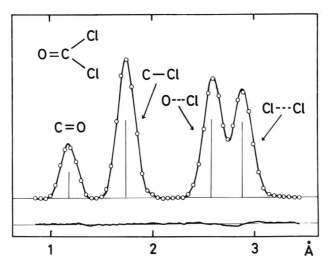

Fig. 3. Radial distribution curves for Cl_2CO (Nakata *et al.*, 1980). Experimental and best fit theoretical values are represented by open circles and the solid curve, respectively. The differences are shown below on the same scale. Vertical bars indicate the positions of average internuclear distances.

The above independent-atom model is shown to be applicable in most cases when nuclear geometry (average internuclear distances) is determined from experiment. A typical application of (20) and its modified Fourier transform (i.e. the radial distribution function, which is the $P(r)/r$ function convoluted with an appropriate experimental aperture function) are shown in Figures 2 and 3 for the Cl_2CO molecule. More sophisticated treatments of the effect of anharmonic thermal vibrations are given in Kuchitsu (1967), Kuchitsu and Cyvin (1972) and Bartell (1980).

3.4. Physical Significance of Geometrical Parameters

The geometrical parameters of polyatomic molecules can be defined in several different ways. The experimental data of gas electron diffraction are so precise and accurate that the derived parameters based on different definitions can be significantly different. Furthermore, it has become more and more common to make a critical comparison of the structures determined by gas electron diffraction with those determined with similar precision by high resolution spectroscopy, crystallography, and theory, or to use data obtained by these methods together with electron-diffraction data to derive the 'best' structure. In such a case, geometrical parameters must be defined clearly and consistently. Otherwise, the analysis may result in a significant systematic error. There are the following three important definitions of molecular geometry (Kuchitsu and Cyvin, 1972):

A. Equilibrium nuclear positions (r_e structure)

This structure corresponds to the nuclear positions of minimum potential energy (a hypothetical state).

B. Average nuclear positions (r_z or r_α structure)

This structure corresponds to the nuclear positions averaged over vibrational motions. Averages over the ground vibrational state i.e. at zero Kelvin, and over thermal equilibrium distributions are denoted as r_z and r_α structures, respectively. The conversion from r_α to r_z can be made by extrapolation to zero Kelvin (usually by a simple theoretical estimation). These structures differ from r_e due to vibrational anharmonicity. The rotational constants determined by high resolution spectroscopy can be used, after small correction for harmonic vibrations, to determine or estimate the r_z structure. Diffraction studies of crystals can also be used to determine thermal average nuclear positions (the r_α structure).

C. Average internuclear distances (r_g structure)

Diffraction studies in the gas, liquid, and amorphous phases give information on the internuclear distances averaged over thermal motions. They are called the r_g

distances. The r_g distance is equal to the centre of gravity of the $P(r)$ function (Bartell, 1955),

$$r_g = \frac{\int_0^\infty r\, P(r)\, dr}{\int_0^\infty P(r)\, dr},$$ (21)

and is related to the r_a distance in (20) by

$$r_g = r_a + l^2/r_a,$$ (22)

and l^2 is the mean square amplitude of vibration of the atoms.

The r_α and r_g structures defined in (b) and (c) are identical only in a diatomic case. In general, for two atoms i and j in a molecule,

$$r_g^{ij} = \langle |r_j - r_i| \rangle$$

in a local Cartesian co-ordinate system, whereas

$$r_\alpha^{ij} = |\langle r_j \rangle - \langle r_i \rangle|.$$

The r_g distance, which allows for correlation between the motions of the pair of atoms, is slightly longer than the r_α and r_z distances for any bonded or nonbonded pair because of the freedom of motion perpendicular to the equilibrium internuclear axis. The r_g and r_α or r_z structures can be calculated from each other if an approximate force field is known.

The r_g distance for a bonded pair is related to the r_e distance as

$$r_g = r_e + \langle \Delta r \rangle$$

where $\langle \Delta r \rangle$ is the displacement of the bond distance averaged over thermal vibration. The $\langle \Delta r \rangle$ term for a bonded pair is positive due to the anharmonic bond stretching vibration. In most reports of gas electron diffraction, the r_g distances for bonded pairs are reported, because r_g bond-distances are convenient and consistent measures for a systematic comparison of bond-lengths in analogous molecules.

The r_g distances for nonbonded atom pairs can also be defined clearly, but they cannot be used with bonded r_g distances to determine well-defined bond angles. This inconsistency in the geometrical conditions of r_g parameters is known as linear (or nonlinear) shrinkage effect. For example, the r_g distance of the O—O pair in CO_2 is significantly smaller than twice the $r_g\,(C—O)$ distance because of the O—C—O bending vibration. The correction for the shrinkage effect is equivalent to the conversion of the r_g structure to the r_z or r_α structure.

3.5. Large Amplitude Motions

Even when the molecule has one or more motions with low frequencies and large amplitudes, (20) is still applicable to calculation of $I_m(s)$ if $P(r)$ is defined. Such calculations have been made for many examples, particularly for molecules with internal rotation (Debye, 1941; Karle, 1954; Bastiansen, Seip and Boggs, 1971; Robiette, 1976; Bastiansen, Kveseth and Møllendal, 1979). However, there still remain many unsolved problems in the method of analysis for deriving accurate information on the structure of such nonrigid systems from gas electron diffraction and spectroscopic data.

3.6. Intramolecular Multiple Scattering

Precise measurements of molecular intensities for molecules containing a heavy atom, such as ReF_6 (Jacob and Bartell, 1970), showed systematic discrepancies between the corresponding intensities calculated by (20). These discrepancies were accounted for in terms of the effect of intramolecular multiple scattering using the Glauber approximation (Miller and Bartell, 1980, Kohl and Arvedson, 1980) (see Bartell's article).

4. EXPERIMENTAL TECHNIQUES

4.1. Standard Apparatus and Procedures

A typical apparatus of gas electron diffraction is composed of the following parts: (a) an electron gun, (b) a simple electron-optical system, (c) a nozzle with a small diameter (e.g., 0.2 mm i.d.) tip, (d) a cold trap to condense the sample gas, (e) a rotating sector, (f) a camera box storing photographic plates, and (g) a fast pumping system. The nozzle is placed perpendicular to the electron beam; the distance between the tip of the nozzle and the beam is about 0.3 mm. The distance from the nozzle to the photographic plate varies from about 10 cm to 30 cm. The sector is a spiral- or heart-shaped thin metal plate, with an angular opening proportional to a regular function of the radius (e.g., an r^3-sector). Examples of the apparatus, the sector, and a diffraction pattern recorded on a photographic plate are shown in Figures 4–6.

The following three points need special remarks: (a) If there is an unknown impurity in the sample gas, it is often left undiscovered in both experiment and analysis, and it can seriously influence the accuracy of the analysis. (b) It is sometimes difficult to define or measure the temperature of the sample, and in some cases the effective vibrational temperature of a certain mode can be much lower than the nozzle temperature, or even be ill-defined because the vibrational populations are non-Boltzmann. However, it is believed that under normal conditions the sample is nearly in thermal equilibrium with the nozzle tip. This makes it much easier to apply gas electron diffraction to thermochemical studies such as gas-phase conformational equilibrium (Bohn, 1977). (c) Corrections for the effect of a finite sample size (scattering volume) are possible if the density distribution of the sample gas is measured. Whether or not the correction is important depends on the nozzle design and the sample pressure.

The scale factor, which defines the s value on the photographic plate, depends on the electron wavelength and the nozzle-to-plate distance. The method used most frequently to determine the scale factor is to analyze a diffraction pattern of a standard gaseous sample, such as CO_2 or benzene.

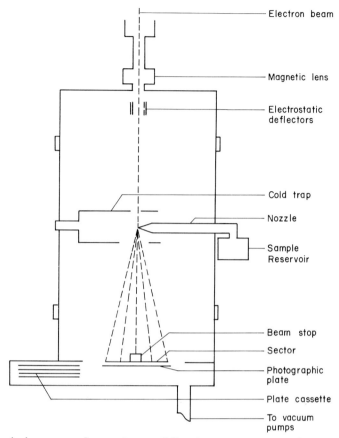

Electron beam

Magnetic lens

Electrostatic
deflectors

Cold trap

Nozzle

Sample
Reservoir

Beam stop

Sector

Photographic
plate

Plate cassette

To vacuum
pumps

Fig. 4.   A typical apparatus for gas electron diffraction (schematic) at Oregon State University
(Hedberg, 1980).

Fig. 5.   A typical r^3-sector. Oregon State University (Hedberg, 1980).

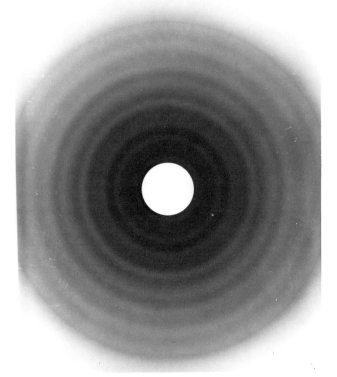

Fig. 6.   A typical photograph taken with an r^3-sector. A diffraction pattern for SiF_4 gas taken at
Oregon State University (Hedberg, 1980).

The photographic density of the plate is measured by a microphotometer.
The plate is usually rotated about the diffraction center during photometry. In
this way the radially-symmetric scattered intensity recorded on all the parts of
the photographic plate is formed into a radial average. Digital microphotometry
accompanied by data processing with a computer is carried out more or less
automatically. The photographic density is converted to relative electron
intensity after correction for nonlinear response, and a smooth background is
drawn through the molecular fluctuations, usually on an empirical basis. It is a
general procedure to average the molecular terms (20) obtained from several
photographic plates for analysis. Other details of the experimental apparatus
and procedure are described in the references in Bartell, Kuchitsu and Seip,
1976.

4.2. New Technical Developments

4.2.1. High temperature studies

Various ovens and high-temperature nozzles have been designed primarily for
(a) achieving high enough vapor pressure, (b) studying the change of structural

parameters, conformational compositions, etc. with temperature. Products of pyrolytic reactions are also being studied. The heating is made by one of the following methods: (a) electric current, either direct (Legett, Kennerly and Kohl, 1974) or indirect (by thermal conduction), (b) circulation of hot gas or fluid, (c) electron bombardment, and (d) laser irradiation (Bartell, Doun and Goates, 1979). Caution is needed to protect the photographic plate from radiation. A nozzle temperature as high as 2500 °C has been reported (Spiridonov and Zasorin, 1979).

4.2.2. High pressure studies

A supersonic nozzle with a stagnation pressure up to several tens of atmospheres has been used to produce atomic or molecular clusters. Average cluster size, approximate atomic arrangements, and lattice parameters can be estimated from the scattering intensity, which shows intermediate patterns between an assembly of free gas molecules and micro-crystals. Detailed studies have been made on the clusters of rare gas atoms and CO_2 molecules (Farges, de Feraudy, Raoult and Torchet, 1975; Bartell, Raoult and Torchet, 1977) and clusters of metal atoms, such as lead (Yokozeki, 1978). In the latter experiments, clusters were produced by cooling the metal vapors emitting from an oven by cold carrier gas.

4.2.3. Nonphotographic methods.

A photographic plate is a very sensitive and reliable detector of electrons in the energy range of about 40 keV. However, there is an insurmountable limit of accuracy in the photographic measurement of electron intensity (of the order of 0.1%), which mainly originates from the irregularities of photographic emulsion and chemical processes of development. Various techniques of direct electron counting have recently been developed in several laboratories. For example, an accuracy of 0.1% has been reported by use of scintillator-photomultiplier systems (Fink, Moore and Gregory, 1979).

5. ANALYSIS

5.1. General Procedure

The analysis of gas electron diffraction is in principle much simpler than a three-dimensional analysis of diffraction data taken from a crystal sample. The molecular term is calculated by (20) and a least squares method is used to obtain the most probable estimate of the structural parameters (Hedberg and Iwasaki, 1964; Morino, Kuchitsu and Murata, 1965). A trial structure is first assumed, and a reasonable number of parameters are refined by iterative fitting of intensity curves such as that defined by (20). Parameters such as the κ parameters and, usually, some of the l as well as corrections for the shrinkage

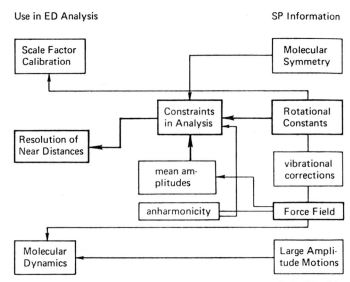

Fig. 7.   Use of spectroscopic data in the electron-diffraction analysis (Kuchitsu, 1974).

effects are fixed to their calculated values. In almost all cases, i.e., except when the molecule is simple and has very high symmetry, there exists correlation among the structural parameters, which can lead to indeterminate or multiple solutions. Much caution has to be taken in such a case on the selection of the parameters to be refined in the analysis (Hilderbrandt and Bonham, 1971; Hedberg, 1974). A conjoint use of accurate experimental data taken from other independent sources, such as rotational constants obtained by high resolution spectroscopy, is often used for reducing or even eliminating correlation and determining a unique structure (Kuchitsu, 1972a; Robiette, 1973, 1976). A diagram showing the use of spectroscopic data in the analysis of gas electron diffraction is shown in Figure 7 (Kuchitsu, 1974).

Random errors can be estimated from the least squares analysis and also from the reproducibility of the structures obtained by repeated measurements of the scattering intensity. In addition, various systematic errors in the measurements of the intensity and the scale factor have to be included in the estimation of total uncertainties in the structural parameters (Kuchitsu, 1972b). In the most favorable case, the total uncertainty limits in bond distances, angles and root mean square amplitudes are about 0.001 Å, 0.1°, and 0.002 Å, respectively.

The time and effort required for an experiment and structure analysis for one molecule depends on various conditions. In a laboratory equipped with a good diffraction apparatus and an automated microphotometer connected to a computer, a routine structure analysis for a simple molecule may take only a day or two. It is not uncommon, however, that a much longer time is spent on the instrumentation and on experimental or analytical procedures.

5.2. Examples of the Analysis

One of the simplest cases, phosgene (Nakata, Kohata, Fukuyama and Kuchitsu,

1980), and a case of medium complexity, 7-thianorbornane (Fukuyama, Oyanagi and Kuchitsu, 1976), are taken as examples.

5.2.1. Phosgene

In this case the electron-diffraction intensity contains sufficient information for determining the three geometrical parameters uniquely. As shown in the radial distribution curve (Figure 3), the peaks are separate and they are very nearly Gaussian. The parameters were determined by a least squares analysis. The rotational constants calculated by use of these parameters agreed with those determined by microwave spectroscopy. Hence, the parameters were further refined to achieve the best fit to both the electron-diffraction intensity and the microwave rotational constants. The following r_z structure was obtained: $r(C=O) = 1.1785 \pm 0.0026$ Å, $r(C-Cl) = 1.7424 \pm 0.0013$ Å, and $\angle Cl-C-Cl = 111.83 \pm 0.11°$, where the errors represent the limits of uncertainty. The geometrical structure having been determined precisely, the rotational constants for several isotopic species were then introduced in order to estimate the isotopic displacements in the average nuclear positions. Hence the degrees of anharmonicity in the $C=O$ and $C-Cl$ bond stretching vibrations and the r_e structure could be estimated.

5.2.2. Thianorbornane

By a least squares analysis of the electron-diffraction intensity alone, the $C-S$ bond distance and the $C-S-C$ valence angle in this molecule (Figure 8) were determined without difficulty. For the C_1-C_2 and C_2-C_3 bonds, however, it was not possible to determine these nearly-equal distances separately, and only their weighted average value, $\frac{2}{3} r(C_1-C_2) + \frac{1}{3} r(C_2-C_3)$, could be determined. Nevertheless, the rotational constants determined by microwave spectroscopy were available as additional experimental data. A combined analysis of the electron-diffraction intensity and the rotational constants resulted in the following r_z structure: $r(C-S) = 1.834 \pm 0.004$ Å, $r(C_1-C_2) = 1.535 \pm 0.006$ Å, $r(C_2-C_3) = 1.557 \pm 0.015$ Å, $r(C-H$, weighted average$) = 1.101 \pm 0.008$ Å, $\angle C-S-C = 80.2 \pm 0.8°$, and $\angle C_1-C_2-C_3 = 105.2 \pm 0.5°$.

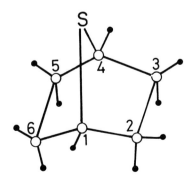

Fig. 8.   7-Thianorbornane, $C_6H_{10}S$.

6. RESULTS

Gas electron diffraction and high resolution (especially microwave) spectro-scopy have been the most important sources of information for determining the geometrical structures of free molecules in the ground electronic state. According to the *Gas Electron Diffraction Catalogue* (see below), the structures of approximately 1250 molecules (900 organic and 350 inorganic) have been determined by gas electron diffraction in the period of 1960–1979 in about 30 research groups in the world. A number of molecules among them have been studied repeatedly. In many instances, particularly for relatively complicated molecules, gas electron diffraction has been the only source of experimental information, although this method cannot always provide sufficient information to determine a complete structure. In a number of other cases, spectroscopic structures have also been determined or estimated. In almost all such cases, the electron diffraction and spectroscopic structures are consistent within experi-mental uncertainties. For a few simple molecules such as CH_4 and NH_3, information on vibrational anharmonicity, in addition to the usual geometrical and vibrational parameters, has been derived by a combined analysis of electron diffraction and spectroscopic data (Kuchitsu, 1972a).

A critical comparison of the structures of free molecules in the gas phase with those of molecules in the crystal phase has been, and still remains, an interesting and important problem of structural chemistry (Beagley, 1975). For example, significant changes in conformations have been observed. A relatively strong intermolecular force in the crystal, such as a hydrogen bond, also influences bond distances and valence angles significantly (e.g., amides, Kitano and Kuchitsu, 1974).

All the studies made by gas electron diffraction and their references are well documented and reviewed (See articles by Sutton and Morino). All the important structural parameters determined by gas electron diffraction and published in the literature are collected into the *Gas Electron Diffraction Catalogue* by B. Starck and her associates in the Sektion für Strukturdokumentation, Universität Ulm. This catalogue was used by the author as a basic source of data when the structures determined by this method were evaluated critically and listed in the *Landolt–Börnstein Tables* (Callomon *et al.*, 1976). *The GED Information Service Letter*, which includes the references of recent publications, has been organized by the Commission on Electron Diffraction, IUCr, and is compiled and distributed regularly to the research groups in this field by B. Starck. Comprehensive and up-to-date review articles appear regularly in *Specialist Periodical Report*, The Chemical Society, London (see Sutton's article), and there are several other recent review articles in *Topics in Current Chemistry*, Springer, Berlin, and in *Diffraction Studies on Non-Crystalline Substances*, Akadémiai Kiadó, Budapest.

7. STUDY OF CHARGE DISTRIBUTIONS

Active studies of the electron density distributions in atoms and molecules by gas electron diffraction started about a decade after pioneering experimental work for the argon atom (Bartell and Brockway, 1953) (see Bartell's article). For electrons of 30–50 keV the first Born approximation provides an appropriate and convenient theoretical basis for this study (Fink *et al.*, 1979). In this approximation the elastic and total scattering intensities from a molecule of N electrons, $I_E(s)$ and $I_T(s)$, respectively, are given in atomic units by Kohl and Bartell (1969)

$$I_E(s) = s^{-4}[\sigma_{nn}(s) + \sigma_{ne}(s) + \langle \int d\mathbf{r}\, j_0\, (sr) \int d\mathbf{r}'\, \rho(\mathbf{r})\rho(\mathbf{r}+\mathbf{r}') \rangle_{\text{vib}} \tag{23}$$

$$I_T(s) = s^{-4}[\sigma_{nn}(s) + \sigma_{ne}(s) + \sigma_{ee}(s) + N] \tag{24}$$

where

$$\sigma_{nn}(s) = \sum_i Z_i^2 + \sum\sum_{i \neq j} Z_i Z_j \langle j_0(sr_{ij}) \rangle_{\text{vib}} \tag{25}$$

$$\sigma_{nn}(s) = -2\sum_i Z_i \langle \int d\mathbf{r}\rho(\mathbf{r}+\mathbf{r}_i) j_0(sr) \rangle_{\text{vib}} \tag{26}$$

$$\sigma_{ee}(s) = \langle \int d\mathbf{r}\rho_c(\mathbf{r}) j_0(sr) \rangle_{\text{vib}} \tag{27}$$

and $\mathbf{r}_i$, r_{ij}, and $\langle\ \rangle_{\text{vib}}$ represent the position of the ith nucleus, the internuclear distance between the nuclei i and j, and the average over the thermal vibrations of the molecule. The one-electron density or distribution function, $\rho(\mathbf{r})$, and the electron-pair-correlation density (2-particle distribution function), $\rho_c(\mathbf{r})$, are defined in terms of the molecular electronic wavefunction, ψ, by

$$\rho(\mathbf{r}) = \sum_{v=1}^{N} \int d\tau_1 \ldots d\tau_N |\psi(\mathbf{r}_1, \cdots \mathbf{r}_N)|^2\, \delta(\mathbf{r} - \mathbf{r}_v) \tag{28}$$

$$\rho_c(\mathbf{r}) = \sum_{v \neq \mu}^{NN} \int d\tau_1 \cdots d\tau_N |\psi(\mathbf{r}_1 \cdots \mathbf{r}_N)|^2\, \delta(\mathbf{r} - \mathbf{r}_v + \mathbf{r}_\mu). \tag{29}$$

Though the three-dimensional function $\rho(\mathbf{r})$ for a molecule cannot be determined uniquely from σ_{ne}, unlike the case of X-ray crystallography, the $I_T(s)$ of gas electron diffraction is an important experimental source of atomic and molecular electronic distributions and wavefunctions (Kohl and Bartell, 1969).

Earlier experimental studies using the sector-microphotometer method were mainly directed to observation of the difference between the experimental $I_T(s)$ and the $I_T(s)$ based on the independent atom model (20) (Bonham, 1969). A simple relationship between this difference and the chemical binding energy of the target molecule (Tavard and Roux, 1965) extended the application of this measurement to studies of electron correlation energy by comparison with thermochemical data. A more quantitative comparison of experimental and theoretical data requires accurate measurements of the electron intensity at small scattering angles ($s \leqslant 10$). Recent measurements using non-photographic methods have reached the required accuracy, and a number of atoms, diatomic and polyatomic molecules are being studied (e.g., Fink *et al.*, 1979).

8. PRESENT AND FUTURE PROBLEMS

In their review article Hilderbrandt and Bonham (1971) pointed out ten possibilities for future developments. Most of the problems predicted by them have since been partially explored, but they still remain important fields of research, as summarized below. Technical improvements and developments are being made in (a) accurate measurement of electron intensity, (b) designs of gas source, and (c) other new aspects of electron scattering by gas molecules.

8.1. Intensity Measurements

Solutions to all the main problems of gas electron diffraction, i.e.:

(a) geometrical structures and intramolecular potential functions of free molecules;

(b) intramolecular motions with large vibrational amplitudes, particularly in systems with weak chemical bonding; and

(c) electronic structures such as effects of chemical bonding (see Section 7),

depend on the accurate measurement of electron intensity over a wide range of scattering angles. Various efforts are being made for this purpose, in both photographic and nonphotographic techniques.

In order to take full advantage of gas electron diffraction, it is almost always necessary to combine accurate data obtained from other experimental or theoretical sources, such as spectroscopy, crystallography, and *ab initio* calculations. Conversely, without electron-diffraction data structural results from these other methods are often incomplete or speculative. By a critical comparison of, and combination with all these data, one can often obtain a complete structure and make a fair estimate of the accuracy of the data obtained.

In addition, structure analysis by use of very low pressures, strong electron beams and highly sensitive electron detectors will probably become increasingly important.

8.2. Gas Source

Efforts are being made to design good high-temperature nozzles (Spiridonov and Zasorin) so as to study geometrical structures of (a) molecules with high boiling (or sublimation) points, (b) molecules in various thermal equilibria, (c) reaction products such as thermal decomposition, and (d) molecules with large thermal motions. Studies are also being made on the production and detection of non-equilibrium vibrational distributions by laser irradiation (Kacner and Bartell, 1979). On the other hand, high-pressure nozzle sources are being used to produce atomic and molecular clusters (van der Waals molecules) (see Section 4.2.2). It is not easy to make any of these studies, and it is highly desirable to combine optical and/or mass spectroscopic techniques to obtain knowledge about the chemical species in the sample and their relative compositions, and if necessary, their quantum states.

8.3. Electron Scattering

As far as gas electron diffraction is concerned, problems related to intraatomic and intramolecular multiple elastic scattering seem to have been understood to a sufficient degree of accuracy (see articles by Schomaker and Bartell). Inelastic electron scattering from gas molecules has been studied extensively by electron spectroscopy (i.e., measurements of kinetic energy of scattered electrons) both for electrons of high and low incident energy (a few tens of eV to tens of keV). However, there still seems to be much to be studied both experimentally and theoretically on various phenomena related to inelastic scattering of 10^4 eV electrons (Bonham, 1979), e.g., the relationship between the inelastically scattered electron intensities at large angles and the electronic structures of the initial and final states of the target molecule.

Electron holography (Bartell, 1979; Bartell and Gignac, 1979) and the reflection of electrons by standing light waves (Bartell, Roskos and Thompson, 1968) may also find interesting applications in future studies (see Bartell's article).

ACKNOWLEDGEMENT

The author wishes to thank Professor K. Hedberg for his valuable comments.

Department of Chemistry
Faculty of Science
The University of Tokyo
Tokyo, Japan

REFERENCES

Adams, W. J., Geise, H. J., and Bartell, L. S.: 1970, *J. Am. Chem. Soc.* **92**, 5013–5019.
Ainsworth, J. and Karle, J.: 1952, *J. Chem. Phys.* **20**, 425–427.
Bartell, L. S.: 1955, *J. Chem. Phys.* **23**, 1219–1222.
Bartell, L. S.: 1979, *J. Chem. Phys.* **70**, 3952–3957.
Bartell, L. S.: 1980, *J. Mol. Struct.* **63**, 259–271.
Bartell, L. S. and Brockway, L. O.: 1953, *Phys. Rev.* **90**, 833–838.
Bartell, L. S., Doun, S. K., and Goates, S. R.: 1979, *J. Chem. Phys.* **70**, 4585–4586.
Bartell, L. S. and Gignac, W. J.: 1979, *J. Chem. Phys.* **70**, 3958–3964.
Bartell, L. S., Kuchitsu, K. and deNeui, R. J.: 1961, *J. Chem. Phys.* **35**, 1211–1218.
Bartell, L. S., Kuchitsu, K., and Seip, H. M.: 1976, *Acta Cryst.* **A32**, 1013–1018.
Bartell, L. S., Raoult, B., and Torchet, G.: 1977, *J. Chem. Phys.* **66**, 5387–5392.
Bartell, L. S., Roskos, R. R., and Thompson, H. W.: 1968, *Phys. Rev.* **166**, 1494.
Bastiansen, O., Hedberg, L., and Hedberg, K.: 1957, *J. Chem. Phys.* **27**, 1311.
Bastiansen, O., Kveseth, K., and Møllendal, H.: 1979, *Topics in Current Chemistry* (ed. F. L. Boschke) Vol. 81, pp. 99–172. Berlin: Springer-Verlag.
Bastiansen, O., Seip, H. M., and Boggs, J. E.: 1971, *Perspectives in Structural Chemistry* (ed. J. D. Dunitz and J. A. Ibers) Vol. 4, pp. 60–165. New York: John Wiley.
Beagley, B.: 1975, *Molecular Structures by Diffraction Methods* (Specialist Periodical Report) (ed. G. A. Sim and L. E. Sutton) Vol. 3, pp. 52–71. London: The Chemical Society.
Bewilogua, L.: 1931, *Physik. Z.* **32**, 265.

Bohn, R. K.: 1977, *Molecular Structures by Diffraction Methods* (Specialist Periodical Report) (ed. G. A. Sim and L. E. Sutton) Vol. 5, pp. 23–94. London: The Chemical Society.

Bonham, R. A.: 1969, *Rec. Chem. Progr.* **30**, 185.

Bonham, R. A.: 1979, *High-Energy Electron Impact Spectroscopy (HEEIS), Electron Spectroscopy: Theory and Technical Applications* (ed. C. R. Brundle and A. D. Baker) Vol. 3, pp. 127–187. London: Academic Press.

Bonham, R. A. and Fink, M.: 1974, *High Energy Electron Scattering* New York: Van Nostrand Reinhold.

Brockway, L. O.: 1936, *Rev. Mod. Phys.* **8**, 231.

Callomon, J. H., Hirota, E., Kuchitsu, K., Lafferty, W. J., Maki, A. G., and Pote, C. S.: 1976, *Landolt-Börnstein*, (ed. K.-H. Hellwege and A. M. Hellwege) Group II, Vol. 7. Berlin: Springer-Verlag.

Cyvin, S. J.: 1968, *Molecular Vibrations and Mean Square Amplitudes* Oslo: Universitetsforlaget, and Amsterdam: Elsevier.

Davis, M. I.: 1971, *Electron Diffraction in Gases* New York: Marcel Dekker.

Debye, P.: 1915, *Ann. Physik* **46**, 809.

Debye, P.: 1930, *Physik. Z.* **31**, 419.

Debye, P.: 1941, *J. Chem. Phys.* **9**, 55.

Debye, P. P.: 1939, *Physik. Z.* **40**, 66; **40**, 404–406.

Debye, P., Bewilogua, L., and Ehrhardt, F.: 1929, *Physik. Z.* **30**, 84.

Ehrenfest, P.: 1915, *Verslag Akad. Wetenschappen*, Amsterdam, **23**, 1132.

Farges, J., de Feraudy, M. F., Raoult, B., and Torchet, G.: 1975, *J. Phys.* (Paris), **36**, 13–17.

Finbak, C., Hassel, O., and Ottar, B.: 1941, *Arch. Math. Naturvidenskab*, Oslo, No. 13, 137.

Fink, M., Moore, P. G., and Gregory, D.: 1979, *J. Chem. Phys.* **71**, 5227.

Fink, M. and Schmiedekamp, C.: 1979, *J. Chem. Phys.* **71**, 5243.

Fink, M., Schmiedekamp, C. W., and Gregory, D.: 1979, *J. Chem. Phys.* **71**, 5238.

Fukuyama, T., Oyanagi, K., and Kuchitsu, K.: 1976, *Bull. Chem. Soc. Japan* **49**, 638–643.

Glauber, R. and Schomaker, V.: 1953, *Phys. Rev.* **89**, 667.

Hedberg, K.: 1974, *Critical Evaluation of Chemical and Physical Structural Information* (ed. D. R. Lide, Jr. and M. A. Paul) pp. 77–93. Washington: National Academy of Sciences.

Hedberg, K.: 1980, Personal communication.

Hedberg, K. and Iwasaki, M.: 1964, *Acta Cryst.* **17**, 529.

Heisenberg, W.: 1932, *Physik. Z.* **32**, 737.

Hilderbrandt, R. L. and Bonham, R. A.: 1971, *Ann. Rev. Phys. Chem.* **22**, 279.

Jacob, E. J. and Bartell, L. S.: 1970, *J. Chem. Phys.* **53**, 2231.

James, R. W.: 1932, *Physik. Z.* **33**, 737.

Kacner, M. A. and Bartell, L. S.: 1979, *J. Chem. Phys.* **71**, 192.

Karle, I. L. and Karle, J.: 1949, *J. Chem. Phys.* **17**, 1052.

Karle, I. L. and Karle, J.: 1950, *J. Chem. Phys.* **18**, 963.

Karle, J.: 1954, *J. Chem. Phys.* **22**, 1242.

Karle, J. and Karle, I. L.: 1950, *J. Chem. Phys.* **18**, 957–962.

Kitano, M. and Kuchitsu, K.: 1974, *Bull. Chem. Soc. Japan* **47**, 67.

Kohl, D. A. and Arvedson, M. M.: 1980, *J. Chem. Phys.* **72**, 1915, 1922.

Kohl, D. A. and Bartell, L. S.: 1969, *J. Chem. Phys.* **51**, 2891, 2896.

Kuchitsu, K.: 1967, *Bull. Chem. Soc. Japan* **40**, 498.

Kuchitsu, K.: 1972a, *MTP International Review of Science* (Phys. Chem. Ser. 1. Vol. 2) (ed. G. Allen) Chap. 6, pp. 203–240. Oxford: Medical and Technical Publ. Co.

Kuchitsu, K.: 1972b, *Molecular Structures and Vibrations* (ed. S. J. Cyvin) pp. 148–170. Amsterdam: Elsevier.

Kuchitsu, K.: 1974, *Critical Evaluation of Chemical and Physical Structural Information* (ed. D. R. Lide, Jr. and M. A. Paul) pp. 132–139. Washington: National Academy of Sciences.

Kuchitsu, K. and Cyvin, S. J.: 1972, *Molecular Structures and Vibrations* (ed. S. J. Cyvin) Chap. 12, pp. 183–211. Amsterdam: Elsevier.

Legett, T. L., Kennerly, R. E., and Kohl, D. A.: 1974, *J. Chem. Phys.* **60**, 3264.

Mark, H. and Wierl, R.: 1930a, *Naturwissenschaften* **18**, 205.

Mark, H. and Wierl, R.: 1930b, *Z. Elektrochem.* **36**, 675.

Maxwell, L. R., Hendricks, S. B., and Moseley, V. M.: 1937, *Phys. Rev.* **52**, 968.

Miller, B. R. and Bartell, L. S.: 1980, *J. Chem. Phys.* **72**, 800.

Morino, Y. and Hirota, E.: 1955, *J. Chem. Phys.* **23**, 737.

Morino, Y., Kuchitsu, K., and Murata, Y.: 1965, *Acta Cryst.* **18**, 549.

Morino, Y., Kuchitsu, K., and Shimanouchi, T.: 1952, *J. Chem. Phys.* **20**, 726.

Morse, P. M.: 1932, *Physik. Z.* **33**, 443.

Nakata, M., Kohata, K., Fukuyama, T., and Kuchitsu, K.: 1980, *J. Mol. Spectrosc.* **83**, 105.

Pauling, L.: 1939, *The Nature of Chemical Bond* Ithaca, New York: Cornell University Press.

Pauling, L. and Brockway, L. O.: 1935, *J. Am. Chem. Soc.* **57**, 2684.

Pirenne, M. H.: 1946, *The Diffraction of X-Rays and Electrons by Free Molecules* London: Cambridge University Press.

Robiette, A. G.: 1973, *Molecular Structures by Diffraction Methods* (Specialist Periodical Report) (ed. G. A. Sim and L. E. Sutton) Vol. 1, pp. 160–197. London: The Chemical Society.

Robiette, A. G.: 1976, *Molecular Structures by Diffraction Methods* (Specialist Periodical Report) (ed. G. A. Sim and L. E. Sutton, Vol. 4, pp. 45–61. London: The Chemical Society.

Schomaker, V. and Glauber, R.: 1952, *Nature* **170**, 290.

Seip, H. M.: 1973, *Molecular Structures by Diffraction Methods* (Specialist Periodical Report) (ed. G. A. Sim and L. E. Sutton) Vol. 1, pp. 7–58. London: The Chemical Society.

Sellers, H. L., Schäfer, L., and Bonham, R. A.: 1978, *J. Mol. Struct.* **49**, 125.

Spiridonov, V. P. and Zasorin, E. Z.: 1979, *Proceedings of the 10th Materials Research Symposium on High Temperature Vapors and Gases*, National Bureau of Standards Special Publication No. 561, pp. 711–755.

Tamagawa, K., Iijima, T., and Kimura, M.: 1976, *J. Mol. Struct.* **30**, 243.

Tavard, C. and Roux, M.: 1965, *Compt. Rend.* **260**, 4460, 4933.

Trendelenburg, F.: 1933, *Naturwissenschaften* **21**, 173.

Wierl, R.: 1930, *Physik. Z.* **31**, 1028.

Wierl, R.: 1931, *Ann. Physik* **8**, 521.

Wierl, R.: 1932, *Ann. Physik* **13**, 453.

Yokozeki, A.: 1978, *J. Chem. Phys.* **68**, 3766.

S. AMELINCKX

III.4. Electron Diffraction in Transmission Electron Microscopy

1. INTRODUCTION

The pioneering work in the field of transmission electron microscopy of thin foils was performed in the late fifties mainly, at the Cavendish Laboratory in Cambridge by P. B. Hirsch and coworkers (see article earlier in this volume). Very soon thereafter the potential of the technique was widely realized, and its application to an extended range of materials and a large variety of problems was developed in a number of laboratories. Although initially applied to metallurgical problems it was soon clear that materials science in general could greatly benefit from the technique.

It is fair to say that at present solid state transmission electron microscopy has become a discipline of its own, based largely on the combined use of microscopy and diffraction, occasionally supplemented by X-ray fluorescence analysis in order to supply chemical information. The fact that these techniques can be applied using the same instrument and on the same specimen, the first two with the specimen kept in the same position and orientation, is an essential feature for crystallographic studies. The additional possibilities such as tilting, heating and cooling have allowed the real-time study of structural features of phase transformations in great detail. Moreover *in situ* studies of deformation, radiation damage and of chemical reactions have lead to considerable progress in all these areas.

In this short survey we shall briefly outline the main results of solid state electron microscopy, emphasizing the role played by electron diffraction in providing the local crystallographic information needed for detailed interpretation of structures, and defect structures.

As opposed to X-ray diffraction, which can only provide information averaged over a large number of unit cells, electron microscopy and electron diffraction allow us to study local structures. In particular, in crystals which are highly fragmented into symmetry related domains, electron diffraction may still provide monodomain diffraction patterns and thus make an unambiguous interpretation of the diffraction pattern possible. Single defects and their effect on the diffraction pattern can be studied.

2. TWO-BEAM AND MULTIPLE-BEAM DYNAMICAL THEORY

Images are often formed under conditions where several beams contribute; the stiuation is then described as a multiple-beam diffraction situation. On the other hand quantitative electron microscopy, especially defect study, is often performed under two-beam conditions. By this is meant that apart from the incident beam only one diffracted beam contributes significantly to the image, the other beams being present, but sufficiently weak so as to be neglected to a good approximation. If under such conditions an image is made with the diffracted beam this is called a two-beam dark field image; similarly one can make two-beam bright field images.

Under two-beam conditions a well defined extinction distance can be attributed to each reflexion. This extinction distance is revealed as the depth period of images if the Bragg condition is exactly satisfied.

The deviation from the exact Bragg condition is measured by s, the excitation error, which is the distance from the excited reciprocal lattice node to Ewald's sphere, measured perpendicular to the foil surface; it is positive if the reciprocal lattice node is inside Ewald's sphere and negative when outside. If $s = 0$ the Bragg condition is exactly satisfied. For large s values, or very thin foils, multiple diffraction can be neglected and the kinematical theory is an acceptable approximation (Hirsch, Howie and Whelan, 1960).

If $s \neq 0$ the extinction distance belonging to the reflection g is given by $(t_g)_{\text{eff}} = t_g/(1 + (s t_g)^2)^{1/2}$. If $s = 0$ the transmitted and scattered beams behave similarly in the crystal, the intensities being complementary at each level; the intensity is transferred periodically with a period equal to the extinction distance from one to the other beam. Depending on the exact crystal thickness either the transmitted or the scattered beam will have maximum intensity (Pendellösung effect: Heidenreich and Sturkey, 1945).

3. CONTRAST AT DEFECTS; THE COLUMN APPROXIMATION

The presence of defects, such as dislocations and stacking faults, is revealed in the image through their effect on the local intensity at the back surface of a crystal. In this respect it is important to remember that the Bragg angles are very small in electron diffraction, leading to a small lateral spread of electrons inside a thin foil. To a good approximation one can say that electrons are travelling along narrow columns of crystal of which the size depends on the thickness and on the Bragg angle, but which is small as compared to the widths of most defect images.

It is therefore in general a good approximation to adopt the so called *column approximation* (Whelan and Hirsch, 1957) when calculating defect images. Only for the smallest defects are noticeable deviations detected with an exact calculation (Howie and Basinski, 1968).

The approximation consists in calculating the intensities of transmitted and scattered beams propagating along a column in which the variation of s with the distance from the front surface of the foil is determined by the displacement field of the defect. The intensity at the back surface of the crystal at a given point is thus identified with the intensity emerging from the column located at that point. The column approximation in fact implies that all parallel columns into which the crystal plate can be subdivided are identical with the considered one, i.e. are deformed in the same way. Defects are thus seen through their displacement field, i.e. through their effect on the local s-value.

It should be noted that the column approximation leads to an exact result in the case of a perfect crystal, but it is an approximation in the case of a deformed crystal. Defects can also be considered as scattering centers causing interbranch scattering (Howie, 1962) and thus creating intensity differences as compared to perfect crystal parts, thereby leading to an image (Hashimoto, Howie and Whelan, 1962).

Even in the two-beam case an exact analytical solution of the system of equations describing dynamical diffraction by a deformed crystal can only be found in the case of planar interfaces, i.e. of a stacking fault, an inversion boundary (Gevers, Van Landuyt and Amelinckx, 1965) and of a coherent twin boundary (Goringe and Valdré, 1966).

4. DISLOCATION CONTRAST

The contrast at dislocations is one-sided for $s \neq 0$, i.e. dislocations are observed as dark lines situated slightly on one side of the actual position of the dislocation core (Hirsch, Howie and Whelan, 1960). The side of the dislocation on which the image line occurs, referred to here as the 'image side', is determined by the sign of the product $p = (\mathbf{g} \cdot \mathbf{b}) \, s$. Accepting the FS/RH convention* for the definition of the Burgers vector and the above-mentioned definition of s and arranging a normal positive print (i.e. one as viewed from below the specimen) so as to define a left and right side of the dislocation, then, for a dislocation line direction defined from bottom to top on the print, the image will be at the right for $p > 0$ and the left for $p < 0$. Similarly, for a closed dislocation loop where the sense of the dislocation line is defined by a clockwise circuit, the image will be inside the loop for $p > 0$ and outside for $p < 0$.

The following intuitive reasoning (Amelinckx, 1964) explains the origin of contrast at dislocations and allows us to determine the 'image side'. Let the foil contain an edge dislocation in E. In the perfect part of the foil the orientation is such that the transmitted and the scattered beams are comparable in intensity for reflection against the lattice planes shown in Figure 1. The line width is a rough measure for the intensity of the beams. The s-value of the active diffraction vector which points to the right is positive.

At the left side of the dislocation in E_1 the considered lattice planes are sloping

---

* FS/RH, or Finish-Start/Right Hand convention refers to the way in which a Burgers circuit is defined around a dislocation line (Bilby, Bullough and Smith, 1955).

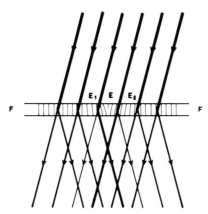

Fig. 1.   Intuitive picture explaining the formation of dislocation images. In E an edge dislocation
is present in the foil.

in such a way that the Bragg condition is locally better satisfied. As a result
locally more electrons will be scattered into the diffracted beam than in the
perfect part and a lack of electrons will be noted all along a line slightly to the left
of the dislocation; i.e. a dark line will be observed in the bright field image. In
this approximation a bright line would be observed in the dark field image. The
opposite would apply to the crystal part slightly right of the dislocation i.e. in E_2
since the lattice planes are sloping in the opposite sense compared to the
orientation in the perfect part. A similar reasoning is applicable to screw
dislocations since lattice planes are also inclined in opposite senses left and right
of the dislocation.

The same intuitive reasoning also demonstrates that Bragg reflection from the
family of lattice planes which are left undeformed by the dislocation will not
reveal the presence of the dislocation. The condition for extinction of the image
is thus $\mathbf{g} \cdot \mathbf{b} = 0$. This is an approximation however; the lattice planes parallel
with the glide plane of an edge dislocation for instance satisfy the extinction
criterion, but nevertheless some contrast is observed, which is due to the slight
deformation of such planes; in the case of an edge dislocation the displacement is
perpendicular to the glide plane. This effect is clearly visible for pure edge
prismatic loops observed with $\mathbf{g}$ in the plane of the loop. In the latter case the
extinction will only be complete along those parts of the loop where the radial
displacement is perpendicular to $\mathbf{g}$ (Howie and Whelan, 1960).

From the one-sided nature of the contrast at dislocations it is possible to
determine the sign of dislocations. The intuitive reasoning easily demonstrates
that changing the sign of the dislocation i.e. of the Burger vector changes the
image side, as well for edge as for screw dislocations; this result also follows from
the change in sign of p as $\mathbf{b}$ changes sign.

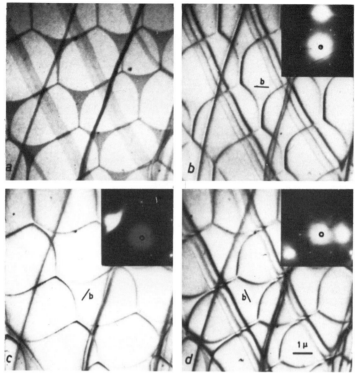

Fig. 2.   Determination of the direction of the Burgers vector of partial dislocations in a network of extended and contracted nodes in graphite. For different active diffraction vectors under two-beam conditions, different partial dislocations are out of contrast. The Burgers vector of the dislocations which are out of contrast are perpendicular to the active diffraction vector and moreover parallel with the foil plane, which is also the glide plane. Faulted areas show up darker (a). The magnitude of the stacking fault energy can be deduced from the widths of single or triple ribbons and from the geometry of the extended nodes. Note also the presence of triple ribbons consisting of three partials with the same Burgers vector. In (c) the three partials are simultaneously out of contrast. Although the three partials in the ribbon have the same Burgers vectors they do not exhibit the same line contrast as is evident from (b) and (d).

The problem of determining whether a prismatic loop is due to the precipitation of vacancies or of interstitials is in fact equivalent to determining the sign of the dislocation bordering the loop. Several practical methods, based on determining the sign of $(\mathbf{g} \cdot \mathbf{b})s$, have been described (Groves and Kelly, 1961; 1962). Such methods have been used extensively in the study of radiation damage, of quench defects in metals and alloys and of loops due to nonstoichiometry.

5. SPECIAL CONTRAST EFFECTS AT DISLOCATIONS

A thin foil containing an edge dislocation undergoes a discontinuous bend at the dislocation; two parts on either side of the dislocation thus differ slightly in

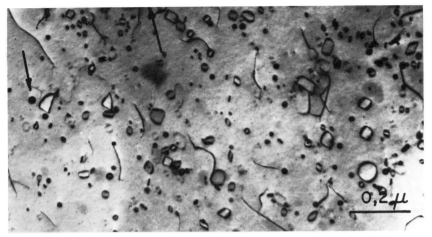

Fig. 3.   Small dislocation loops produced by neutron irradiation in platinum.

orientation. This orientation difference depends on the position of the dislocation with respect to the foil centre; it is maximum when the dislocation is in the center of the foil. This orientation difference is revealed in the image as an intensity difference in the crystal parts on either side of the dislocation. The sense of bending and the orientation difference can be determined from the difference in position of the Kikuchi lines produced by the crystal parts on either side of the dislocation. The so-determined sense can be correlated with the sign of the dislocation determined from the image side. The sign of a dislocation can thus be

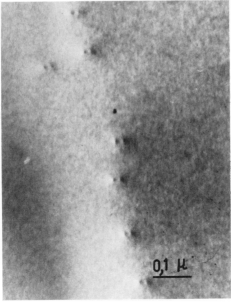

Fig. 4.   The dislocations are out of contrast, only the lattice relaxation at the surface produces contrast.

determined by two independent methods (Siems, Delavignette and Amelinckx, 1962).

For small s values dislocations which are not parallel with the foil plane produce an image which changes periodically with the level in the foil, the period being the effective extinction distance. This is called oscillating or dotted contrast. The oscillations in bright and dark field images are similar (i.e. in phase) at the top of the foil and complementary at the bottom of the foil (i.e. in anti-phase) (Howie and Whelan, 1961; 1962).

Small dislocation loops or very small precipitates may produce black or bright dots as images (Ashby & Brown, 1963). Dislocations seen end on (Tunstall, Hirsch & Steeds, 1964) produce characteristic contrast effects which are to a large extent due to surface relaxation along their emergence points. In cases where the dislocation image itself is extinct, the emergence points may still produce contrast.

6. COMPUTER SIMULATION OF DISLOCATION IMAGES

Computer programs have been developed to simulate two-beam dislocation images. Identification of the characteristics of dislocations proceeds through comparison of observed and simulated images (Head, Humble, Clarebrough, Morton & Forwood, 1973). The strength of dislocation contrast depends on $n = \mathbf{g} \cdot \mathbf{b}$, which is a small integer for perfect dislocations ($n = 1, 2, 3$) and on the diffraction variables. For partial dislocations n adopts no integral values; for $n = \frac{2}{3}$ the dislocation will usually still be visible, whereas for $n = \frac{1}{3}$ the visibility becomes questionable.

It is possible to determine the magnitude of the Burgers vector from the knowledge of n, $\mathbf{g}$ and the direction of $\mathbf{b}$ (Amelinckx, 1967; 1974).

The images of closely neighbouring parallel dislocations with the same Burgers vector may be quite different; this is due to the fact that the combined displacement field of the two dislocations produces the contrast. The image is therefore not the superposition of the two images that would be produced by two isolated single dislocations (Holland, Lindenmeyer, Trivedi and Amelinckx, 1965). This 'vicinity' effect is especially striking in layer structures where it extends far along the layer planes as a result of the elastic anisotropy of such materials. In graphite for instance triple ribbons containing three partial dislocations with the same Burgers vector are frequently observed (DeLavignette, Trivedi, Gevers and Amelinckx, 1966). Nevertheless the three partials produce quite different images; the image of these ribbons is furthermore strongly dependent on the sign of s. An analytical theory, based on the kinematical approximation, accounts for these observations.

7. THE WEAK BEAM METHOD

The images of dislocations become finer and finer lines as the magnitude of s increases; this effect is systematically exploited in the *weak beam technique*

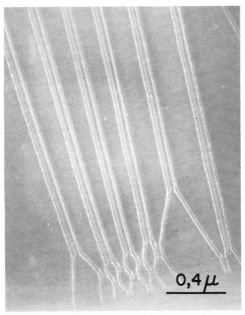

Fig. 5. Weak beam image of dislocation ribbons in RhSe$_2$; such images are used to determine stacking fault energies.

(Cockayne, Ray and Whelan, 1969). Weak beam images are mainly used to study the fine structure of dislocations, i.e. to study the splitting of perfect dislocations into multiribbons of partials in complicated crystals (Cockayne, Jenkins and Ray, 1971).

The separation of partials is determined, among other factors, by the magnitude of the stacking fault energy; weak beam images therefore offer a unique method for the quantitative determination of stacking fault energies.

For large s-values the effective extinction distance becomes smaller; under such conditions planar interfaces produce many fringes, which may sometimes be useful, such as for the study of anti-phase boundaries in ordered alloys. The extinction distances corresponding with the superstructure reflections which are needed to image anti-phase boundaries are in general large, thus producing a small number of fringes only, unless large s-values are used.

8. PHASE TRANSFORMATIONS

Displacive phase transitions are almost invariably accompanied by a reduction in symmetry of the structures generated on cooling. Usually a number of orientation variants (or twins) are formed, which are related by the symmetry operations lost during the transformation (e.g.: Amelinckx, 1975). As a result a single crystal of the high temperature phase becomes fragmented into symmetry related domains on transforming into the low temperature phase. Such domain structures produce diffraction patterns of which the spots are broken up into a

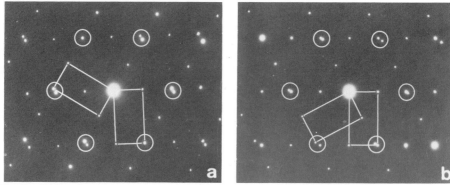

Fig. 6. Diffraction patterns (a) and (b) due to two types of reflection twins in molybdenum ditelluride. The row of unsplit spots has a different orientation in the two cases. Note that the spot splitting is in both cases parallel with the row of unsplit spots.

number of components. Selected area diffraction patterns taken across a single domain boundary allow one to determine the symmetry operation relating the two domains; in particular one can distinguish reflection twins and rotation twins by tilting experiments, i.e. by studying several sections of reciprocal space (Boulesteix, Van Landuyt and Amelinckx, 1976).

9. INVERSION BOUNDARIES

Interfaces separating domains of which the structures are related by an inversion operation, can be identified in dark field images taken under multiple beam conditions, along a zone-axis which does not produce a centre of symmetry in projection. Under such conditions Frieldel's law is violated and inversion-related domains show up with a different intensity in dark field images, but not in the bright field image. The boundaries themselves are imaged by fringes,

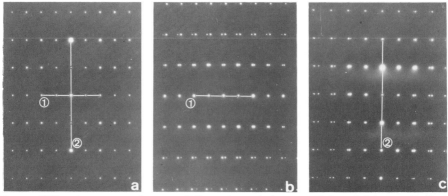

Fig. 7. Diffraction patterns due to a 180° rotation twin in monoclinic molybdenum ditelluride. The identification requires exploring reciprocal space in a tilting experiment. In (a) the reciprocal plane of unsplit spots is revealed; tilting about the axis 1 and 2 produces respectively the pattern in (b) and (c); which proves the presence of a single plane of unsplit spots.

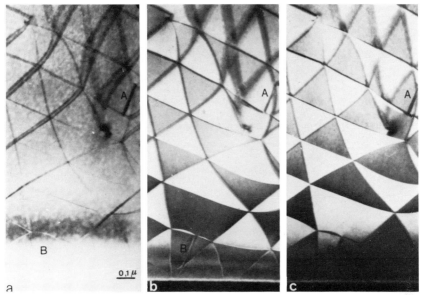

Fig. 8. Triangular domain structure of inversion boundaries in a χ-phase alloy. (a) The fringes reveal the inversion boundaries, which intersect by three along a common line; (b) In a multiple beam situation with the incident beam along a non-centrosymmetrical zone the two families of inversion domains show up with different intensities as a results of the violation of Friedel's law; (c) The contrast can be reversed by changing the diffraction conditions.

which have properties similar to those produced at stacking faults (Serneels, Snykers, Delavignette, Gevers and Amelinckx, 1973; also, Morton, 1974, gives experimental results).

10. DOMAIN STRUCTURES

By the choice of suitable diffraction conditions one can in general reveal any domain structure; it is sufficient to excite a reciprocal lattice node for which the s-values are somewhat different, in different domains. If the reciprocal lattices coincide for all domains, the structure amplitudes may still have a different value for different domains and thus give rise to a difference in transmitted or scattered intensity. This method was e.g. used to reveal Dauphiné twins in quartz (Van Tendeloo, Van Landuyt and Amelinckx, 1976).

If a phase transition proceeds through the propagation of a sharp transformation front, this can be revealed in a dark field image made in a diffraction spot belonging to only one of the two phases. The phase from which the selected diffracted beam originates will show up bright in such a dark field image (Van Tendeloo, Van Landuyt and Amelinckx, 1976).

If on transformation the unit cell size increases, i.e. if there is a loss of translation symmetry, in general translation variants are formed (Van Landuyt, Remaut & Amelinckx, 1970; Manolikas & Amelinckx, 1980) separated by

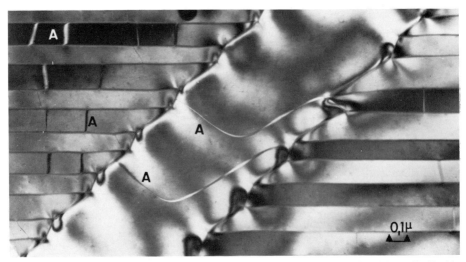

Fig. 9. Domain structures in the monoclinic low temperature phase of niobium ditelluride. Three orientation variants differing by 120° in orientation result from the phase transition. Since the unit cell size increases by a factor of three also three translation variants can be formed. The image shows orientation domains of two types, as well as antiphase boundaries (A) separating translation variants.

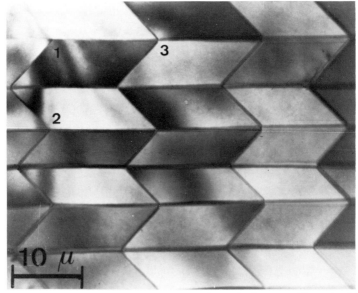

Fig. 10.   Domain structures in a rhombohedrally deformed γ-bronze with composition Al_8Mn_4. This domain configuration minimizes the strain energy; it contains four orientation variants.

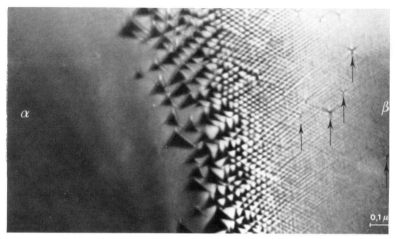

Fig. 11. Network of triangular areas of Dauphiné twins in a natural crystal of α- quartz kept close to the α–β transition temperature. The image is produced by the 3031 reflection which coincides for the two variants. However the structure amplitudes and their phases are different for the two variants and the contrast is thus structure factor contrast. There is a temperature gradient across the specimen in such a manner that α and β are present simultaneously in different parts of the specimen.

anti-phase boundaries. Translation variants produce the same contrast under all diffraction conditions in bright field as well as in dark field images; moreover they usually do not produce a visible fragmentation of the diffraction spots. The displacement vector $\mathbf{R}$ of the translation interfaces can be deduced modulo a lattice vector, from the extinction criterium $\mathbf{g} \cdot \mathbf{R} =$ integer. (i.e. a vector $\mathbf{R}_1$ cannot be distinguished from $\mathbf{R}_1 + n \cdot \mathbf{g}$, n an integer).

11. INTERFACE MODULATED STRUCTURES

Electron diffraction is very well suited to study *modulated structures* generated from some basic structure, either by the periodic insertion of planar interfaces such as stacking faults, anti-phase boundaries, twin interfaces or inversion boundaries, by the periodic distortion of the structure or by periodic changes in composition (Amelinckx, 1979). One dimensional long period structures formed by the periodic introduction of identical translation interfaces with a displacement vector $\mathbf{R}$ and a spacing d produce a diffraction pattern of which the spot positions are given by the relation

$$\mathbf{g} = \mathbf{H} + (1/d)(m - \mathbf{H} \cdot \mathbf{R})\mathbf{e}_u$$

where $\mathbf{e}_u$ is the unit vector perpendicular to the interfaces, $\mathbf{g}$ and $\mathbf{H}$ are the diffraction vectors respectively of the superlattice and of the basic lattice and m is an integer (Van Landuyt, De Ridder, Gevers and Amelinckx, 1970). The intensity of the spots decreases with increasing m. The diffraction pattern of the superstructure thus consists of linear arrays of superlattice spots $\mathbf{g}$ with a spacing

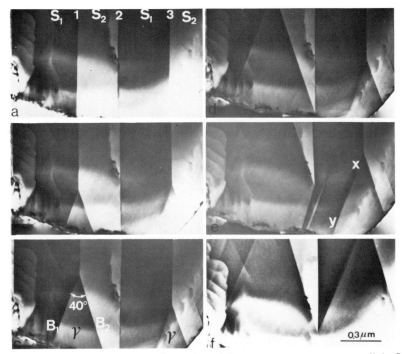

Fig. 12.   Successive stages in the transformation of the room temperature monoclinic β -phase which is fragmented in ferroelastic domains, into the high temperature ($\sim$ 120 °C) hexagonal prototype phase of lead orthovanadate. The different β- domains show up a in a different intensity; the γ-phase consists of a single domain. The contact plane between β and γ is well defined crystallographically; it is determined by the condition of lattice compatibility between β and γ along the contact plane.

$1/d$, clustered around each of the basic spots $\mathbf{H}$ and shifted with respect to these basic spots by a fraction $\mathbf{H} \cdot \mathbf{R}$ of the interspot distance. From the fractional shift the dot product can be deduced and hence $\mathbf{R}$ determined provided two independent sections of reciprocal space are available. This of course implies the knowledge of the positions of the basic spots $\mathbf{H}$. The fractional shift can be deduced directly from selected area diffraction patterns of crystal parts containing both basic and modulated structure. Under such conditions one can also make a microprobe analysis of the faulted and of the unfaulted areas separately and determine whether or not composition differences are present and therefore decide as to the conservative or nonconservative nature of the interfaces (Van Tendeloo, Van Landuyt, Delavignette and Amelinckx, 1974).

The diffraction patterns due to crystals consisting of periodic poly-synthetic twins, present similar characteristics. Arrays of superlattice spots are now associated with the spot positions due to the two twin variants. From the intensities information can be obtained concerning the relative widths of the two twin strips (Van Dyck, Colaitis and Amelinckx, 1980).

The microscope mode allows us to image the periodic interfaces directly and to observe deviations from the ideally equidistant array. In the diffraction pattern such irregularities lead to streaking along arrays of superlattice reflections. Regular mixtures of two (or more) spacings, leading to a fractional mean spacing, may give rise to spacing anomalies in the diffraction pattern, i.e. arrays of superlattice spots belonging to different basic spots do not match where they meet. Systematic ledging of the periodic interfaces may furthermore give rise to orientation anomalies, i.e. the arrays of superlattice spots may have directions which are not simply related to the directions of the basic lattice. The origin of these diffraction effects can unambiguously be established by imaging the interfaces (Van Tendeloo and Amelinckx, 1977).

12. DEFORMATION MODULATED STRUCTURES

Deformation modulated structures produce diffraction patterns consisting of spots due to the basic structure, each such spot being surrounded by an array of satellite spots. The vectors joining the basic spot to its satellites are the wave vectors of the Fourier components of the deformation wave (see *Modulated Structures*, AIP Conf. Proc., 1971). According to the kinematical theory the intensities of the satellites associated with basic spots far from the origin will be relatively stronger than those associated with basic spots close to the origin; this feature can be used to identify the nature of the modulation.

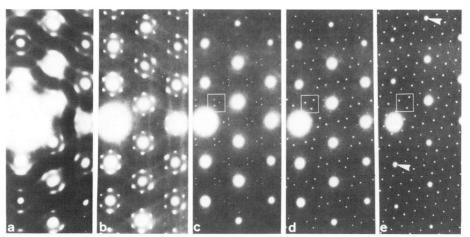

Fig. 13. Evolution of the diffraction pattern of the octahedrally coordinated form of tantalum disulfide on cooling through the different phase transitions which are generally assumed to be related to charge density wave formation. (a) At room temperature diffuse scattering is present along circles centered on the basic diffraction spots; (b) On cooling satellites develop which reveal an octahedral complex of points in reciprocal space. The wave vectors of the charge density waves are obtained by joining the basic spots to their satellites; (c) A first incommensurate phase is formed; note the triangular configuration of weak spots; (d) A second incommensurate phase formed discontinuously from the first; (e) On further cooling this second incommensurate phase gradually evolves into a commensurate superstructure. The distance between the two indicated basic spots is divided in 13 equal intervals by superlattice spots.

Deformation modulated structures, attributed to charge density waves, have been studied in several transition metal dichalcogenides (e.g. Van Landuyt, Wiegers and Amelinckx, 1978). The wave vectors of the modulation waves are in a number of cases incommensurate with the lattice of the basic structure; a commensurate phase may be formed at low temperature however. In transition metal trichalcogenides one-dimensional charge density waves have been found by means of electron diffraction (Van Tendeloo, Van Landuyt and Amelinckx, 1977).

13. COMPOSITION MODULATED STRUCTURES

The diffraction patterns of certain types of composition modulated structures exhibit characteristic features which have been studied in detail and compared with observations in Ni_2Te_{3+x} and Cu_2Te_{3+x} (Colaitis, Delavignette, Van Dyck and Amelinckx, 1979; 1980).

14. HIGH VOLTAGE ELECTRON MICROSCOPY

Electron microscopes with an accelerating voltage significantly higher than the conventional 100 kV have come into use over the latest decade. They offer the possibility of a greater penetrating power and hence the use of thicker specimens which are more representative of the bulk. Due to the shorter wave length of the electrons used they also allow better instrumental resolution to be achieved. Moreover the contrast transfer function of the lens system can be designed so as to produce roughly the same phase shift for a larger angular range of beams and hence produce a more faithful representation of crystal structures than with 100 kV microscopes. The use of high resolution, high voltage microscopy offers at present good perspectives for the direct study of crystal structures; striking examples have been produced already (see for example: Bredero and Van Landuyt, 1980). However, the displacement and ionization damage produced by electrons of high energy constitute an intrinsic limitation which restricts the observation time. High resolution, medium voltage electron microscopes may turn out to be the best compromise for a number of applications.

15. ATOM DISPLACEMENTS

The controlled displacement of atoms by high energy electrons can be used to study radiation damage in metals and alloys (Makin, 1968; Thomas, 1970). The displacement threshold can be determined in this manner. The possibility of following *in situ* the change in the diffraction pattern of alloys allows the real time observation of both radiation ordering (Van Tendeloo, Van Landuyt and Amelinckx, 1979) and radiation disordering (Howe, Rainville and Schulson, 1974).

16. STRUCTURE IMAGES; STRUCTURE DETERMINATION

As the instrumental resolution of microscopes improves and reaches interatomic distances (< 2 Å), structure determination by direct imaging becomes feasible. Images directly interpretable in terms of crystal structures have been obtained mainly in oxides (Iijima and Allpress, 1973) and silicate minerals (Iijima and Buseck, 1976) but more recently also in ordered alloys (Van Tendeloo and Amelinckx, 1978). Detailed imaging of the structure is not possible in practice because of the finite aperture of the lenses limiting necessarily the number of beams (Fourier components) contributing to the image and also because of phase changes introduced by the lens aberrations (Fejes, 1977).

The intensity distribution in the image corresponding to a given structure can be calculated numerically for a crystal of given thickness and orientation, using multiple beam dynamical theories, for an instrument with given aberrations and knowing the amount of defocus. Several computer programs have been developed for this purpose (e.g.: Lynch and O'Keefe, 1972; Van Dyck, 1975). The inverse problem i.e. deducing the structure from an observed intensity distribution is a more difficult problem comparable in nature with the phase problem in X-ray diffraction.

Different empirical approaches have been used. When knowing the structures and having corresponding images of a few compounds in a homologous series, the images can in a sense be 'calibrated' and the structure of other unknown

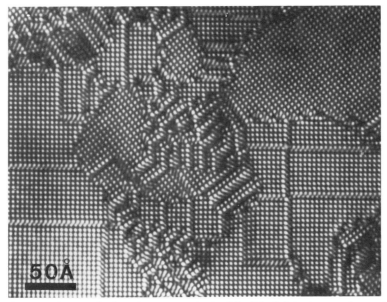

Fig. 14. Structure image of a domain structure in Au_4Mn. The bright dots can be considered as imaging the manganese columns in the structure. All super-lattice spots contained in one mesh of the reciprocal lattice of the basic structure were used for imaging. Orientation variants as well as translation variants can be seen clearly.

Fig. 15.   High resolution image of a dislocation in germanium. A Burgers circuit has been drawn
around the dislocation.

members of the series deduced (Van Landuyt, Amelinckx, Kohn and Eckart, 1974; Van Dyck, Van Landuyt and Amelinckx *et al.*, 1976).

For some as yet not completely understood reasons, recognizable representations of the projected structure are sometimes obtained by chance. This is more often the case the thinner the foil, but it sometimes also occurs in thicker foils. It is clear that further development of the method of interpretation is needed.

The usual approach in structure determination from an image is the equivalent of the 'trial and error' approach in X-ray structure determination, i.e. a series of images of a trial structure is simulated for different thicknesses and different defocus values and fitted with the observed image.

Structure images have also been useful for the study of local deviations from strict periodicity, such as antiphase boundaries, stacking faults and dislocations. A good example of a lattice-dislocation image is shown in Figure 15.

From the experience accumulated so far, one can conclude that structure images can with a certain degree of confidence be interpreted as representing the projected 'charge density' or atomic density'; regions of the unit cell containing heavy atoms being represented as dark on a positive print and regions of empty space (tunnels) or of light atoms being represented as lighter areas. However this seems to be justified only if certain conditions are met: the foil has to be very thin

(< 100 Å); the instrumental resolution of the microscope has to be good enough (~ 2 Å) and the contrast transfer function should be reasonably flat up to the instrumental resolution under the focussing conditions used. Crystal alignment, reflecting the full crystal symmetry along the zone-axis of the diffraction pattern, and a critical amount of underfocus (~ 1000 Å) with respect to the Gaussian focus, are essential.

Most structure images have been obtained so far by collecting as many beams as are compatible with the instrumental resolution in the bright field mode. However, use has been made of the dark field mode as well. Twinning on the unit cell level can be detected by using non-central rows of diffraction spots (e.g.: Bourret and Desseau, 1979). Ordered alloys have been studied by collecting superlattice reflections only (Van Sande, Van Tendeloo, Amelinckx & Airo, 1979).

ACKNOWLEDGEMENTS

It is a pleasure to acknowledge the courtesy of my colleagues and friends in providing the following illustrations: A. Bourret (Figure 15), P. Delavignette (Figures 2, 3, 4), C. Manolikas (Figures 5, 6, 7, 12), G. Remaut (Figure 9), E. Reudl (Figures 3, 4), M. Snykers (Figure 8), J. Van Landuyt (Figures 9, 10, 11, 13), M. Van Sande (Figure 10), G. Van Tendeloo (Figures 11, 13, 14).

Universiteit Antwerpen (RUCA)
Antwerp, Belgium, and
S.C.K.-Mol. Belgium

REFERENCES

Amelinckx, S.: 1964, *The Direct Observations of Dislocations* (ed F. Seitz and D. Turnbull) p. 127. Academic Press, N.Y.
Amelinckx, S.: 1967, *Trace characterization – Chemical and Physical* (ed. W. Meinke and B. Sribner) p. 427, NBS, Monograph 100, Washington, USA.
Amelinckx, S.: 1974, *J. Cryst. Growth*, **24/25**, 6.
Amelinckx, S.: 1975, *Frontiers in Materials Science*. M. Dekker, N.Y.
Amelinckx, S.: 1979, In AIP Conf. Proc. No. 53, p. 102.
Ashby, M. F. and Brown, L. M.: 1963, *Phil. Mag.* **8**, 1083.
Bilby, B. A., Bullough, R., and Smith, E.: 1955, *Proc. Roy. Soc.* A**231**, 263.
Boulesteix, C., Van Landuyt, J., and Amelinckx, S.: 1976, *Phys. Stat. Sol.* (a)**33**, 595.
Bourret, A. and Desseau, J.: 1979, *Phil. Mag.* **39**, 405; *ibid*, p. 419.
Bredero, P. and Van Landuyt, J. (eds.): 1980, *Proceedings of the International Conference on HVEM*, Antwerp.
Cockayne, D. J. H., Jenkins, M. L., and Ray, I. L. F.: 1971, *Phil Mag.* **24**, 1383.
Cockayne, D. J. H., Ray, I. L. F. and Whelan, M. J.: 1969, *Phil. Mag.* **20**, 1265.
Colaitis, D., Delavignette, P., Van Dyck, D. and Amelinckx, S.: 1979, *Phys. Stat. Sol.* (a)**51**, 657; *ibid* (1979) **53**, 105 423; *ibid* (1980) **58**, 271.
Delavignette, P., Trivedi, R., Gevers, R., and Amelinckx, S.: 1966, *Phys. Stat. Sol.* **17**, 221.
Fejes, P. L.: 1977, *Acta Cryst.* A**33**, 109.

Gevers, R., Van Landuyt, J., and Amelinckx, S.: 1965, *Phys. Stat. Sol.* **11**, 689.

Goringe, M. J. and Valdré, U.: 1966, *Proc. Roy. Soc.* A**295**, 192.

Groves, G. W. and Kelly, A.: 1961, *Phil. Mag.* **6**, 1527; *ibid.* (1962) **7**, 892.

Hashimoto, H., Howie, A., and Whelan, M. J.: 1962, *Proc. Roy. Soc.* A**269**, 80.

Heidenreich, R. D. and Sturkey, L.: 1945, *J. App. Phys.* **16,** 97.

Head, A. K., Humble, P., Clarebrough, L. M., Morton, A. J., and Forwood, C. T.: 1973, *Computed Electron Micrographs and Defect Identification*. North-Holland, Amsterdam.

Hirsch, P. B., Howie, A., and Whelan, M. J.: 1960, *Phil. Trans.* A**252**, 499.

Holland, V. F., Lindenmeyer, P. H., Trivedi, R. and Amelinckx, S.: 1965, *Phys. Stat. Sol.* **10**, 543.

Howe, L. M., Rainville, M., and Schulson, E. M.: 1974, *J. Nucl. Mat.* **50**, 139.

Howie, A.: 1962, 5th Int. Congr. on Electron Microscopy, Philadel. (Academic Press, New York) AA-10.

Howie, A. and Basinski, Z. S.: 1968, *Phil. Mag.* **17**, 1039.

Howie, A. and Whelan, M. J.: 1960, Proc. European Reg. Conf. on Electron Microscopy Delft, Vol. 1, p. 194.

Howie, A. and Whelan, M. J.: 1961, *Proc. Roy. Soc.* A**263**, 217; *ibid* (1962) A**267**, 206.

Iijima, S. and Allpress, J. G.: 1973, *J. Solid State Chem.* **7**, 94.

Iijima, S. and Buseck, P. R.: 1976, *Application of Electron Microscopy in Mineralogy* (ed. H. R. Wenk) Berlin, Springer, p. 319.

Lynch, D. F. and O'Keefe, M. A.: 1972, *Acta Cryst.* A**28**, 536.

Makin, M. J.: 1968, *Phil. Mag.* **18**, 637.

Malonikas, C. and Amelinckx, S.: 1980, *Physica* **99B**, 31.

Morton, H. J.: 1974, *Phys. Stat. Sol.* (a)**23**, 275.

Serneels, R., Snykers, M., Delavignette, P., Gevers, R., and Amelinckx, S.: 1973, *Phys. Stat. Sol.* (b)**58**, 277.

Siems, R., Delavignette, P., and Amelinckx, S.: 1962, *Phys. Stat. Sol.* **2**, 421.

Thomas, L. E.: 1970, *Radiation Effects* **5**, 183.

Tunstall, W. J., Hirsch, P. B., and Steeds, J.: 1964, *Phil. Mag.* **9**, 99.

Van Dyck, D.: 1975, *Phys. Stat. Sol.* (b)**72**, 321.

Van Dyck, D., Colaitis, D., and Amelinckx, S., to be published.

Van Dyck, D., Van Landuyt, J., Amelinckx, S., Nguyen en Huy-Dung, and Dragon, C.: 1976, *J. Solid State Chem.* **19**, 179.

Van Landuyt, J., Amelinckx, S., Kohn, J. A., and Eckart, D. W.: 1974, *J. Solid State Chem.* **9**, 103.

Van Landuyt, J., De Ridder, R., Gevers, R., and Amelinckx, S.: 1970, *Mat. Res. Bull.* **5**, 353.

Van Landuyt, J., Remaut, G., and Amelinckx, S.: 1970, *Phys. Stat. Sol.* **41**, 271.

Van Landuyt, J., Wiegers, G. A., and Amelinckx, S. 1978, *Phys. Stat. Sol.* (a)**46**, 179.

Van Tendeloo, C. and Amelinckx, S.: 1977, *Phys. Stat. Sol.* (a)**43**, 553.

Van Tendeloo, G. and Amelinckx, S.: 1978, *Phys. Stat. Sol.* (a)**50**, 53.

Van Tendeloo, G., Van Landuyt, J., and Amelinckx, S.: 1976, *Phys. Stat. Sol.* (a)**33**, 723.

Van Tendeloo, G., Van Landuyt, J. and Amelinckx, S.: 1977, *Phys. Stat. Sol.* (a)**43**, K 137.

Van Tendeloo, G., Van Landuyt, J. and Amelinckx, S.: 1979, *Rad. Effects* **41**, 179.

Van Tendeloo, C., Van Landuyt, J., Delavignette, P., and Amelinckx, S.: 1974, *Phys. Stat. Sol.* (a)**25**, 697.

Van Sande, M., Van Tendeloo, G., Amelinckx, S., and Airo, P.: 1979, *Phys. Stat. Sol.* (a) 449.

Whelan, M. J. and Hirsch, P. B.: 1957, *Phil Mag.* **2**, 1121.

III.5. *Inelastic Electron Scattering*

1. HISTORICAL REMARKS

An understanding of the nature of energy losses suffered by electrons passing through dense material, as solids or liquids, has developed rather slowly. Its quantum character was demonstrated in gases by the experiment of Franck and Hertz (1914), but it took a long time for the situation to be as well understood in solids. Important contributions were the reflection experiments of Rudberg (1936- with low energy electrons and the transmission experiments with fast electrons of Ruthemann (1942). The different discrete energy losses obtained with low energy electrons have been ascribed to excitations between levels of the bandstructure of the solid, whereas the discrete losses in the transmission experiment, e.g. the 15 eV loss and its multiples in aluminium, could not be explained in 1942. The concept of oscillations of the electron density in a metal with a quantum energy $\hbar\omega_p$ (where $\omega_p \sim \sqrt{n}$, n being the electron density) has been introduced by Bohm, Gross and Pines (1949, 1951). It took a rather long time for the interpretation of the discrete losses in metals as plasmons to be accepted. The follow-up to this went in two directions: on the one side was the investigation of the physics of plasmons, their properties and their theoretical description, on the other side their role in electron diffraction studies and electron microscope work. This is still the stiutation today.

2. ELASTIC AND INELASTIC SCATTERING

If electrons pass through a crystal, elastic and inelastic interaction takes place. The elastic interaction is important for electron diffraction and imaging: the electrons transfer no energy to the crystal since the crystal mass is very high compared to that of the electron. However the electrons lose and gain momentum $\hbar\Delta\mathbf{K} = \hbar\,\mathbf{K}^{cl} - \hbar\mathbf{K}_0^{cl}$, where $\mathbf{K}^{cl}$ and $\mathbf{K}_0^{cl}$ are the wave vectors of the scattered and incoming electrons respectively. Due to the translational periodicity multiples of $\hbar\mathbf{g} = \hbar|\Sigma_i h_i\mathbf{b}_i|$ with $\mathbf{g}$ the reciprocal lattice vector, can be transferred to the electron which leads to the Laue/Bragg diffraction given by $\hbar\Delta\mathbf{K} = \hbar\mathbf{g}$.

Inelastic scattering takes place, if energy is transferred to the elements of the crystal for example as vibrations of the ions or phonons with an energy of some 10 meV, excitation of plasmons: volume and surface oscillations with an energy of about 10 eV, and excitation of interband transitions which vary between 1 eV up to several keV. The corresponding momentum transfer $\hbar\Delta\mathbf{K}$ is governed by the dispersion relation of the process involved e.g. the phonon dispersion relation $\omega(\Delta K)$ or the volume plasmon dispersion relation $\omega_{vol}(\Delta K)$, or that of the surface plasmons. In case of interband transition non-vertical (i.e. momentum change), direct, transitions take place, i.e. in contrast to the situation for photon scattering, so that the excited electron has obtained a momentum $\hbar\Delta\mathbf{K}$ with which it moves in the higher energy level. At first we consider mainly volume plasmons and interbandtransitions. Of these, the probability of producing plasmons is higher than that of exciting the deeper atomic levels. Further, it is interesting that the angular intensity distribution of the inelastic scattering is rather different from the elastic one. The probability for an electron being scattered into the solid angle element $d\Omega$ and suffering an energy loss ΔE is given by

$$\frac{\partial^3 W}{\partial \Delta E\, \partial^2 \Omega} = \text{const.} \times D/q^2 \times \text{Im}[-1/\varepsilon(\omega)] \tag{1}$$

where $\varepsilon(\omega)$ is the dielectric function, and $\text{Im}[-1/\varepsilon(\omega)]$ is the 'loss function' or the imaginary part of $[\varepsilon(\omega)]^{-1}$ determining the energy ($\hbar\omega$) dependence of the scattering probability. Of the remaining terms, D is the thickness of the transmitting film, W is the specific stopping power, and q is identical with the ΔK mentioned above, and can be written

$$q^2 = (K_0^{cl}\,\theta)^2 + \left(K_0^{cl}\frac{\Delta E}{2E_0}\right)^2,$$

where θ is the scattering angle and E_0 is the kinetic energy of the primary electrons. This relation shows that the intensity of electrons having lost ΔE has decreased to half of its value at $^H\theta_{incl} = \Delta E/2E_0$, which is a very small angle; for 50 keV electrons and $\Delta E = 10$ eV we obtain $^H\theta_{incl} = 10^{-4}$ rad. This means that the inelastic intensity is concentrated in the forward direction. In contrast, the corresponding angular half-width for intensity elastically scattered by a Gaussian-shaped atom of radius a is (by fourier transformation)

$$^H\theta_{cl} = 1/K^{cl}a$$

i.e. very much larger than $^H\theta_{incl}$; it has a value for 50 keV electrons of about 0.01 rad ($\approx 1°$), almost independent of atomic number Z (a is a very slowly-varying function of Z). On the other hand the angular width of a diffracted beam is determined by the size of the crystal; assuming a linear dimension of Nd, the angular width of the beam is given by $^H\theta_{cl} \approx (\lambda/d)(1/N)$ and thus nearly comparable with $^H\theta_{incl}$; this is verified in the experiments, see below.

3. EXPERIMENTAL DEVICES AND RESULTS

Two types of instrumentation for the study of energy losses of electrons have been developed: the transmission type and the reflection type. The first has the

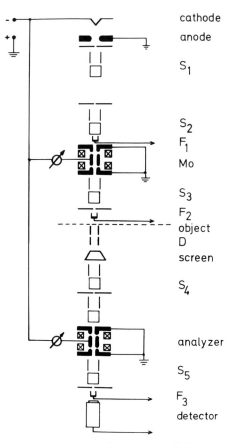

cathode

anode

S_1

S_2
F_1

Mo

S_3
F_2
object
D
screen

S_4

analyzer

S_5

F_3
detector

Fig. 1.   Scheme of an energy loss spectrometer for electrons of about 50–100 keV in transmission.

advantage that the penetration depth is given by the thickness of the transmitted film; it can be chosen to be of the order of the mean free path length for inelastic collision ($\sim 10^3$ Å) so that the elementary excitation can be studied. Further, the observation of the angular distribution (θ)dependence) allows one to measure the momentum dependence or the dispersion relation of the process, which is of basic interest.

The reflection of electrons allows one to investigate the properties of surfaces. If low-energy electrons (1–100 eV) are used, the scattering takes place at the very first layers of the crystal surface; the analysis of the loss spectrum thus gives important information on the nature of the surface structure.

A typical instrument for transmission is shown in Figure 1: the electrons accelerated to energies of 50–100 keV are monochromatised in the spectrometer-lens Mo, so that their energy spread is reduced from ~ 1 eV to 0.1 eV or 0.01 eV. After having passed the object they enter an analyser, which may be one of several designs: e.g. a $127°$ electrostatic cylinder, or a modified filter lens (Hartl lens). A typical loss spectrum is reproduced in Figure 2; it represents the loss function which is the key term in the differential probability function mentioned

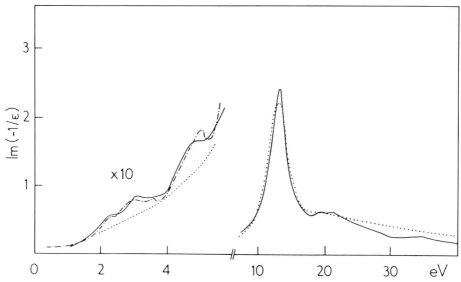

Fig. 2.   Loss function $\mathrm{Im}[-1/\varepsilon(\omega)]$ of InSb. *Full Line*: monocrystalline *Dash-dotted line*: polycrystalline. *Dotted line*: amorphous state.

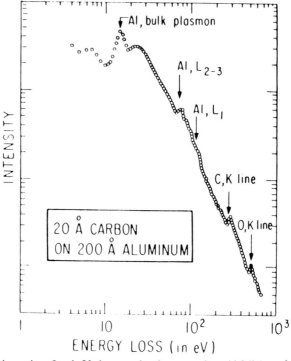

Fig. 3   Inelastic intensity of 25 keV electrons having passed an Al foil (200 Å thick) on a carbon substrate (20 Å thick) in the energy range of 10 eV up to 1000 eV. Double logarithmic scale. In addition to the strong volume plasmon a number of K and L edges of Al, C and O (oxide of Al) are observed.

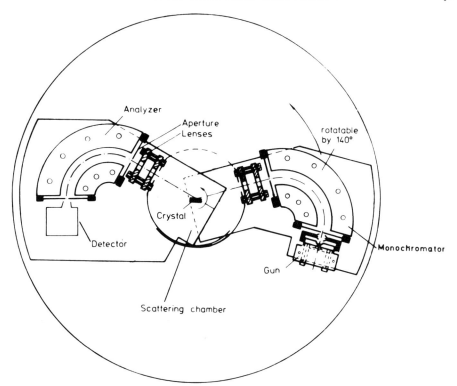

Fig. 4. Schematic drawing of a high energy-resolution spectrometer for reflected, low energy electrons.

above, for InSb. One recognizes the strong volumen plasmon at 12.8 eV and nearby some low energy interband transitions. The full line is obtained from single crystal material, the dashed line from polycrystalline substances. The dotted line stems from amorphous InSb. Extending the analysis to higher energy losses one records the excitation of the deeper levels of the atom as the $K L M \ldots$ shells. An example is given in Figure 3 which displays the loss spectrum of Al between about 10 eV to 1000 eV. The volume plasmon loss at 15 eV can be seen as well as those for the different shells.

The devices for reflection experiments are simpler, (see Figure 4) Electrons of E_0 between 1 and 100 eV are used. Having passed a monochromator, which is here a cylindrical electrostatic deflection system, they hit the crystal surface. The reflected electrons enter the analyser which is very similar in construction to the monochromator. Due to the low energy of the electrons a rather small energy-width of the beam can be realised. In case of fast electrons a similar monochromator type yields an energy resolution of $E_0/\Delta E \sim 10^4$. Reducing E_0 to 10 eV, one obtains ~ 1 meV in a device as shown in Figure 4.

If the primary energy E_0 amounts to some 10 eV plasmons can be excited – volume as well as surface plasmons. With electrons of less than ~ 10 eV plasmons cannot in general be excited, but losses due to surface phonons can be detected which have an energy of some 10 meV. If molecules are adsorbed at the crystal surface, vibrations of these adsorbates are excited which have an energy of this order. The study of these losses allows one to draw conclusions about the position of the adsorbed molecules on the crystal surface. The method has become a technique of surface physics providing information in many cases parallel to that from infrared spectroscopy.

The reflection method can also be applied to fast electrons if grazing incidence ($\sim 1°$) is used. Under these conditions fast electrons penetrate the boundary very little – less than 10 Å – and are specularly reflected. This happens if the mean roughness δ is small compared to the wavelength of the electrons, more exactly if $2\phi\delta \ll \lambda$, where ϕ is the grazing angle (~ 0.01 rad); the value of δ comes out (with 50 keV electrons) to < 2 Å. In other words, the degree of smoothness required is that of a stationary liquid surface. As we see the inelastic processes are of two types: we have the so-called collective excitations like plasmons in which a high number of particles participate, and the single particle excitation, where the exterior field acts only on very few particles, in general one. This behaviour is described by the equation (1): if the dielectric function $\varepsilon(\omega) = \varepsilon'(\omega) + i\varepsilon''(\omega)$ has at a certain energy $\hbar\omega$:

$$\varepsilon' = 0 \quad \text{with} \quad \varepsilon'' \ll 1$$

the loss function passes a maximum at this energy. This is characteristic for a collective behaviour. At higher energies ΔE the value of $\varepsilon'(\omega)$ approaches the value of 1 and $\varepsilon''(\omega)$ becomes small. Thus the loss function $\mathrm{Im}[-1/\varepsilon(\omega)] = \varepsilon''/\varepsilon'^2 + \varepsilon''^2$ can be written $\mathrm{Im}[\varepsilon(\omega)] = \varepsilon''(\omega)$, which is correlated with interband transitions. This means that with increasing ΔE we approach the single particle excitation.

4. APPLICATIONS

A. Electron Diffraction and Electron Microscopy

In an ordinary electron diffraction diagram elastic and inelastic intensity are mixed. Due to the high angular concentration of the inelastic intensity in the forward direction the angular width of the diffracted beam is in general not much changed and the diffracted intensity is given by the sum of elastic and inelastic intensities. In order to see the role of the inelastic part, it can be separated by a counterfield which cuts off all those electrons having lost more than 2.5 eV. A comparison between the filtered and unfiltered powder diagram for Al appears in Figure 5. The striking difference of both is the reduced background; maxima became visible which came from amorphous Al_2O_3. Detailed study indicates also an influence of the inelastic intensity on the line width.

This type of equipment has been developed into a scanning electron diffraction unit which allows one to register diffraction patterns in a short time and to see their changes (Denbich and Grigson, 1965). Energy filters have also been installed in an electron microscope to study the influence of the inelastic electrons. The Wien filter is well known. Another device, the Ω filter, has been

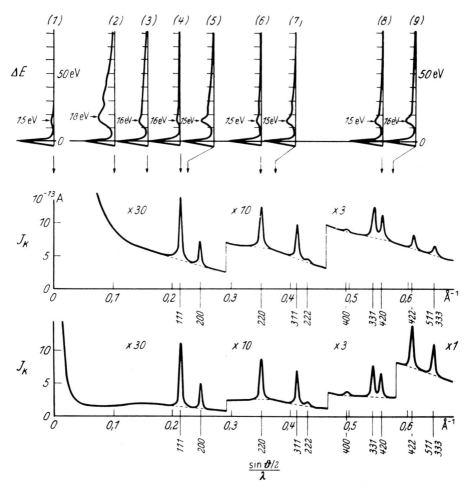

Fig. 5. Electron Diffraction pattern of 200 Å aluminium film on a 20 Å carbon substrate in logarithmic scale. Comparison of unfiltered (middle) and filtered (low) powder diagrams of Al. Crystal size 240 Å, $E_0 = 48$ keV. *Above*: at the indicated $\sin \frac{1}{2}\theta/\lambda$ values. The crystal size is so small that the inelastic electrons do not noticeably enlarge the beam. In the reduced background of the filtered diagram the flat maxima of an amorphous Al_2O_3 coating of the Al film can be recognized.

(Reproduced from Horstmann and Meyer (1961): *Z. f. Phys.* **159**.)

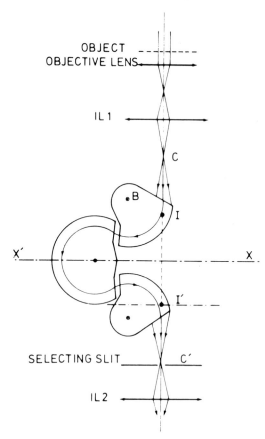

Fig. 6.   Ω filter built up of three magnetic prisms used in a high voltage electron microscope
of 1 MeV.

designed which is of particular importance for high-voltage electron microscopy
and is displayed in Figure 6. It is built up of three magnetic prisms which bend
the beam back into the original direction and select a certain energy range from
the spectrum. It is thus possible to produce an image with electrons of just this
energy range. Figure 7 shows how the mean free path length λ_p for plasmon
excitation depends on the energy of the primary electrons. This increase of λ_p is
of interest in metal physics.

B. Microanalysis

As we have seen in Figure 3, edges in the loss spectrum due to the excitation of
deeper levels of the atom of the solid are rather well defined, and can be used to
identify the irradiated atoms. The height of the step in the continuously
decreasing intensity can be measured and with the knowledge of the cross section
for this process the number of atoms in the irradiated region can be calculated.
Since the number of inelastically, forward-scattered electrons is counted (and

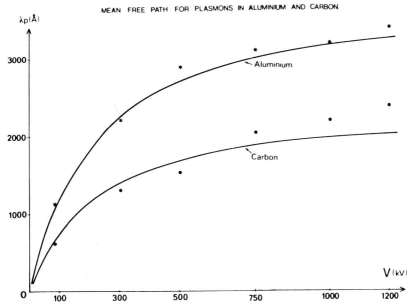

Fig. 7. Mean free path length for volume plasmons in Al and carbon as a function of the electron energy. *Full line*: calculated values. *Points*: measured values.

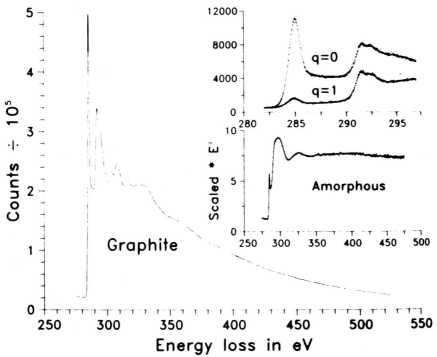

Fig. 8. Energy loss spectrum of crystalline (left) and amorphous (lower inset) graphite. The EXAFS of the crystalline state is much more pronounced than the fine structure of the amorphous state indicating a more ordered neighbourhood for the excited carbon atom in the crystalline solid.

not the fluorescence radiation scattered into the solid angle of 4π), this method is extremely sensitive. By reducing the beam diameter of the primary electrons this procedure can be used as a microanalytical tool. Up to now quantities of some 10^{-19} g have been detected. (Isaacson and Johnson, 1975; Hren, Goldstein and Joy, 1979).

C. EXAFS

The extended X-ray-absorption fine-structure is detected if one looks at the edges of Figure 3 with a better energy resolution. Going to smaller wavelengths the absorption makes a steep rise due to the excitation of photoelectrons. It follows a region of fluctuating, slowly decreasing intensity up to several 100 eV above the edge. If instead of X-rays electrons are used, and the intensity of the electrons is measured as function of their energy loss, the same fluctuations are observed. This fine structure is produced because the photoelectron wave leaving the deep shell is partially back-reflected from the atoms of the neighborhood, which leads to a modulation of the excitation-probability of photoelectrons. This modulation depends on the wavelength of the photoelectrons or on ΔE. Figure 8 demonstrates the environmental dependence of the fine-structure of the K edge of carbon, using a graphite sample. Evaluation of such an EXAFS gives interesting information on the distance and number of nearest neighbours and on the chemical nature of the absorbing atom. e.g. experiments on Cu give the distance of the atoms equal to the values obtained by X-ray diffraction within 0.02 Å. If one compares relative distances the accuracy increases by about a factor of 10.

5. CONCLUSION

This short survey on the present state of inelastic scattering of electrons in solid materials demonstrates an interesting field of physics. In addition, the application of electron energy loss spectroscopy to surface physics as well as to microanalysis has so far brought many important new results. It is furthermore to be considered as a subject still is its infancy.

Institut für Angewandte Physik
Hamburg, West Germany

REFERENCES

Additional references to the topics discussed in this article are collected in the book by the present author 'Excitation of Plasmons and Interband Transitions by Electrons': Springer Tracts in Modern Physics Vol. 88 (1980).

Bohm, D. and Gross, E. P.: 1949, *Phys. Rev.* **75**, 1864.
Bohm, D. and Pines, D.: 1951, *Phys. Rev.* **82**, 625.
Denbich, P. N. and Grigson, C. W.: 1965, *J. Sci. Inst.* **42**, 305.
Franck, J. and Hertz, G.: 1914, *Phys. Ges. Verhandlungen* **16**, 512.

Hren, J. J., Goldstein, J. I., and Joy, D. C. (eds.): 1979, *Introduction to Analytical Microscopy*, Plenum Press, New York.: Chapters 7, 8, 9.
Isaacson, M. and Johnson, D.: 1975, *Ultramicroscopy* **1**, 33.
Rudberg, E.: 1936, *Phys. Rev.* **50**, 138.
Ruthemann, G.: 1942, *Naturwiss.* **30**, 145.

JON GJØNNES

III.6. Structure Determination by Electron Diffraction

1. INTRODUCTION

Electron diffraction is the most sensitive of the three diffraction methods commonly used in crystal structure analysis. The interaction with matter is 100 to 1000 times stronger than for X-rays, hence the scattering from a solid specimen is appreciable after a thickness of a few nm. In the first years of electron diffraction this sensitivity may have appeared a difficulty to be overcome, rather than an advantage to be exploited. It called for experimental techniques capable of handling very thin crystals and a theoretical apparatus which could treat the strong interaction. Both of which took time to be developed.

It also appears that the type of problems in which electrons could offer specific advantages were relatively rare in a period when determination of the idealized, average structure was very much the focus of structure research. For that kind of structure analysis, X-ray diffraction is usually a far more accurate and reliable method – provided that suitable single crystals can be produced – which is not always the case. Quite early on, therefore, electron diffraction became an alternative when only very small crystals were available. This is not just a question of preparation technique; many man-made materials as well as natural objects consist of small crystals. Often the crystallographic relations between the crystals are as important for the understanding of properties and previous history as the average structure of the individual grain.

The sensitivity of electron diffraction, combined with the image-forming properties of the electrons, has made electron diffraction/electron microscopy an indispensable tool for structure analysis of such fine-grained materials. Hence electron diffraction has become very much a practical method, devoted to the study of actual materials rather than the idealized, average structure which is the aim of the X-ray crystallographer.

There may be an apparent contradiction between this development and the considerable effort directed towards theory and experiment in scattering from perfect, rather simple crystals. But we find the advances in practical application of electron diffraction often linked with these theoretical studies. It is fruitful, therefore, to see the history of electron diffraction in terms of such a dual approach, even if this, to some extent may overemphasize a planned, or strategic element in the development of the field.

2. SOME RESULTS FROM THEORY

The development of the dynamical theory of electron diffraction is presented in the article by A. F. Moodie in this volume. Let us here refer to only some results which are essential to our discussion. In the kinematical approximation which forms the basis for virtually all structure analysis with X-rays and neutrons, the scattering amplitude is given by the Fourier transform of a distribution which characterizes the object. The intensity distribution around each reciprocal lattice point h is given by the structure factor F_{hkl} and the shape transform S which is the same around every reflection:

$$I_h(s)^{\text{kinematical}} = F^2{}_{hkl} S^2\,(s,h) \tag{1}$$

where s is the scattering variable. In X-ray and neutron diffraction the variation of I_h with s, the rocking curve, is rarely used. Normally the measurement includes an integration of I_h over s; the integrated intensity is taken as a measure of F^2_{hkl}, which is the starting point for further structure analysis.

In electron diffraction the kinematical approximation is usually not valid and there is no such simple scheme for obtaining the structure factors, which we here may call V_{hkl} – the Fourier components of the time-averaged potential distribution. This is not just because the dynamical theory leads to a more complex calculation of the scattered intensity and its dependence on thickness, z,

$$I_h^{\text{dynamical}} = I(V_h, V_g, \ldots, s_h, s_g, \ldots, z) \tag{2}$$

where the structure factors V_g and excitation errors, s_g, of several reflections contribute. It also means that the simple concept of an integrated intensity for the reflection h breaks down. The intensities of the reflection spots must be calculated from an expression like (2) and averaged over the distribution of orientation and thickness appropriate to the object – a distribution which is usually not known. Only in the simple two-beam case when there is just one strong reflection, can a treatment similar to the kinematical case be applied. The two-beam expression

$$I_h^{\text{2-beam}} = \frac{V^2{}_h}{V^2{}_h + (s_h/2)^2}\sin^2[V^2{}_h + (\tfrac{1}{2}S_h)^2]^{\frac{1}{2}}z \tag{3}$$

can be integrated over excitation errors and thicknesses to yield:

$$I_h^{\text{2-beam, integr.}} \propto V_{hkl} \tag{4}$$

The effect of other beams can, to some extent, be included by the so-called Bethe potentials (Bethe, 1928):

$$V_h^{\text{Bethe}} = V_h + \sum_g V_g V_{h-g}/2k s_g \tag{5}$$

The second term expresses the effect of 'umweganregung' via the beam g into the beam h and is sometimes useful for the discussion of three-beam effects, which have been studied from Shinohara's (1932) early work onwards.

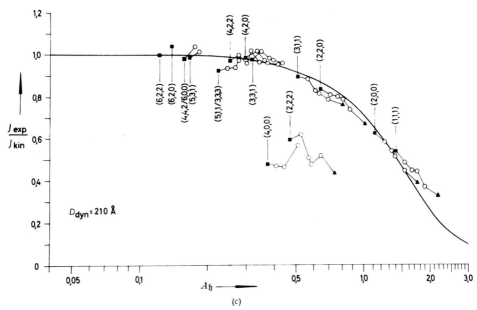

Fig. 1. The relation $I_{exp}/I_{kinematical}$ for polycrystalline Al, with average crystal thickness 21 nm. Full line curve calculated for two-beam dynamical theory (Horstmann and Meyer, 1962).

The range of validity of kinematical, two-beam or other dynamical expression has been the subject of many investigations. It seems that the weaker reflections in a powder pattern can be described reasonably well by a kinematical expression, as seen in Figure 1 (Horstmann and Meyer, 1962) – the strongest are often close to a two-beam value. But as was emphasized, particularly by Cowley: *the main advantage of electron diffraction is the possibility of obtaining single crystal patterns from small individual crystals or domains* – powder patterns may just as well be taken with X-rays.

Partly for this reason, Cowley and Moodie (1957) developed the thin phase grating expression as an approximate theory for electrons incident along a zone axis. It seems particularly useful in high resolution microscopy.

N-beam dynamical solutions can, of course, be worked out for any given diffraction condition and crystal thickness, using slice formulation or diagonalization procedures. But the problem remains: How can this be translated into a set of intensities for a spot pattern from a crystal which, due to mosaic structure or bending, will have a range of orientations and thicknesses?

As is explained below, the answer came along two main lines: To proceed from the spot pattern to the rocking curve of the convergent beam pattern, or to use spot intensities mainly qualitatively.

3. THE EXPERIMENTAL WORK FROM POWDERS TO CONVERGENT BEAM

The early experimental work, e.g. the extensive studies by the group in London

(Finch and Wilman, 1937) was to a large extent concerned with the theory and method. Much of the work was done on powders, which were convenient for the theory (Blackman, 1939). Systematic and extensive crystal structure studies with electrons were taken up especially by the Soviet school of electron diffractionists, Z. G. Pinsker, B. Vainshtein and their coworkers. They used both powder and single crystal spot patterns, but put special emphasis on the development of the oblique texture patterns. These are obtained from oriented samples formed when thin crystals are deposited on a grid or supporting film. When such a specimen is inclined to the beam, reflections from different layers ($hk0$, 1, 2, . . .) are found on separate layer lines. The overlap of reflections is thus much less than in the powder pattern, whereas one may hope that most of the intensities can still be represented by a kinematical expression.

From such data the Soviet school carried out many structure determinations, largely using Fourier methods similar to the X-ray work in the same period, and to comparable accuracy. From a few hundred reflections for a structure with twenty or so independent structure parameters they reached R-values of about 0.20 (for a review, see Pinsker, Zvyagin, and Semiletov, this volume). The substances covered many fields of chemistry; salts, intermetallic compounds, carbides and nitrides and also organic structure, like diketopiperazine where Vainshtein (1955) used electron diffraction to determine position of hydrogen, utilizing the somewhat better ratio between scattering amplitudes for light and heavy atoms in the electron case. In his well known studies of clay minerals, Zvyagin (1967) demonstrated the application to substances where X-ray diffraction often produces little but diffuse powder patterns.

Judged by contemporary standards, these studies can be said to represent highlights in the use of electron diffraction intensities. But, as it turned out, there was too little scope for improvement in comparison with the tremendous advance made in the techniques of X-ray determination of crystal structure in the age of automatic diffractometers.

The Japanese effort in electron diffraction covered almost every aspect of the field; theory, methods, experimental development and structure determination. The emphasis was very much on theoretical and experimental work on the diffraction effects resulting from dynamical interactions. Their structure studies were quite early directed towards single crystal work such as the investigations of superlattices in alloys, which were taken up by Ogawa and Watanabe (1952).

J. M. Cowley may have had the clearest concept of electron diffraction as a tool in structure research. With his coworkers in Melbourne he directed a broad effort towards the use of single crystal spot patterns in crystallography, including theory, experimental development as well as typical applications. It may be that these single crystal studies, (see e.g. Cowley, 1953, 1956) produced relatively few structures and that the expectations for the method based upon spot patterns were not met. Nevertheless, this emphasis on single crystal work was highly significant for the development which followed, a development which turned electron diffraction into a practical method in structure research through the modern electron microscope.

Selected area diffraction (SAD) in the electron microscope had already been demonstrated by Boersch (1936). It was introduced in commercial electron microscopes after World War II, but its full potential was realized only towards 1960. The introduction of movable apertures, goniometer stages and illumination tilt made the electron microscope a unique crystallographic instrument with the capability to reveal crystal structure and morphology and the interplay between them in fine-grained materials.

The impact was immediate in metallurgy, where the study of defects and submicroscopic particles opened up almost a new world to the metallographer, and at the same time making crystallography visual, not just something from a textbook. Through this development, which is described in the article by Amelinckx, electron diffraction also took on a more qualitative and practical aspect. The single crystal spot pattern, taken in SAD-mode, is usually not a step in structure determination in the conventional X-ray sense. It is more likely to be taken as a preliminary record of crystal type or orientation, e.g. in connection with a study of crystal defects or morphology as described in specialist textbooks (e.g. Thomas and Goringe, 1979, or Edington, 1975) (see however section 7 below).

But parallel to this qualitative line, which is more concerned with the investigations of materials than with structure of crystals, new efforts were also made towards more accurate electron diffraction experiments on crystals.

Convergent beam diffraction patterns (Figures 3 and 4) are obtained when a fine electron beam is focussed near the specimen (Kossel and Möllenstedt 1939 and earlier article, this volume). If this is a flat and perfect single crystal, one obtains disks of two-dimensional rocking curves instead of the spot pattern. The intensity variations due to dynamical interactions were used already in 1940 by McGillavry to deduce structure factors for mica.

Goodman and Lehmpfuhl (1964) revived the technique and developed it into a very sensitive method for the study of symmetry and determination of structure factors. Other methods for structure-factor determination from dynamical effects were also introduced, e.g. the critical voltage method (Watanabe, Uyeda and Kogiso, 1968).

The introduction of analytical equipment, expecially X-ray energy dispersive spectrometers and, later, electron energy loss spectrometers as attachments to the electron microscope, was of more immediate practical value.

The combination of microscopy, diffraction and chemical point analysis on a micron-scale or finer further increased the potential of the electron microscope in the study of fine-grained material. The rapid development of electron microscopy in mineralogy (e.g. Wenk, 1976) offered many examples.

By this experimental development through the last twenty years electron diffraction has become ever more part of a complex of techniques based upon the electron beam, and has thereby diverged very much from the tradition of X-ray crystallography. It is evident that this also has led to a different approach to structure analysis.

4. THE STRUCTURE DETERMINATION

The determination of crystal structure in the conventional, X-ray sense is a well-defined process which can be divided into main steps: The establishment of the unit cell dimensions; the derivation of space group symmetry; the determination and refinement of atomic coordinates from the intensities of reflections to a certain accuracy which should be attained before the structure can be regarded as 'determined'. If this process is not completed, the result is rather a structure proposal.

Electron diffraction is much less frequently completely carried through such a scheme. The reason is not just that the accuracy may be hard to attain, but also that the aim of the study is wider and often related to other questions than atomic coordinates and bond lengths and angles. In a sense the very concept of structure is looser than in X-ray diffraction. Bearing this in mind, we now proceed to the various steps in structure studies.

5. LATTICE PARAMETER AND UNIT CELL

The high monochromaticity and short wavelength of electrons permit very high precision in the determination of lattice parameters. But this resolution is partly sacrificed in the selected area technique because of the action of a strong lens after the specimen. Even with careful calibration and focusing, the accuracy may not be much better than one percent. Therefore X-ray powder lines are often used, instead, if a bulk value is sufficient.

If accurate measurements from small areas are required, the Kikuchi pattern can be used. Methods which are quite similar to the use of Kossel lines in the X-ray case are based upon measurement of dimensions in the small triangle formed by three nearly intersecting Kikuchi lines, not belonging to the same zone. Høier (1969) showed how an accuracy of 10^{-3} is readily achieved for a cubic substance. The extension to crystals of lower symmetry was studied by Olsen (1976). He determined lattice parameters for an olivine and a feldspar from measurements of some twenty intersections. The result was uncertain when all six parameters in the triclinic case were varied in the least square fitting procedure. But if one or two parameters could be fixed, or determined by other methods, an accuracy close to 10^{-3} was again obtained.

The Kikuchi pattern can also be used to determine small orientational differences (see e.g. Johari and Thomas, 1969). When these occur across coherent boundaries they can be used to assist in lattice parameter determination. Olsen (1977) applied several such methods together with X-ray microanalysis in the microscope to the investigation of local composition variations and site occupancies in labradorite feldspars. The most difficult and time-consuming element in such use of Kikuchi-patterns from low-symmetry structures is the indexing. This is simplified considerably by computer generation of the pattern (Pirouz & Boswara 1974).

Solid state transformations are often accompanied by changes in unit cell dimensions and symmetry, as in the cubic to tetragonal or orthorhombic transition which occurs upon ordering in many alloys, defect oxides, carbides, nitrides, etc. The associated changes in the dimensions of the fundamental

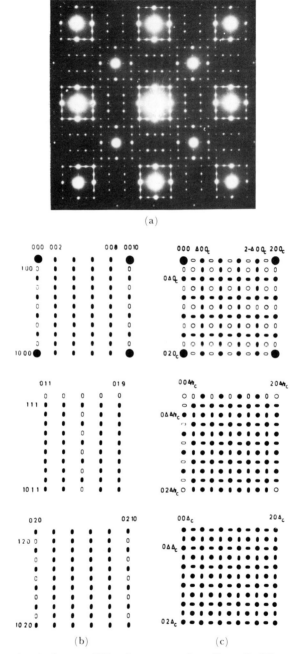

(a)

(b)   (c)

Fig. 2.   (a) Above:(001)$_c$ electron diffraction pattern from Fe$_{0.89}$O. (The strong 110$_c$ spots are due to Fe$_3$O$_4$.) (b) Lower left: Schematic representation of reciprocal lattice for one of six orthorhombic contributions for (h0l), (h1l) and (h2l): (c) Lower right: The same layers with all six contributions added (Andersson and Sletnes, 1977).

sub-cell are often small and hard to detect in diffraction patterns, which will be a superposition of patterns from equivalent orientations. The existence of equivalent orientations is frequently revealed by the resulting domain structure; in order to ascertain the true unit cell, dark field micrographs have to be taken. A typical example is the defect iron monoxide, which in X-ray studies (Koch and Cohen, 1969) appeared to be cubic. But the pattern of Figure 2 is the superposition of six equivalent orientations of an orthorhombic pattern, shown by Andersson and Sletnes (1977) to have the space group $ABm2$ and a periodic antiphase relation. This diffraction information led to a new defect model, and a structure proposal which could be checked against the reported X-ray intensities. The lattice imaging study by Iijima (1974) is consistent with the electron diffraction result, but could not reveal the three-dimensional structure.

6. SPACE GROUP SYMMETRY

In kinematical diffraction the space group of a structure is deduced from Laue symmetry and the systematic absences of reflections. It was recognized quite early that kinematically forbidden spots frequently appear in electron diffraction through 'unweganregung'. In spot patterns one can often avoid this by finding projections which do not include the possibility of 'umweganregung' paths, but this is not always easy.

Cowley and Moodie (1959) pointed out that the kinematically forbidden reflection should remain extinct under dynamic conditions when the beam is entering along the zone axis of the projection. This argument, which essentially was confined to the two-dimensional space group, was extended by Gjønnes and Moodie (1965) to three dimensions and general directions. They showed that a 2-fold screw axis leads to an extinct band at the Bragg condition for the forbidden reflection, whereas the glide plane leads to dynamical extinction when the incident beam is parallel to the glide plane. If interactions only in the projection is considered, both extinctions apply, producing a dark cross through the kinematically forbidden reflection, as was demonstrated by Goodman and Lehmpfuhl (1964). No similar effects are associated with the 4- and 6-fold screw axes. A practical example of the use of dynamical extinction is provided in Fig. 3.

The effect of dynamical interactions on the symmetry in diffraction patterns was pointed out by Miyake and Uyeda (1950), who demonstrated the failure of Friedels law for non-centrosymmetrical crystals, i.e. the failure of reversal symmetry between the reflections h and $\bar{h}$ when the crystal or beam is rotated 180° about an axis in the plane h. This effect can be used to distinguish between different planar space groups, e.g. pm and $p2m$ (Goodman and Lehmpfuhl 1968).

In a more comprehensive study, Goodman (1975) described the procedure for complete space group determination by means of the convergent beam technique and applied this to the monoclinic mineral $2M$-biotite, which was shown to belong to the space group $C2$, rather than $C2/c$ which had previously been allocated from X-ray data. Another approach was that of Tanaka *et al.*

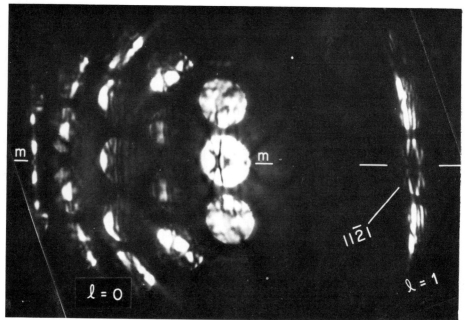

Fig. 3.   An illustration of dynamical extinction being used to identify a 2-fold screw axis in GaS. The crystal is oriented to give exact symmetrical orientation of the 1 1 2̄ 1 reflection (from Goodman and Whitfield, 1980).

Table I. From the symmetries in bright field, dark field and between the $+G$ and $-G$ dark fields an unequivocal allocation of space group could be made. A thorough discussion of symmetry in electron diffraction, including the effect of surfaces is given by Buxton et al. (1976).

Convergent beam patterns can also be used to distinguish between enantio-morphous space groups (determination of handedness). Goodman and Secomb (1977) developed a procedure analogous to the classical X-ray method by Bijvoet (1949); instead of the anomalous scattering effect in the X-ray case, the N-beam interactions were exploited. The use of selected four-beam interactions (Goodman and Johnson 1977) may be seen as a more elegant solution to the problem.

TABLE I

Symmetries of CBED patterns expected for the three space groups $C2/c$, Cc and $C2$ (Tanaka et al. (1980)).

Space group	Point group	Diffraction group	Bright field	Whole pattern	Dark field	$\pm G$
$C2/c$	$2/m$	21_g	2	2	2	21_g
Cc	m	1_g	2	1	2	1
$C2$	2	2	2	2	1	2

The diffraction groups described in column 3 are expected for the case of the [010] electron incidence.

These studies show in a convincing way that the convergent beam electron diffraction is a powerful method for symmetry determination, surpassing that of X-rays in several ways: The easy identification of a symmetry centre, distinguishing between the members of enantiomorphous pairs and not least, the possibility of recording the symmetry with high spatial resolution down to a few tens of a nanometer.

With recent improvement in probe resolution, and the use of low temperatures, it has been possible to isolate high-symmetry phases in so-called 'solid solutions' of geological-type samples. (Johnson, personal communication: Figure 4).

The determination of symmetry has for some time been of increasing interest in electron diffraction studies of structures. The fine grains or domains in the

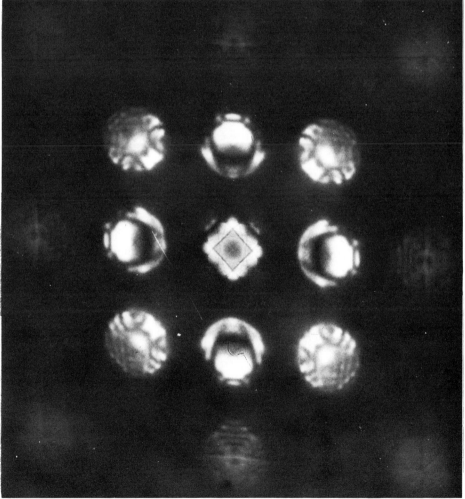

Fig. 4.   A phase of $Fd3m$ symmetry in a spinel of bulk composition $Mg_{0.36}Al_{2.43}O_4$ (A. W. S. Johnson, personal communication).

material under investigation are very often the product of a phase transformation involving a change in symmetry. Many electron diffraction studies exemplify this, notably in connection with superstructures. In such cases the determination of unit cell and space group may be sufficient for arriving at a structure proposal, sometimes with the support of qualitative comparison with the observed spot intensities. In several such studies X-ray powder intensities have been used to augment the electron diffraction information. A model study was the investigation of titanium monoxide by Watanabe et al. (1967). Previous X-ray work on the composition had not produced sufficient information to establish the superlattice formed by the ordering of titanium and oxygen vacancies in the structure. Through electron diffraction a superlattice structure was proposed and refined using X-ray powder intensities. It may be added, however, that recent single crystal studies (Terauchi et al., 1978) led to some minor modifications of the structure which had been derived. Recent studies of phase transitions in organic molecular crystals (e.g. Rees et al., 1980) show that similar problems may await investigation in such systems, as the difficulties arising from beam damage are overcome. But through the study of more complex structures than the simple ordering on ideal sites, the need for reliable intensity data may also increase.

7. SPOT PATTERN INTENSITIES AND THEIR USE

In X-ray crystallography the collection of large numbers of fairly accurate intensities of single crystal reflections has permitted the determination of atomic coordinates from ever larger and more complicated structures. The simple concept of the mosaic crystal, where kinematical scattering takes place within the individual block forms the basis of the X-ray work. The lack of a similar simple relation between intensities of spots and Fourier components of the structure has been a major limitation in electron diffraction. No practical scheme for averaging the n-beam dynamical calculation over an undetermined distribution of orientation and thickness has as yet been devised. The simple kinematical expression $F^{hkl} \propto \sqrt{I^{hkl}/d}$ used by e.g. Harada and Kitamura (1964) or Udalova and Pinsker (1964) with dynamical corrections only for the strongest reflections is, at best a rough approximation. It is therefore not strange that the recent trend has been to avoid or circumvent this difficulty, either by turning to problems where quantitative structure factors are not so important or where sufficient intensity data can be extracted from X-ray powder patterns. Rejecting the idea of electron diffraction powder patterns, the oblique texture pattern may appear a possible compromise, in cases where such a preferred orientation can be produced. But even this can be strongly influenced by dynamical interactions, as pointed out by Turner and Cowley (1969). Another possibility is to turn to more refined diffraction techniques, like the convergent beam where no average over orientation is required.

Even so, spot (SAD) intensities from mosaic crystals with a spread of orientations can sometimes be used, even when the scattering is clearly dynamic.

In his study of lower vanadium oxides, Andersson (1975) determined the unit cell and space group of a monoclinic oxide which occurs near the composition V_2O. Due to the overlap of many lines in his X-ray powder pattern he deemed this insufficient for obtaining a structure proposal. The electron diffraction patterns although clearly influenced by N-beam dynamical interactions, appeared to contain considerable structure information. It was also clear that even a qualitative description of the structure would depend upon some knowledge of atomic displacements away from special positions. Using 120 reflections from three selected projections, which were chosen so as to be sensitive to the structure elements under consideration and not too severely influenced by dynamical interactions, it was possible to distinguish between the nine different structure models which were consistent with the space group, and the possible compositions $V_{14}O_8$, $V_{12}O_8$, $V_{14}O_6$, $V_{12}O_6$. Calculated intensities were based upon dynamical diffraction, but with an empirical angular correlation factor instead of the proper average over orientations. Ten coordinates for five independent atoms were determined, with an overall R-factor of 0.15.

It is of interest to compare this result with the independent study by Hiraga and Hirabayashi (1975), which was based on more extensive X-ray data. They assumed a slightly different composition, $V_{14}O_6$, but with extra oxygen statistically distributed on the remaining oxygen positions. The metal coordinates were the same, however, within the standard deviation, as shown in Table II.

Used with care there is no doubt that dynamically affected spot pattern intensities can give useful information, maybe at a level which can best be compared with X-ray crystallography in the days of visual intensities. Despite the quite extensive work in the sixties e.g. Cowley and Kuwubara (1962), there may still be scope for work aimed at obtaining better insight into the uncertainties and error sources (Table III).

Furthermore, there may be much to be gained by going to thinner crystals and light elements. Several recent studies represent attempts to treat spot patterns from organic crystals by kinematical theory (Dorset and Hauptmann 1976; Claffey et al. 1974). The validity of this for thin organic specimens or the thin phase grating approximation has been investigated by Jap and Glaeser (1980) who derived thickness limits at different voltages for the somewhat severe demand of $R = 0.5$.

TABLE II

Comparison between metal coordinates in $V_{14}O_{6+2n}$ obtained by electron spot pattern and X-ray powder.

	Andersson (1975)			Hiraga and Hirabayashi (1975)		
	x	y	z	x	y	z
V (1)	0.274	0	0.179	0.272	0	0.183
V (2)	0.561	0	0.316	0.560	0	0.320
V (3)	0.845	0	0.463	0.848	0	0.457

TABLE III

Crystal structure factors of Si by X-ray pendellösung and electron diffraction methods. R: Rocking curve, CV: Critical voltage, ET: Equal thickness fringes, CB: Convergent beam, IKL: intersecting Kikuchi lines.

	F_{111}	F_{222}	F_{220}
X-ray Pendellösung			
Tanemura and Kato (1972)	60.658 ± 0.012		69.74 ± 0.01
Aldred and Hart (1973)	60.726 ± 0.034	1.42 ± 0.04	69.176 ± 0.072
Fehlman and Fujimoto (1975)		1.724 ± 0.036	
Electron diffraction			
Kreutle and Meyer-Ehmsen (1971) R	60.66 ± 0.05	-1.7 ± 0.2	
Hewat and Humphreys (1974) CV	60.590 ± 0.068		
Ando, Ichimiya and Uyeda (1974) ET	60.57 ± 0.47	-1.94 ± 0.89	
Smith and Lehmpfuhl (1975) CB	60.87 ± 0.10	-1.49 ± 0.42	
Terasaki, Watanabe and Gjønnes (1979) IKL			
	60.81 ± 0.28	3.28 ± 1.36	
	60.58 ± 0.23	2.24 ± 1.12	
	60.98 ± 0.28	0.32 ± 1.12	

8. THE CONVERGENT BEAM PATTERN

Goodman and Lehmpfuhl (1967) demonstrated that structure factors can be obtained with far better accuracy from convergent beam electron diffraction than from spot patterns or powder patterns. In their study of systematic *hoo* reflections from thin MgO crystals, the first order structure factor V_{200} was determined from the photometered rocking curves to about one percent – which was also considerably better than had been obtained with X-rays. There are rather few uncertain factors in these measurements: The scale of the rocking curve is fixed by fine details due to high-order reflections, the thickness can be determined quite accurately by oscillations in the outer (s_h large) part of the rocking curve, whereas the value near $s_h = 0$ is sensitive to $V_h z$.

The need for a parallel plate crystal of uniform orientation is clearly a limitation. But with finer electron probes in modern instruments employing brighter sources and having much lower contamination rates, this is easier to obtain.

The early convergent beam experiments were performed on relatively thick crystals; pronounced dynamical effects were exploited in the measurements. The accuracy is then very good for the strong reflections. But there are clear disadvantages: An appreciable background mainly due to inelastic scattering has to be subtracted. This background cannot be readily calculated and is not featureless, but quite strongly affected by Bragg interactions, as Kikuchi bands etc. Another difficulty is due to the N-beam effects on the weaker, outer reflections. The accuracy in the determination of structure factors decreases rapidly when one proceeds to the weaker reflections.

There are two ways of overcoming these difficulties. One is to search for ever

Fig. 5. An illustration of thin-crystal convergent beam patterns being used for quantitative data collection: (a) shows the hexagonal zone axis from thin graphite; (b) shows the alignment for exact symmetrical excitation of the 110 reflection; the small inner discs visible indicate the regions used for quantitative intensity measurement (Goodman, 1976).

more details which are sensitive to the structure factors or parameters to be determined. The other is to go back to situations which are less dynamical, that is to thinner crystals. This course was explored in recent studies of MoO_3 by Bursill et al. (1977) and of graphite (Goodman, 1976). The latter investigation is particularly interesting, since the structure problem was partly connected with symmetry, partly with bonding effects in inner reflections and partly to local variations in structure, associated with defects. From a number of convergent beam photographs, taken from different areas Johnson (1972) had already shown that the symmetry is indeed hexagonal as in Bernal's classical structure and that regions of trigonal or orthorhombic symmetry are associated with defects. The determination of inner reflections confirmed the anomalous ratio between F_{100} and F_{110}, which is attributed to bonding effects (see Figure 5).

In these studies the intensity variations within each disk were used to define the orientation and determine thickness. Accurate intensity measurements were taken at corresponding, special points in each disk in much the same way as spot intensities. To a certain extent this is returning to a nearly kinematical case, but in a situation which allows corrections to the kinematical intensities to be calculated quite precisely. The inelastic background is, of course, greatly reduced. Although no systematic study involving a large number of reflections seems to have been performed as yet, this way of extracting precise single crystal data appears to offer considerable possibilities, available in modern electron microscopes with STEM optics. The need for a special habit may limit the amount of three-dimensional data which can be collected, however.

9. STRUCTURE FACTOR DETERMINATION FROM OTHER DYNAMICAL EFFECTS

The convergent beam determination of structure factors using a systematic row of reflections from a thick crystal is one of several methods which utilize simple, essentially two-beam, dynamical effects. Even in the presence of several reflections, the intensity distribution will contain terms of the type in equation (3); quite often one such term will dominate. The sinusoidal variation can be studied in equal thickness fringes at constant s_h, preferably $s_h = 0$, or as a function of s_h for constant thickness, (rocking curve). The effective V_h which is thus obtained is proportional to the gap at the dispersion surface and can be calculated from N-beam dynamical expressions.

The first such method, the beam splitting at a crystal wedge, was introduced by Honjo and Mihama (1954), Molière and Niehrs (1954) and Cowley, Goodman and Rees (1957). The angular separation between the two waves producing the sinusoidal term in (3), was measured in the split spots appearing in diffraction rings from magnesium oxide smoke. Lehmpfuhl (1972) improved the method considerably by using a macroscopic crystal wedge, which was rotated in a controlled way. Since perfect crystal wedges are needed, however, the application has so far only been to a few substances.

Essentially the same magnitude is measured in the equal thickness fringe method. If the wedge angle is known and also the magnification of the microscope (which is not trivial), the fringe period in thickness can be determined and again the gap at the dispersion surface is involved. Due to difficulties in maintaining the predetermined diffraction condition, the method was subject to some experimental uncertainties. These appear to have been overcome through the work by Uyeda and Nonoyama (1965) and Ichimiya *et al.* (1973); by using different diffraction conditions and/or voltages, uncertainties in magnification, wedge angles etc. can be eliminated.

In the intersecting Kikuchi line method (Gjønnes and Høier, 1971: Figure 6) the gap at the dispersion surface is sampled in quite different ways. When a weak Kikuchi line crosses a strong line, the former is split into two segments. The magnitude of the separation is proportional to the gap at the dispersion surface for the strong reflection and can therefore be used to determine the structure factor. Høier and Andersson (1974) used the method to determine the metal interstitial concentration in $VO_{1.25}$. As with most of these methods, the accuracy can be improved considerably by employing higher voltage (see Figure 7).

The other and more well known Kikuchi line method is the critical voltage, (Watanabe, Uyeda and Kogiso, 1968). This is rather a three-beam effect: the effective potential can become zero if the 'umweganregung' term cancels the leading term in equation (5). This is reflected in a vanishing gap at the dispersion surface and hence vanishing contrast of the Kikuchi line. Such a condition can be obtained for a second-order reflection in a controlled fashion by

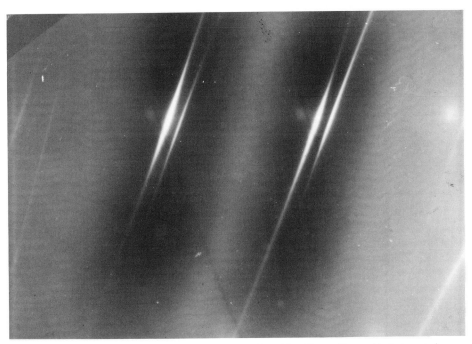

Fig. 6.   Kikuchi line intersections with 220 band in silicon (Terasaki *et al.*, 1979).

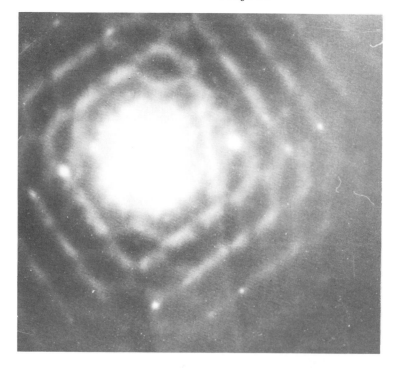

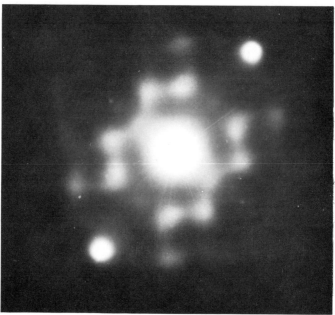

Fig. 7.   (a) Diffuse scattering from VO$_{1.23}$ 411-projection, used to obtain order parameters for inter-cluster ordering (Andersson *et al.*, 1974). (b) Pattern taken at 1 MeV, beam near [001] direction; the outer part of the pattern is interpreted in terms of displacement parameters (Andersson, 1979).

changing the beam voltage in a high-voltage microscope, and thereby the effective mass of the electron. At a certain critical voltage the gap becomes zero; whence, approximately:

$$V_{2h} \approx \frac{V^2_{k}}{2ks_h} \tag{6}$$

It seems well established that zero gap is the condition for vanishing Kikuchi line contrast and also that in the systematic case this can be calculated by using the Bethe potential.

It is also possible to vary the excitation error s_h for the strong reflection in order to obtain the extinction of contrast. This means measuring the distance of an extinction point from the strong band edge, as was done by Gjønnes and Høier (1971). There seems to be rather few cases suitable for accurate measurements of such 'critical excitation error', however.

The critical voltage measurement does not depend upon a special crystal shape and is thus relatively easy to apply when the condition (6) can be met, usually for only one or two reflections. Many such measurements have been performed, accuracies of about 0.5% have been attained (Hewat and Humphreys 1974). Sometimes the measurements have been dedicated to special structure problems; a bonding state, ionicity or site occupancy. Smart and Humphreys (1980) have used the critical voltage structure factors together with other data to construct bonding charts for several metals.

The limit to accuracy seems to be associated with absorption, which limits the number of equal thickness fringes and hence the accuracy in measurement of fringe period. This will also be reflected in the sharpness of the dispersion surface and in the gap – hence also in the ultimate sharpness of Kikuchi lines.

A more serious practical limitation is the rather small number of structure factors which can be determined by these methods. Frequently even these will depend, to some extent, upon knowledge of other structure factors. An obvious answer to this is to combine several methods, especially the critical voltage with one of the two-beam methods (Watanabe et al., 1974; Shishido and Tanaka, 1976). But even this will be limited to a few inner reflections, which may be useful in special studies of bonding etc. but of limited application in ordinary structure determinations.

Another way of extending the measurements may be indicated by the principles involved, especially in the Kikuchi-line methods. Here special, sensitive, details are sought out. A more systematic and comprehensive search for such details has not been attempted, although the use of qualitative comparisons by Steeds and Jones (1975) is a legitimate step in this direction.

Another important possibility of improving accuracy should be mentioned. Through the use of current recording, and filtering of inelastic scattering, Kreutle and Meyer-Ehmsen (1971) achieved higher accuracy in their rocking-curve measurement than had been obtained by photographic recording. The modern, combined TEM/STEM instruments with the possibility of including velocity analyzers will definitely increase the use of such techniques beyond present recognition.

In addition to the application of convergent beam electron diffraction this may be the most promising way to obtain more quantitative data for structure determination. There is doubtless much to be gained by more automated recording, using STEM systems, comparison with simulated patterns and so on. But there is also a long way to go until even a small part of the data which are obtained as a routine in X-ray diffraction can be collected.

10. DETERMINATION OF PHASES — AND OF POSITIONS

The dynamical effects involving more than two beams are sensitive to the sign or phase of structure factors. Thence Kambe (1957) and others have pointed out that three-beam effects can reveal the sign of the structure factor product $U_g U_h U_{h-g}$. So far little use has been made of that possibility, although the more systematic study of qualitative effects in Kikuchi patterns, convergent beam and zone axis patterns may incorporate also these effects.

Another phase or position sensitive effect is the channelling-like effects on the emission of secondary radiation, which is shown in the X-ray emission by Taftø (1979). But it should be remembered that the most effective way of extracting phase information is to combine several beams to form an image as in the electron microscope.

11. DIFFUSE SCATTERING

Patterns of diffuse scattering from local order, thermal motion etc. are usually easier to obtain with electrons than with other diffraction methods. But the interpretation of the patterns has mainly been qualitative; quantitative studies in terms of order parameters etc. has usually been left to X-ray diffraction. The main reason for this is again the stronger interaction and the more complicated theory. Although the individual Fourier components responsible for diffuse scattering may be quite weak, scattering in directions differing by a reciprocal lattice vector will be coupled through Bragg diffraction. This leads for example to excess and deficient Kikuchi lines and other effects. The theory for the influence of Bragg scattering on the diffuse scattering has been developed by Kainuma, 1955; Gjønnes, 1966 and others.

It follows from these theories that diffuse scattering which is repeated from one Brillouin zone to the next, as for short range order scattering from binary alloys, is not modified in any drastic way by the dynamical interactions. Hence they need not be considered in qualitative interpretation, like the study of Fermi surface effects in alloys. These are a reflection of the long-range forces due to conduction electrons. As was shown by Krivoglaz (1969) and Moss (1969), flat parts of the Fermi surface appear as maxima in the short-range order scattering (see Moss, earlier this volume). Such maxima are easily identified in electron diffraction patterns (Ohshima and Watanabe, 1977). The position agrees well with theoretical calculations.

Other examples of qualitative interpretation of diffuse scattering are the studies of thermal diffuse streaks by Komatsu and Teramoto (1966), the calculation of order parameters from the shape of diffuse scattering maxima by Sauvage and Parthe (1972) and the cluster models proposed by de Ridder *et al.* (1976).

In recent years diffuse scattering of electrons from defect oxides, carbides and similar compounds has attracted much interest.

In the disordered rock-salt type vanadium monoxide the defects include metal and oxygen vacancies, metal interstitials as well as correlated atomic displacements. Andersson *et al.* (1974) interpreted the central part of the diffuse scattering pattern from the oxygen rich $VO_{1.23}$ in terms of order parameters describing the local order of tetrahedral defect cluster. Projections were selected so as to minimize dynamical interactions, which were of only moderate influence to the results. In a later study of scattering at higher angles, Andersson (1979) showed how the displacement order parameters (Borie and Sparks, 1971) could be extracted from high angle data obtained by high voltage electron diffraction; the large range of scattering variable thus available was of considerable value for the separation of the two types of scattering.

The theory of dynamical interactions was used in these studies partly as a guide to conditions where they could be minimized, but also as correction to the kinematical scattering which was essential in the latter study. The most serious difficulty appears to be associated with the background, however. Due to multiple scattering and dynamical effects in the inelastic scattering there is no way of calculating a theoretical background, a problem which can be quite serious also in the study of the continuous scattering from amorphous films.

12. PAST AND FUTURE APPLICATIONS — CONCLUDING REMARKS

Considerable efforts in electron diffraction have gone into investigations of silicon, magnesium oxide, calcium flourite etc. largely because of the attraction of perfect crystal when studying the theory of scattering. But the application of electrons in structure research has become ever more associated with the study of fine grained material, domains and local variations in structure.

The examples of such systems are increasing, one of the first were the superstructures found in substitutional alloys. Although the first structure work was done with X-rays in the 1930s, we owe most of our knowledge of these structures to electron diffraction studies, by Ogawa and others. These structures which are determined mainly by long-range forces are typically an electron diffraction speciality.

Another large field concerns the phases occurring in metallurgical specimens, precipitates, martensite plates etc. These frequently metastable structures are often accessible only to electron diffraction because they occur within a surrounding matrix. Crystal structure determination may be relatively rare in such systems, the metallurgical interest being connected with defects and relations between crystals more than with the ideal structure of each phase. The

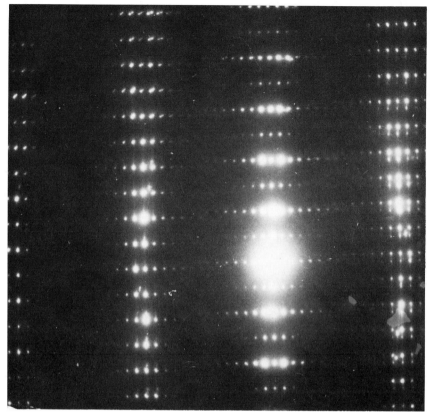

Fig. 8.   Electron diffraction pattern from a 10-layer antigorite with approximately 2550 atoms in
the unit cell (A. Olsen).

many texts on electron metallography published in the past decade reveal this
very clearly.

The mineral world has become an important growth area for applications of
microscopy and diffraction, with studies rather similar to the metallurgical ones.
Here the crystal structures are generally more complex and of more chemical
interest. The progress from rather well defined intergrowths with distinct
boundaries to the more complex compositional and structural variations in
labradorites may be examples of problems where electron diffraction enters as
one of several investigation methods (for a review see Wenk, 1976).

Through the studies of geological specimens many complicated and poorly
known structures become the subject of electron diffraction studies – sometimes
quite complex ones, as the antigorite studied by Olsen (1980). From electron
diffraction patterns (Figure 8) and electron micrographs taken at moderate
resolution, he was able to propose a structure model including 2550 atoms
within a monoclinic (pseudoorthorhombic) cell.

The field of spatial variations in structure is evidently a field with increasing
potential and interest. Many of the more recent electron diffraction studies

Fig. 9. Rocking beam methods illustrated by patterns from 111 Ge film: (a) without signal processing; (b) with signal processing; and (c) modified, or 'LACBED' pattern giving a wide-angle scan, unrestricted by neighbouring reflections (pictures by courtesy of M. Tanaka).

demonstrate this. Johnson's study of minerals (Figure 4) shows the application of the convergent beam method to structure within small regions; the studies of ordering in nonstoichiometric oxides reveal composition fluctuations to be an important element in what are normally taken to be single phase structures. There may also be a need to supplement the high resolution electron microscopy investigations of defects with diffraction information from small regions. Fields and Cowley (1978) have calculated the diffraction patterns to be expected from isolated defects. Experimentally, diffraction patterns from regions down to 40 Å have been reported by Geiss (1975).

The possibility of obtaining patterns from small regions is one reason for applying the method to organic substances, as was shown already by Honjo and Watanabe's (1958) patterns from cellulose. The application to phase transformations in organic solids can be studied by using low beam currents and cooled specimens (Rees *et al.*, 1977).

An obvious field of application for electron diffraction is, of course, to particles sampled from the environment or extracted from biological materials.

As has already been emphasized, the trend has been to use electron diffraction together with other techniques, microscopy and microanalysis, especially in complex systems. But the application to crystallographically more complex structures may also lead to an increased need for developing the quantitative side of electron diffraction. There are several lines of development in that direction. Typical examples are convergent beam diffraction from thin crystals and the systematic comparison of qualitative features in experimental and calculated patterns as shown by Steeds *et al.* (1976) in zone axis patterns. The more systematic use of the recording possibilities offered by modern STEM systems may enhance such development. Striking results have most recently been obtained by Tanaka *et al.*. (1980) using an automated beam-rocking technique. In this way he can not only invoke signal processing, such as d.c. filtering, and, in principle, electron-energy filtering, but can also, by a modification of the technique, collect any diffracted order over a large angular range, thus removing a key limitation to the use of conventional convergent beam diffraction in the presence of a superlattice or closely spaced orders. This technique is illustrated in Figure 9.

Finally, the more common and systematic use of computed patterns and facilities for storing data and controlling experiments will undoubtedly prevail. This will increase the power of the method in dealing with structure problems. But there is also reason to believe that a main feature of electron diffraction will remain the unique possibility to see the crystalline nature of small objects in a very direct and quick way, which has been a main practical advantage of electron diffraction for many years – due to its unique sensitivity as a diffraction method.

Department of Physics
University of Oslo, Norway

REFERENCES

Aldred, P. J. E. and Hart, M.: 1973, *Proc. Roy. Soc.* **A332**, 233.

Andersson, B., Gjønnes, J., and Taftø, J.: 1974, *Acta Cryst.* **A30**, 216.

Andersson, B.: 1979, *Acta Cryst.* **A35**, 718.

Andersson, B. and Sletnes, J. O.: 1977, *Acta Cryst.* **A33**, 268.

Andersson, B.: 1979, *Acta Cryst.* **A35**, 718.

Ando, Y., Ichimiya, A., and Uyeda, R.: 1974, *Acta Cryst.* **A30**, 600.

Bethe, H.: 1928, *Ann. Phys. Lpz.* **87**, 55.

Bijvoet, J. M.: 1949, *Proc. Roy. Soc. Amsterdam* **52**, 313.

Blackman, M.: 1939, *Proc. Roy. Soc.* **A173**, 68.

Boersch, H.: 1936, *Ann. Phys. Lpz.* **26**, 631.

Borie, B. and Sparks, C. J.: 1971, *Acta. Cryst.* **A27**, 198.

Bursill, L. A., Dowell. W. C. T., Goodman, P. and Tate, N.: 1977, *Acta Cryst.* **A34**, 296.

Buxton, B. F., Eades, J. A., Steeds, J. W., and Rackham, G. M.: 1976, *Phil. Trans. R. Soc. London* **281**, 171.

Claffey, W., Gardner, K., Blackwell, J., Lando, J., and Geil, P. H.: 1974, *Phil. Mag.* **30**, 1223.

Cowley, J. M.: 1953, *Acta Cryst.* **6**, 522.

Cowley, J. M.: 1956, *Acta Cryst.* **9**, 391.

Cowley, J. M. and Moodie, A. F.: 1957, *Acta Cryst.* **10**, 609.

Cowley, J. M. and Moodie, A. F.: 1959, *Acta Cryst.* **12**, 360.

Cowley, J. M., Goodman, P., and Rees, A. L. G.: 1957, *Acta Cryst.* **10**, 19.

Cowley, J. M. Kuwabara, S.: 1962, *Acta Cryst* **A34**, 556.

Dorset, D. L. and Hauptmann, H. A.: 1976, *Ultramicroscopy* **1**, 195.

Edington, J. W.: 1975, *Practical Electron Microscopy in Materials Science* Vols. 1–4, Philips Technical Library, London: MacMillan.

Fehlmann, M. and Fujimoto, I.: 1974, *Diffraction Studies of Real Atoms and Real Crystals* A.

Fields, P. M. and Cowley, J. M.: 1978, *Acta Cryst.* **A34**, 103.

Finch, G. J. and Wilman, H.: 1937, *Ergeb. Exakten Naturwiss.* **16**, 353.

Geiss, R. H.: 1975, *Appl. Phys. Lett.* **27**, 174.

Gjønnes, J.: 1966, *Acta Cryst.* **20**, 240.

Gjønnes, J. and Moodie, A. F.: 1965, *Acta Cryst.* **19**, 65–67.

Gjønnes, J. and Høier, R.: 1971, *Acta Cryst.* **A27**, 313.

Goodman, P.: 1975, *Acta Cryst.* **A31**, 804.

Goodman, P.: 1976, *Acta Cryst.* **A32**, 793.

Goodman, P. and Johnson, A. W. S.: 1977, *Acta Cryst.* **A33**, 997.

Goodman, P. and Lehmpfuhl, G.: 1964, *Z. Naturforsch.* **19**(a), 818.

Goodman, P. and Lehmpfuhl, G.: 1967, *Acta Cryst.* **A23**; 1968, *Acta Cryst.* A **24**, 339.

Goodman, P. & Secomb, T. W.: 1977, *Acta Cryst.* **A33**, 126.

Goodman, P. and Whitfield, H. J. 1980, *Acta Cryst.* **A36**, 219.

Harada, J. and Kitamura, M.: 1964, *J. Phys. Soc. Japan* **19**, 328.

Hewat, E. A. and Humphreys, C. J.: 1974, *High Voltage Microscopy* (ed. P. R. Swann) London: Academic Press pp. 42–56.

Hiraga, K. and Hirabayashi, M.: 1975, *J. Solid State Chemistry* **14**, 219.

Honjo, G. and Mihama, K.: 1954, *J. Phys. Soc. Japan* **9**, 184.

Honjo, G. and Watanabe, D.: 1958, *Nature* **181**, 326.

Horstman, M. and Mayer, G.: 1962, *Acta Cryst.* **15**, 271.

Høier, R.: 1969, *Acta Cryst.* **A25**, 516.

Høier, R. and Andersson, B: 1974, *Acta Cryst.* **A30**, 93.

Ichimiya, A., Arii, T., Uyeda, R., and Fukugara, A.: 1973, *Acta Cryst.* **A29**, 724.

Iijima, S.: 1974, *Diffraction Studies of Real Atoms and Real Crystals*, Australian Academy of Science, pp. 217–218.

Jap, B K. and Glaeser, R. M.: 1980, *Acta Cryst.* **A36**, 57.

Johari, O. and Thomas, G.: 1969, *The Stereographic Projection and its Applications.* Techniques of Metals Research Vol IIa, New York: Interscience.

Johnson, A. W. S. : 1972, *Acta Cryst.* **A28**, 89.

Kainuma, Y.: 1955, *Acta Cryst.* **8**, 247.

Kambe, K.: 1957, *J. Phys. Soc. Japan* **12**, 13.

Koch, F. and Cohen, J. B.: 1969, *Acta Cryst.* B**25**, 275.

Komatsu, K. and Teramoto, K.: 1966, *J. Phys. Soc. Japan* **21**, 1152.

Kossel, W. and Möllenstedt, G.: 1939, *Ann. Phys. Lpz.* **36**, 113.

Kreutle, M. A. and Meyer-Ehmsen, G.: 1971, *Phys. Stat. Sol.* (a)**8**, 111.

Krivoglaz, M. A.: 1969, *Theory of X-ray and Thermal Neutron Scattering by Crystals.* New York: Plenum. pp 50–55.

Lehmpfuhl, G.: 1972, *Z. Naturforsch.* **27**(a), 425.

MacGillavry, C. H.: 1940, *Physica* **7**, 329.

Miyake, S. and Uyeda, R.: 1950, *Acta Cryst.* **3**, 314.

Molière, K. and Niehrs, H.: 1954, *Z. Phys.* **137**, 445.

Moss, C. S.: 1969, *Phys. Rev. Lett.* **22**, 1108.

Ogawa, S. and Watanabe, D.: 1952, *J. Phys. Soc. Japan,* **7**, 36.

Ohshima, K. and Watanabe, D.: 1977, *Acta Cryst.* A**33**, 784.

Olsen, A.: 1976, *J. Appl. Phys.* **9**, 9.

Olsen, A.: 1977, *Acta Cryst.* A**33**, 706.

Pirouz, P. and Boswara, I. M.: 1974, *Phys. Stat. Sol.* (a)**26**, 407.

de Ridder, R., van Tendeloo, G., and Amelinckx, S.: 1976, *Acta Cryst.* A**32**, 216.

Rees, W. L., Goringe, M. J., Thomas, J. M. Jones, W.: 1980, *Electron Microscopy 1980.*Proc. 7th. European Cong. The Hague.

Sauvage, M. and Parthe, E.: 1972, *Acta Cryst.* A**29**, 520.

Shinohara, K.: 1932, *Sci. Pap. Inst. Phys. Chem. Res. Tokyo* **20**, 39.

Shishido, T. and Tanaka, M.: 1976, *Phys. Stat. Sol.* (a)**38**, 453.

Smart, D. J. and Humphreys, C. J.: 1980, *Electron Microscopy and Analysis 1979,* Int. Phys. Conf. ser. No. 52, 1980. London: The Institute of Physics.

Smith, P. J. and Lehmpfuhl, G.: 1975, *Collected Abstracts, 10th International Congress of Crystallography,* I.U.Cr., Amsterdam, 220.

Steeds, J. W. and Jones, P. M.: 1975, *Collected Abstracts, 10th International Congress of Crystallography,* I.U.Cr. Amsterdam, 252.

Steeds, J. W., Jones, P. M., Rackham, G. M., and Shannon, M. D.: 1976, *EMAG 75 Development in Electron Microscopy and Analysis.* London: Academic Press pp. 351–356.

Taftø, J.: 1979, *Z. Naturforsch.* **34**A, 452.

Tanaka, M., Katuyoshi, U., and Hirata, Y.: 1980, *Japanese J. Appl. Phys.* **19**, L201.

Tanaka, M., Saito, R., Ueno, K., Hirata, Y., and Harada, Y.: 1980, *Electron Microscopy 1980,* (proceedings of the 7th European Congress on Electron Microscopy, The Hague),

Tanaka, M. Saito, R. and Watanabe, D.: 1980, *Acta Cryst.* A**36**, 350.

Tanemura, S. and Kato, N.: 1972, *Acta Cryst.* **A28**, 69.

Terasaki, O., Watanabe, D. and Gjønnes, J.: 1979, *Acta Cryst.* A**35**, 895.

Terauchi, H., Cohen, J. B., and Reed, T. B.: 1978, *Acta Cryst.* A**34**, 556.

Thomas, G. T. and Goringe, M. J.: 1979, *Transmission Electron Microscopy of Materials.* New York: Wiley.

Turner, P. and Cowley, J. M.: 1969, *Acta Cryst.* A**25**, 475.

Vainshtein, B. K.: 1955, *J. Phys. Chem. Moscow* **29**, 327.

Udalova, V. V. and Pinsker, Z. G,: 1963, *Sov. Phys. Cryst.* **8**, 433.

Uyeda, R. and Nonoyama, M.: 1965, *Jap. J. Appl. Phys.* **4**, 489.

Watanabe, D., Andersson, B., Gjønnes, J., and Terasaki, O.: 1974, *Acta Cryst.* A**30**, 772.

Watanabe, D., Castles. J. R., Jostens, A., and Malen, A. S.: 1967, *Acta Cryst.* **23**, 307.

Watanabe, D., Uyeda, R., and Kogiso, M.: 1968, *Acta Cryst.* A**24**, 249.

Wenk, H. R. (Ed): 1976, *Electron Microscopy in Mineralogy.* Berlin: Springer.

Zvyagin, B. B.: 1967, *Electron Diffraction Analysis of Clay Mineral Structures.* New York: Monographs in Geoscience, New York: Plenum.

Name Index

Chief Topic Index